Mastering Retrieval-Augmented Generation

Advanced Techniques and Production-Ready Solutions for Enterprise AI

Ranajoy Bose

Apress®

Mastering Retrieval-Augmented Generation: Advanced Techniques and Production-Ready Solutions for Enterprise AI

Ranajoy Bose
Bangalore, Karnataka, India

ISBN-13 (pbk): 979-8-8688-1807-3   ISBN-13 (electronic): 979-8-8688-1808-0
https://doi.org/10.1007/979-8-8688-1808-0

Managing Director, Apress Media LLC: Welmoed Spahr
Acquisitions Editor: Celestin Suresh John
Development Editor: Laura Berendson
Coordinating Editor: Gryffin Winkler

Distributed to the book trade worldwide by Springer Science+Business Media New York, 1 New York Plaza, New York, NY 10004. Phone 1-800-SPRINGER, fax (201) 348-4505, e-mail orders-ny@springer-sbm.com, or visit www.springeronline.com. Apress Media, LLC is a Delaware LLC and the sole member (owner) is Springer Science + Business Media Finance Inc (SSBM Finance Inc). SSBM Finance Inc is a **Delaware** corporation.

For information on translations, please e-mail booktranslations@springernature.com; for reprint, paperback, or audio rights, please e-mail bookpermissions@springernature.com.

Apress titles may be purchased in bulk for academic, corporate, or promotional use. eBook versions and licenses are also available for most titles. For more information, reference our Print and eBook Bulk Sales web page at http://www.apress.com/bulk-sales.

Any source code or other supplementary material referenced by the author in this book is available to readers on GitHub (https://github.com/Apress). For more detailed information, please visit https://www.apress.com/gp/services/source-code.

If disposing of this product, please recycle the paper

To the innovators and builders who dare to transform ideas into reality.

This book is dedicated to every engineer, data scientist, and technologist who believes that artificial intelligence should serve humanity's greatest challenges. Your relentless pursuit of solutions that matter inspires the future we're building together.

To my family and friends, whose unwavering support makes every ambitious journey possible.

And to the next generation of AI practitioners—may you build systems that not only demonstrate technical excellence but also create meaningful impact in the world.

Table of Contents

TABLE OF CONTENTS

TABLE OF CONTENTS

About the Author

Ranajoy Bose is a technologist, entrepreneur, and thought leader in the fields of Generative AI, MLOps, and enterprise data systems. As Co-founder and Global Head of Engineering at Morfius, he is at the helm of building cutting-edge AI solutions that power real-world transformation through Retrieval-Augmented Generation (RAG) and large-scale language models.

Before Morfius, Ranajoy held leadership roles at Oracle, where he led the Cloud Engineering organization for North America. His work was instrumental in advancing the adoption of data lakehouse architectures, modern analytics, AI/ML platforms, and cloud-native services for Fortune 500 clients.

Recognized as a 40-under-40 Data Scientist, Ranajoy also led a team ranked among *Analytics India Magazine*'s Top 10 data science workplaces. Beyond his corporate leadership, he remains a committed advocate for innovation and learning—frequently speaking at global conferences, contributing to academic and industry forums, and mentoring the next generation of AI practitioners.

Driven by curiosity and purpose, Ranajoy continues to push the boundaries of enterprise AI, translating complex technology into impactful solutions for the modern world.

About the Technical Reviewer

Dr. Swapan Ghosh is an Assistant Professor of Management at the Eli Broad College of Business, Michigan State University, with over 25 years of executive experience in technology firms such as Oracle, GE Digital, and VMware. A specialist in digital transformation and technology strategy, he has led global initiatives in artificial intelligence, industrial IoT, blockchain, and big data analytics across the high-tech, manufacturing, and life sciences sectors. Dr. Ghosh earned his Ph.D. in Management from Loughborough University, UK, where his research focused on digital transformative capabilities in industrial businesses. His academic expertise integrates dynamic capabilities theory, the resource-based view, and platform-based innovation to explain how firms adapt and compete in turbulent, tech-intensive environments.

Dr. Ghosh's research has been published in leading journals such as *Technovation, Journal of Engineering and Technology Management*, and *Business Horizons*, and he has contributed chapters to scholarly volumes on AI-driven business model innovation and digital entrepreneurship. His work examines how digital technologies, particularly AI, digital twins, and integrated PLM systems, enable strategic renewal and servitization in B2B manufacturing. He is also an active reviewer for journals, including *Technovation* and *California Management Review*, and participates in scholarly communities such as the Academy of Management and the Strategic Management Society.

As an educator, Dr. Ghosh has taught strategic management, innovation, and digital transformation at various universities in the United States and Mexico. Known for blending academic rigor with industry relevance, he draws on extensive field experience to bring real-world insights into the classroom. He has also founded and led technology startups, authored influential industry white papers and blogs, and organized executive seminars on AI-led strategy. His current research explores self-organizing ecosystems, generative AI, and the strategic role of integrated digital capabilities in industrial innovation.

Acknowledgments

Writing this book has been a journey that would not have been possible without the support, guidance, and collaboration of many remarkable individuals and organizations who have shaped both my understanding of RAG technology and my ability to share that knowledge effectively.

First and foremost, I want to thank the entire team at Morfius for their unwavering support throughout this writing process. Building cutting-edge RAG solutions together has provided the real-world insights and practical experiences that form the backbone of this book. Your dedication to pushing the boundaries of what's possible with AI continues to inspire me daily.

My deep appreciation extends to Dr. Swapan Ghosh, who served as the technical reviewer for this book. His meticulous review, insightful feedback, and expert guidance helped identify gaps, clarify complex concepts, and ensure that the content serves its intended audience effectively. His contributions have significantly enhanced the quality and accuracy of this work.

My sincere gratitude goes to the Apress team, particularly Celestin and the editorial staff, who believed in this project from the beginning and provided invaluable guidance in shaping the manuscript. Your expertise in technical publishing and commitment to quality have elevated this work significantly.

To the clients and partners who trusted us with their most challenging AI implementation projects: thank you for providing the real-world testing grounds where these concepts were refined and validated. Your willingness to push boundaries and demand excellence has shaped every recommendation in this book.

ACKNOWLEDGMENTS

Special recognition goes to my mentors throughout my career who emphasized the importance of translating complex technology into practical solutions that create real value. Your guidance continues to influence how I approach both technology leadership and knowledge sharing.

I am deeply grateful to the vibrant AI and machine learning community—researchers, practitioners, and thought leaders who continuously share their knowledge through open source contributions, research papers, and community discussions. The collaborative spirit of this community has accelerated innovation in RAG technology and made it possible to build upon a foundation of collective wisdom.

I want to acknowledge the academic researchers and industry pioneers who laid the groundwork for RAG technology. While this book focuses on practical implementation, it builds upon years of foundational research and theoretical advances that made these applications possible.

Finally, I want to thank my family and friends for their patience and encouragement during the intensive writing process. Your support made it possible to balance the demands of running a growing company while dedicating the time and focus necessary to create a comprehensive resource for the AI community.

This book represents not just individual effort but the collective wisdom of an entire community working to advance the state of artificial intelligence. To everyone who contributed directly or indirectly to this work: thank you for helping bring these ideas to life.

The future of RAG technology depends on continued collaboration, innovation, and knowledge sharing. I hope this book contributes to that collective effort and helps the next generation of AI practitioners build systems that truly make a difference in the world.

Introduction

Retrieval-Augmented Generation (RAG) represents one of the most significant breakthroughs in artificial intelligence, fundamentally changing how we build intelligent systems that can access, understand, and reason with vast amounts of information. While large language models have demonstrated remarkable capabilities, they face inherent limitations: knowledge cutoffs, hallucinations, and inability to access real-time or proprietary information. RAG solves these challenges by combining the reasoning power of language models with dynamic information retrieval, creating systems that are both knowledgeable and grounded in factual data.

This book emerges from years of hands-on experience building production RAG systems for enterprise clients across diverse industries. Having witnessed the evolution from experimental prototypes to mission-critical applications serving millions of users, I've seen firsthand both the tremendous potential and the practical challenges of implementing RAG at scale. The techniques, patterns, and strategies presented here have been battle-tested in real-world environments where performance, reliability, and accuracy are non-negotiable.

What This Book Covers

Mastering Retrieval-Augmented Generation provides a comprehensive journey from foundational concepts to advanced production deployment. Rather than focusing purely on theory, this book emphasizes practical implementation with enterprise-grade considerations throughout.

Part I: Foundations establishes the theoretical groundwork and introduces core RAG concepts. You'll understand why RAG matters, how it works, and build your first functional system. This foundation prepares you for the more sophisticated implementations that follow.

Part II: Core Components dives deep into each element of the RAG pipeline. From document loading and text processing to embedding generation and vector storage, you'll master the technical building blocks that determine system performance and reliability.

Part III: Advanced Implementation explores cutting-edge RAG patterns including structured data integration, graph-based approaches, and agentic systems. These chapters prepare you to tackle complex real-world scenarios that basic RAG approaches cannot handle.

Part IV: Production and Evaluation focuses on the critical aspects of deploying RAG systems at scale. Evaluation frameworks, production deployment strategies, and security considerations ensure your systems perform reliably in enterprise environments.

Who This Book Is For

This book targets senior AI/ML engineers, data scientists, and technical architects who need to build production-ready RAG systems. If you're responsible for implementing AI solutions that deliver measurable business value, this book provides the expertise you need.

Secondary audiences include engineering managers and technical leads planning next-generation AI capabilities, as well as AI researchers working with large-scale language models and information retrieval systems.

You should have intermediate Python programming skills, basic understanding of machine learning concepts, and familiarity with natural language processing fundamentals. While prior RAG experience isn't required, you'll benefit most if you're already working with language models or information retrieval systems.

How This Book Is Structured

Each chapter builds upon previous concepts while remaining accessible as stand-alone reference material. Code examples progress from simple illustrations to production-ready implementations, with emphasis on patterns you can adapt to your specific use cases.

Throughout the book, you'll find real-world case studies demonstrating how leading organizations have successfully implemented RAG systems. These examples illustrate not just technical implementation but also business impact and lessons learned from production deployments.

The book balances breadth with depth—covering the full spectrum of RAG implementation while providing sufficient detail for practical application. Whether you're building your first RAG system or optimizing existing implementations, you'll find actionable insights that directly apply to your work.

What Makes This Book Different

Unlike purely academic treatments of RAG, this book maintains relentless focus on production readiness. Every technique, pattern, and recommendation has been validated in enterprise environments where reliability and performance matter.

The book also addresses the complete life cycle of RAG systems—from initial conception through ongoing optimization and maintenance. You'll learn not just how to build RAG systems but how to build them right: scalable, secure, and maintainable.

Most importantly, this book recognizes that RAG is not just a technical challenge but a business enabler. The goal isn't just to implement RAG—it's to create systems that deliver measurable value to users and organizations.

Welcome to the comprehensive guide for mastering Retrieval-Augmented Generation. Let's build intelligent systems that truly make a difference.

PART I

Foundations

CHAPTER 1

Introduction to Retrieval-Augmented Generation (RAG)

We live in an age where the world is captivated by the boundless potential of generative AI. From startups to global enterprises, we are beginning to grasp how this revolutionary technology can redefine productivity and creativity. Generative AI is no longer just a buzzword; it's a game-changer with tangible impacts across industries.

Consider its versatility: Generative AI can process vast amounts of information, spanning text, images, audio, and even video. It can help developers write and debug code, assist analysts by generating insights for market research, and speed up the creation of professional presentations and documents. Marketing teams are using it to craft engaging campaigns, while content creators are producing art, stories, and designs with an ease that was unimaginable a few years ago.

In short, generative AI is reshaping everything from content creation to data-driven decision-making, driving efficiency, personalization, and engagement like never before. And this is just the beginning.

© Ranajoy Bose 2025
R. Bose, *Mastering Retrieval-Augmented Generation*,
https://doi.org/10.1007/979-8-8688-1808-0_1

Opportunities and Challenges: The Dual Face of Gen AI

However, the journey of generative AI has not been without its share of challenges. Organizations experimenting with and adopting these technologies encounter complex questions about responsible use, ethical implications, and operational efficiency.

On the one hand, generative AI offers unprecedented customization and adaptability. AI models have evolved rapidly in sophistication, enabling solutions tailored to specific industries and user needs. These advancements unlock opportunities across diverse sectors—healthcare, finance, education, entertainment, and beyond.

On the other hand, this evolution comes with heightened expectations. Stakeholders are demanding transparency, reliability, and ethical practices. Leaders must navigate a delicate balancing act: embracing innovation without compromising accountability or security. For example:

- **Transparency:** How can organizations ensure that AI-generated content is verifiable and trustworthy?

- **Bias and Ethics:** How do we prevent generative AI from perpetuating biases or producing harmful content?

- **Security:** What safeguards are needed to protect sensitive data and ensure the integrity of AI systems?

This push-and-pull dynamic forces companies to rethink workflows, reevaluate data practices, and adopt strategies that balance bold innovation with ethical and secure implementation.

What Are LLMs?

LLMs, or **large language models**, are advanced artificial intelligence systems designed to understand, interpret, and generate human language.

Think of them as incredibly skilled language experts trained on massive amounts of text data, such as books, articles, and websites. These models rely on deep neural networks with billions—or even trillions—of parameters, enabling them to process language in a remarkably sophisticated way.

Key Characteristics of LLMs

LLMs are the backbone of many modern AI applications because of their ability to perform a wide range of language-based tasks. Their defining features include

1. **Language Understanding**

   LLMs excel at interpreting and analyzing text, whether a simple question or a complex instruction. They can decode grammar, tone, and intent nuances, allowing them to provide meaningful responses.

2. **Text Generation**

   One of their standout abilities is generating coherent, human-like text. This makes them ideal for tasks like content creation, chatbot interactions, translations, summaries, etc.

3. **Context Awareness**

   LLMs can grasp the context of conversations or documents. For example, they understand that "Paris" in one sentence might refer to the capital of France and, in another to the famous personality Paris Hilton. This contextual awareness ensures their outputs are relevant and coherent.

Examples of LLMs

Several LLMs dominate the landscape today, each with unique strengths:

- **GPT (OpenAI):** Known for its versatility and wide adoption across industries

- **Gemini (Google):** An evolving platform with deep integration into Google's ecosystem

- **LLaMA (Meta):** Focused on accessibility and open development for research and experimentation

- **Claude (Anthropic):** Designed with safety and alignment as a priority

- **Mistral and Cohere:** Emerging players offering lightweight and efficient solutions for specific applications

- **DeepSeek R1 (DeepSeek AI):** An advanced open-weight LLM built to support code and reasoning tasks optimized for efficiency and scalability

These models are revolutionizing industries such as customer service, education, healthcare, and research by identifying patterns in language to perform tasks once thought to require human expertise.

Limitations of LLMs

While LLMs are powerful, they are far from perfect. Understanding their limitations is crucial for leveraging them effectively:

Knowledge Cutoff and Static Training Data

- **Outdated Information:** LLMs are trained on datasets with a fixed cutoff date, meaning they can't know anything that happened after their training.

- **Static Learning:** They don't automatically update their knowledge, making them less useful for fast-evolving fields like medicine or technology.

Unpredictability and Hallucination

- **Fabricated Facts:** LLMs sometimes "hallucinate," confidently producing responses that seem plausible but are entirely incorrect or made up.

- **Lack of Groundedness:** Their outputs might lack reliability since they don't verify real-time data.

Sensitivity to Prompts

- **Inconsistent Responses:** Slight changes in phrasing can lead to wildly different answers, which can frustrate users.

- **Prompt Crafting:** Getting the desired output often requires expertise in crafting the proper prompts, a skill known as "prompt engineering."

Limited Reasoning and Context Understanding

- **Shallow Understanding:** LLMs operate on patterns and statistics, not proper comprehension or reasoning. They excel at mimicking knowledge but don't genuinely "know" anything.

- **Contextual Limitations:** They may struggle with long or complex inputs, constrained by their "token window" (the maximum length of text they can process at once).

Bias and Ethical Concerns

- **Embedded Biases:** LLMs can unintentionally reflect and amplify biases in their training data, leading to unfair or inappropriate outputs.

- **Ethical Risks:** Poorly designed or unchecked use of LLMs can result in harmful, offensive, or unethical content.

Lack of Explainability and Transparency

- **Opaque Decision-Making:** It's often unclear how an LLM arrived at a particular answer, which can undermine trust in critical applications.

- **Limited Control:** Fine-tuning specific behaviors in an LLM can be challenging, making them unpredictable in some contexts.

A Relatable Perspective

Imagine an LLM as an over-enthusiastic new team member who confidently answers every question but doesn't keep up with the latest developments. While their enthusiasm is impressive, their unwavering assurance—even when wrong—can erode trust. This unpredictability is something businesses need to be wary of, especially in customer-facing applications.

Why RAG Is the Solution

One way to address many of these challenges is through **Retrieval-Augmented Generation (RAG)**. RAG combines the language capabilities of LLMs with reliable, real-time data retrieval from external sources. This grounds the responses in verified information and allows for updates and context-specific answers, bridging the gap between raw language modeling and practical, trustworthy applications.

As we dive deeper into this book, we'll explore how RAG complements LLMs, making them smarter, more reliable, and better suited to tackle real-world problems.

What Is RAG?

Retrieval-Augmented Generation, commonly known as RAG, is an innovative approach that combines traditional information retrieval techniques with generative AI models. This method leverages the strengths of retrieving factual and contextually relevant information from large datasets while utilizing language models' ability to generate coherent and natural-sounding text. As a result, RAG systems can produce more accurate, informative, and context-aware responses than those generated by purely generative or retrieval-based models alone.

At its core, Retrieval-Augmented Generation (RAG) operates by first searching for relevant data within a collection of information, including text, documents, or other structured and unstructured content. A retriever performs this process. The retrieved data is then input into a generative model that utilizes this information to construct a response. By combining these two processes, RAG systems can answer complex questions, generate detailed responses, and assist users when accuracy and specificity are crucial.

For example, consider a user asking a RAG (Retrieval-Augmented Generation) system about the latest advancements in AI research. A traditional generative model might respond based on the data it was trained on, which could be outdated. In contrast, an RAG system would search for the most recent research papers, articles, and other relevant information from reliable sources. It would then retrieve this information and use it to generate a well-informed and up-to-date answer. This makes RAG particularly valuable for applications that require real-time or current information, such as news aggregation, customer support, and technical assistance (see Figure 1-1).

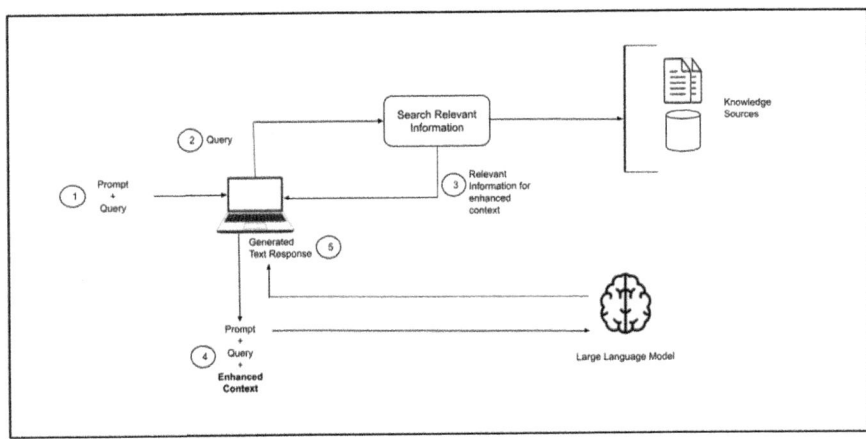

Figure 1-1. *Conceptual flow diagram of RAG*

Overview of RAG Architecture

A Retrieval-Augmented Generation (RAG) system is broadly designed to combine the best of two worlds: retrieving relevant information from external sources and using that information to generate insightful, context-aware responses. This process happens in three key stages: **Indexing**, **Retrieval**, and **Generation**. Let's break it down in simple terms.

Indexing: Organizing Your Knowledge Library

Imagine you have a vast library filled with books, documents, and articles. If these materials were thrown into a big pile, finding the right piece of information would be like looking for a needle in a haystack. The indexing phase of RAG solves this problem by organizing all this knowledge so the system can find what it needs quickly.

Here's how it works:

- **Preparing the Information:** The system processes raw data (such as text documents, reports, or even FAQs) and organizes it, breaking it down into smaller, manageable sections. Think of it as arranging your library into chapters and sections instead of keeping it as whole books. This step is also referred to as chunking.

- **Turning Text into Numbers:** Computers can't read text like we do—they need numbers. So, each piece of text is converted into a unique set of numbers called **embeddings**. These numbers represent the meaning of the text in a way the computer understands.

- **Storing It Smartly:** Instead of stuffing this knowledge into a typical bookshelf (or database), RAG uses something more advanced: **vector databases.** These are special tools designed to handle the embeddings and allow for superfast searches.

This step sets up the foundation for the system to work efficiently. It ensures the "library" is ready for action whenever a question or query comes in.

Retrieval: Finding the Right Answer

Now that the library is organized, the next step is **retrieval**—finding the right book (or chapter or sentence) that contains the answer to your query. Here's how this works:

- **Understanding the Question:** When a user asks a question, the system translates it into numbers (an embedding), like how it processed the library earlier. This ensures the question and the stored information "speak the same language."

- **Finding the Match:** The system then searches the library for pieces of information that closely match the question's meaning. This isn't about finding exact words—it's about finding ideas and concepts that align with what you're asking. This step is also known as similarity search.

- **Picking the Best Bits:** Once it finds relevant information, the system ranks and filters the results, like a librarian handing you the most valuable books instead of dumping a pile on your desk.

This stage ensures the system has the right information to work with before generating a response.

Generation: Crafting the Answer

Now comes the magic: **generation.** The system combines your question with the retrieved information to create a clear, thoughtful answer.

Here's what happens:

- **Putting it Together:** The system pairs the retrieved information with your question to form a detailed input. For example, if you ask about the features of a new phone, the system combines your query with the retrieved specifications.

- **Generating the Answer:** A large language model (LLM)—like OpenAI's GPT or Google's Gemini— takes this combined input and crafts a response. It's like having a knowledgeable assistant who reads the relevant books for you and explains them in plain language.

- **Refine the Response:** The generated response might undergo a final check to clean up any unnecessary details, ensure clarity, or even summarize the information if it's too long.

The generation step is where everything comes together, providing a response that feels human-like, relevant, and to the point.

Bringing It All Together

Think of a RAG system like a well-trained librarian who

1. Organizes the library (Indexing)

2. Finds the exact book or section you need (Retrieval)

3. Explains the content to you in simple terms (Generation)

This three-step process—**indexing, Retrieval,** and **Generation**—ensures that RAG systems can not only generate fluent responses but also ground them in real, trustworthy information (see Figure 1-2). Whether helping a customer find answers, assisting researchers with insights, or powering intelligent chatbots, this pipeline makes RAG a powerful tool for tackling a wide range of real-world problems.

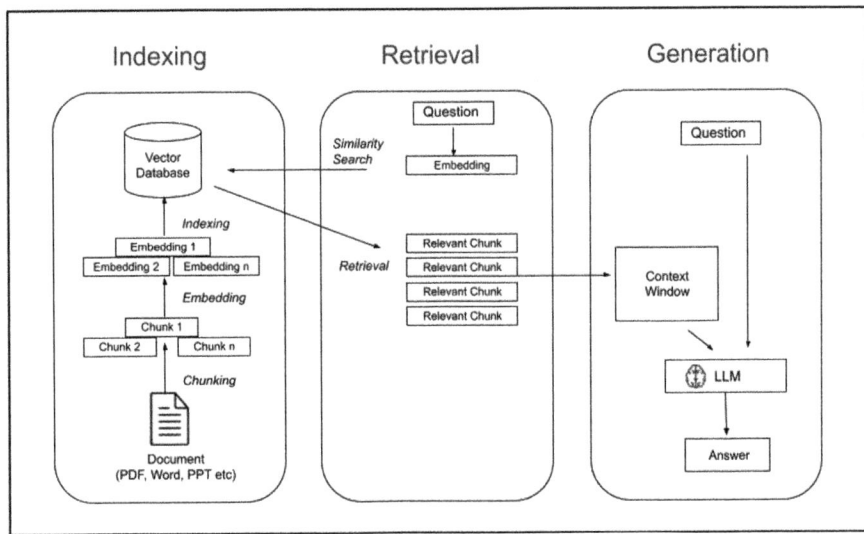

Figure 1-2. *Overview of RAG architecture*

Differences Between Traditional Generative Models and Retrieval-Augmented Systems

The way traditional generative models and Retrieval-Augmented Generation (RAG) systems process and deliver information diverges fundamentally, reflecting their strengths and limitations. Here's a detailed comparison:

Knowledge Source and Grounding

- **Generative Models:**

  Traditional generative models like GPT or LLaMA generate responses by drawing from patterns embedded in the static datasets on which they were trained. This training gives them a broad knowledge base, but their understanding is confined to the scope and time frame of their training data. Consequently, they lack mechanisms to verify the factual accuracy or relevance of their outputs in real time.

  Example: A generative model asked about the latest advancements in AI might provide outdated or fabricated information if it falls outside its training knowledge.

- **RAG Systems:**

  RAG systems augment generative capabilities by incorporating a retrieval step. They pull relevant, up-to-date information from external databases or knowledge repositories before generating a response, ensuring outputs are both contextually accurate and grounded in reliable sources.

Example: When queried about the latest AI developments, an RAG system retrieves information from recent research papers or articles, producing a response that reflects current trends.

Handling of Dynamic Information

- **Generative Models:**

  Once trained, traditional generative models are static and do not automatically update as new information becomes available. They require retraining to incorporate updated knowledge, which can be resource-intensive and time-consuming.

- **RAG Systems:**

  RAG systems are inherently dynamic, accessing external knowledge bases to retrieve the latest data on demand. This makes them ideal for applications requiring real-time updates, such as news aggregation, regulatory compliance, or product support.

  Example: A customer support system powered by RAG can instantly provide guidance for a newly released product, leveraging updated FAQs or documentation without retraining the underlying model.

Accuracy and Hallucination

- **Generative Models:**

  Generative models often "hallucinate," producing confident but incorrect or fabricated information.

This can undermine their reliability in domains where precision is crucial, such as healthcare, legal, or technical domains.

- **RAG Systems:**

  RAG systems significantly reduce hallucination risks by retrieving factual information from trusted sources. The retrieval mechanism provides a factual grounding layer, ensuring more accurate and verifiable responses.

  Example: A legal assistant powered by RAG can pull statutes or case laws directly from legal databases, ensuring its advice is grounded in actual regulations.

Context and Personalization

- **Generative Models:**

  Generative models are adept at understanding language and generating coherent responses but often struggle with specificity and contextual nuances. Their outputs may lack the depth or personalization required for complex queries.

- **RAG Systems:**

  RAG enhances context awareness by retrieving information tailored to the user's specific query. This enables the generation of more relevant and personalized responses, especially for applications like customer service or technical support.

  Example: A RAG-powered troubleshooting assistant retrieves relevant device-specific guides to deliver precise step-by-step instructions for resolving issues.

Scalability and Adaptability

Let's look at generative models vs. RAG (see Table 1-1).

- **Generative Models:**

  Incorporating new knowledge into traditional models requires retraining, making them less adaptable to rapidly evolving data landscapes.

- **RAG Systems:**

  RAG systems can integrate new knowledge sources dynamically without retraining, allowing organizations to scale and adapt effortlessly.

  Example: An e-commerce platform using RAG can seamlessly update its chatbot to provide details about newly added products by linking to the latest catalog.

Table 1-1. *Traditional Generative Models vs. RAG*

Aspect	Generative Models	Retrieval-Augmented Systems (RAG)
Knowledge base	Static, limited to training data	Dynamic, pulls real-time external data
Accuracy	Prone to hallucination	Grounded in factual, retrieved information
Adaptability	Requires retraining for updates	Integrates new data sources dynamically
Context awareness	Generic, limited personalization	Specific, tailored responses
Use cases	Broad, generalized applications	Specialized, accuracy-critical applications

Benefits of RAG Over Traditional Generative Models

Generative AI systems like GPT, LLaMa, and other large language models (LLMs) have showcased remarkable capabilities in understanding and generating human-like language. However, they often stumble when providing accurate, current, and contextually relevant responses. This is where Retrieval-Augmented Generation (RAG) shines, combining the creative prowess of generative models with the precision of retrieval mechanisms. Let's explore how RAG addresses these challenges and offers distinct advantages over traditional generative approaches.

Incorporating Proprietary and Domain-Specific Knowledge

LLMs, by design, are trained on publicly available datasets, which often don't include proprietary or industry-specific information. This limitation can make them inadequate for addressing niche queries or providing insights based on confidential data.

RAG Advantage: RAG allows systems to integrate proprietary or domain-specific information, such as internal memos, product design documents, or technical manuals, into their responses. For example, a tech company's customer support RAG system can retrieve detailed troubleshooting steps for proprietary devices directly from internal knowledge bases, ensuring accurate and specialized support.

Real-Time and Up-to-Date Information

Traditional generative models operate based on static training datasets, meaning their knowledge is frozen at a specific time. They can't adapt to newly released products, changing regulations, or breaking news without extensive retraining.

19

RAG Advantage: RAG systems bypass this limitation by retrieving the most current data from external sources. For instance, if a company releases a new version of a software tool, a RAG-powered support assistant can immediately provide guidance specific to the latest release by pulling information from updated documentation or FAQs. This dynamic capability makes RAG systems invaluable for industries that rely on timely and accurate information, such as finance, healthcare, and news aggregation.

Enhanced Accuracy and Reliability

One of the most notorious challenges with LLMs is their tendency to "hallucinate"—confidently producing plausible but entirely incorrect information. This unreliability can be problematic in high-stakes environments like healthcare, legal, or financial services.

RAG Advantage: RAG mitigates this risk by grounding responses in retrieved data from verified sources. For example, an RAG system assisting in legal research can pull clauses directly from legislation or case law databases, ensuring that its answers are factual and trustworthy. Additionally, some RAG implementations can even include citations of sources, enabling human verification of the generated responses.

Context-Aware and Personalized Responses

While LLMs can produce grammatically correct and coherent text, their responses often lack the depth or specificity needed for complex queries. This is because generative models work from a general knowledge base and can struggle with understanding nuanced or context-specific requirements.

RAG Advantage: By retrieving information directly related to a user's query, RAG ensures that responses are tailored to the context. For example, in a customer service scenario, a RAG system troubleshooting a specific smartphone model will provide step-by-step guidance tailored to that model rather than offering vague or generic advice.

Secure and Controlled Data Access

Once trained, LLMs treat their knowledge as a monolithic block, lacking the ability to differentiate between user permissions or access levels. This can lead to privacy and security concerns when sensitive information is involved.

RAG Advantage: RAG systems can be designed to incorporate fine-grained access controls, retrieving only the data that a user is authorized to see. For example, an employee using a RAG-based HR assistant could retrieve personalized information about their benefits package. At the same time, the same system would deny access to confidential organizational policies meant for executives.

Scalability and Adaptability

Traditional generative models require retraining to incorporate new knowledge, a computationally expensive and time-consuming process.

RAG Advantage: RAG systems are highly adaptable and can seamlessly integrate new knowledge sources without modifying the underlying generative model. For example, an e-commerce chatbot powered by RAG can instantly respond to queries about newly added products by retrieving details from the updated product database. This flexibility makes RAG systems particularly useful for dynamic and evolving environments.

Efficiency in Information Retrieval

LLMs often provide verbose responses or include irrelevant details, leaving users to sift through unnecessary information.

RAG Advantage: RAG systems streamline responses by retrieving only the most relevant data, ensuring concise and actionable outputs. This is critical in call centers, where quick and accurate answers directly impact customer satisfaction.

Conclusion

RAG represents a paradigm shift in natural language processing, overcoming the inherent limitations of traditional generative models by combining retrieval and generation. It ensures responses are accurate, timely, secure, and context-aware—qualities essential for today's complex and data-driven applications.

Whether accessing proprietary knowledge, delivering real-time updates, or enhancing the trustworthiness of AI-generated content, RAG is setting a new standard for intelligent systems across industries. As demand for smarter, more adaptable AI solutions continues to grow, RAG's unique capabilities make it a game-changer in generative AI.

History and Evolution of RAG

The development of Retrieval-Augmented Generation (RAG) marks a transformative milestone in the fields of natural language processing (NLP) and artificial intelligence (AI). Its journey, shaped by decades of innovation in information retrieval (IR) and generative modeling, has culminated in a powerful hybrid approach capable of producing fluent, contextually rich, and factually grounded responses.

To appreciate RAG's emergence and significance, it's essential to explore its historical roots and the key advancements that have shaped its evolution.

From Information Retrieval to RAG

The Foundations: Information Retrieval Systems

The journey began in the mid-20th century with the development of **information retrieval (IR) systems** designed to locate relevant data in large document collections. Early IR systems relied on techniques like term frequency-inverse document frequency (**TF-IDF**) and **Boolean operators**, enabling keyword-based searches. While revolutionary at the time, these systems often returned results that lacked nuance, requiring users to sift through documents for specific details manually.

The explosion of digital content in the following decades demanded more sophisticated retrieval mechanisms. Modern **search engines** emerged, leveraging algorithms that rank documents based on user intent, relevance, and context. Yet, these systems still had limitations—they provided exact matches rather than synthesized answers, leaving gaps in usability and efficiency.

Parallel Advances in Generative Models

At the same time, NLP researchers were pushing the boundaries of language understanding and generation. Early generative models were rudimentary, relying on rule-based logic and statistical methods, which struggled with coherence and context.

The introduction of **deep learning** in the 2010s catalyzed dramatic advancements in generative modeling. Transformer-based architectures like **BERT** and **GPT** brought unprecedented fluency and contextual understanding, enabling models to generate human-like text. However, these generative systems faced inherent limitations:

- Their knowledge was static and bound by their training data.

- They lacked mechanisms to ensure factual accuracy, leading to issues like "hallucination" (producing plausible but incorrect information).

These challenges highlighted the need for a system that could dynamically retrieve and integrate real-time information into generative processes—paving the way for RAG.

The Emergence of RAG

A Hybrid Approach

RAG was born out of the necessity to bridge the gap between **retrieval** and **generation**. It combines the strengths of both domains:

- **Retrieval systems** locate accurate, relevant information from external sources.

- **Generative models** synthesize this information into fluent and contextually rich responses.

This synergy resolves the limitations of traditional generative models by grounding outputs in reliable data, enabling RAG systems to deliver accurate and context-aware results.

Milestones in RAG Development

Dense Embeddings and Semantic Search (2016–2018)

The shift from sparse, keyword-based search to **dense embeddings** marked a turning point in IR. Dense embeddings, powered by neural networks, enabled **semantic search**, where retrieval systems understood the meaning behind queries rather than relying on exact term matches. This innovation formed the foundation of modern retrieval systems, which underpin RAG's capabilities.

Transformer Revolution (2017 Onwards)

The release of the **transformer architecture** revolutionized NLP, leading to models like **BERT** (Bidirectional Encoder Representations from Transformers) and **GPT** (Generative Pre-trained Transformer). These models introduced parallel processing, making them highly efficient for complex language tasks. Their contextual depth set the stage for integrating retrieval mechanisms, enhancing their accuracy and utility.

Birth of RAG (2020)

The concept of **Retrieval-Augmented Generation** was formally introduced in 2020. Early RAG implementations combined retrieval systems with generative models to address hallucination issues, improve accuracy, and enhance response relevance. These models demonstrated the potential of using external knowledge to ground generative outputs in factual data.

Hybrid and Multimodal RAG (2021–2023)

Building on initial successes, researchers explored **hybrid RAG systems** that combined structured and unstructured data sources and **multimodal RAG systems** that processed diverse inputs like images, audio, and video. These innovations expanded RAG's applicability to domains requiring complex, multimodal understanding, such as virtual assistants, technical support, and creative content generation.

Industry Adoption and Open-Source Frameworks (2023 Onwards)

By 2023, RAG transitioned from academic exploration to enterprise adoption. Organizations began leveraging RAG systems to enhance customer support, improve technical documentation, and streamline legal research. The availability of open-source frameworks like **LangChain** and tools from **Hugging Face** accelerated the democratization of RAG, enabling widespread experimentation and innovation.

The history of RAG illustrates the power of combining complementary technologies to overcome individual limitations. From its roots in keyword-based retrieval and statistical language models to today's sophisticated systems, RAG has evolved into a cornerstone of modern AI. RAG has become an essential tool for addressing complex real-world challenges by ensuring responses are accurate, contextually relevant, and human-like. The future of RAG promises continued innovation, including deeper integration with multimodal systems and further enhancements in scalability and precision.

Why RAG Matters Today

In a world driven by data, information, and rapid decision-making, the importance of **Retrieval-Augmented Generation (RAG)** has grown exponentially. Traditional generative models are powerful but often struggle with accuracy, timeliness, and contextual relevance. RAG addresses these shortcomings by marrying the strengths of generative language models with the precision of information retrieval systems. This hybrid approach makes RAG an indispensable tool across various industries.

Real-World Relevance and Use Cases Across Industries

Healthcare

1. **Clinical Decision Support**

   - Healthcare professionals rely on accurate, real-time data to diagnose conditions, recommend treatments, and manage patient care.

- RAG systems retrieve the latest clinical guidelines, research articles, and case studies to provide precise, evidence-based recommendations, enhancing decision-making and patient outcomes.

- Example: Suggesting a treatment protocol based on the latest clinical trial data for a rare disease.

2. **Medical Documentation**

- Automating the generation of discharge summaries, patient histories, and reports by pulling data from electronic health records (EHRs).

- It saves clinicians time, reduces errors, and ensures compliance with documentation standards.

Legal Services

1. **Legal Research**

- RAG systems retrieve relevant case law, statutes, and precedents, synthesizing them into concise, usable summaries.

- Speeds up research and ensures lawyers have access to up-to-date and contextually appropriate information.

- Example: Summarizing rulings from recent high-profile cases relevant to an ongoing trial.

2. **Contract Review and Analysis**

- Automating the extraction of key clauses, compliance checks, and risk assessments in large volumes of contracts

- Ensures consistency, reduces time spent on manual reviews, and highlights potential risks

Finance and Banking

1. **Financial Analysis and Reporting**

- Analysts synthesize vast amounts of data from news, financial statements, and reports.

- RAG systems fetch the latest data, identify trends, and generate insights, enabling quicker and more informed decision-making.

- Example: Generating a market performance report integrating live stock market data with historical trends.

2. **Fraud Detection**

- RAG models can identify suspicious behavior by retrieving historical transaction patterns and correlating them with real-time activity.

- Example: Generating alerts for abnormal activity in financial transactions based on historical fraud patterns.

Customer Service and Support

1. **Intelligent Customer Support Agents**

- RAG-powered chatbots retrieve relevant information from product manuals, FAQs, and troubleshooting guides to provide accurate and context-aware answers.

- Enhances user satisfaction by resolving complex queries efficiently.

- Example: A telecom company using RAG to provide real-time solutions for network issues customers report.

2. **Personalized Product Recommendations**

- In e-commerce, RAG systems leverage user preferences, purchase history, and product data to deliver tailored recommendations.

- Boosts sales, improves user experience, and builds customer loyalty.

Education and Content Creation

1. **Adaptive Learning Systems**

- RAG systems create personalized learning experiences by retrieving and generating content based on textbooks, online courses, and academic journals.

- Provides students with tailored study guides, answers to queries, and detailed explanations.

- Example: Generating a customized study plan for a student preparing for competitive exams.

2. **Automated Content Generation**

- Content creators often need to gather information from multiple sources. RAG systems streamline this process by retrieving relevant data and generating structured drafts.

- Saves time, ensures accuracy, and enhances productivity in journalism and marketing.

- Example: Creating a blog post summarizing the latest industry trends using live and historical data.

The Game-Changer: Why Enterprises Embrace RAG

RAG systems are reshaping how industries interact with data and knowledge by addressing critical challenges such as:

- **Accuracy:** Grounding outputs in verified, up-to-date information

- **Efficiency:** Reducing the time required for manual research and data synthesis

- **Adaptability:** Scaling across domains, from legal and medical to e-commerce and education

In a landscape where businesses and consumers demand **contextually aware, reliable, and actionable insights**, RAG emerges as a cornerstone technology, transforming operations and elevating user experiences.

As adoption grows, RAG will continue to redefine AI's possibilities, setting new benchmarks for how systems retrieve, synthesize, and generate information in real time.

Summary and Next Steps

Throughout this introductory chapter on Retrieval-Augmented Generation (RAG), we've explored the transformative potential of combining generative AI with dynamic knowledge retrieval. Beginning with the broader landscape of generative AI's opportunities and challenges, we examined the limitations of traditional large language models (LLMs)—static knowledge, hallucinations, and ethical risks—and how RAG addresses these gaps by grounding responses in real-time, verifiable data.

Key Takeaways

Let's recap the key themes we've covered:

- The rise of generative AI and its dual promise of innovation and challenges like bias and transparency

- Core limitations of LLMs, including outdated knowledge and lack of explainability

- How RAG systems bridge these gaps by integrating retrieval mechanisms with generative models

- The three-stage RAG architecture (indexing, retrieval, generation) and its real-world advantages

- Industry applications of RAG, from healthcare to customer support, highlighting its versatility

What makes RAG revolutionary is its ability to marry the creativity of LLMs with the precision of curated knowledge sources. While we've used conceptual examples here, the principles apply universally—whether you're working with open-source tools or enterprise-grade systems.

Next Steps
Looking Ahead

As you progress in your RAG journey, consider

- Reflecting on how RAG could solve specific pain points in your domain (e.g., outdated documentation, customer query resolution)

- Experimenting with small-scale prototypes to see retrieval-augmented responses in action

- Exploring the technical components introduced here (like vector databases) in later chapters

The RAG paradigm we've introduced is just the foundation. In Chapter 2, we'll dive deeper into its core mechanisms, such as vector search and hybrid architectures, while Chapter 3 will guide you through hands-on implementation. Like all AI systems, RAG thrives on iteration—start with these concepts, test them in practice, and refine as you go.

CHAPTER 2

Core Concepts of Retrieval-Augmented Generation (RAG)

Retrieval-Augmented Generation (RAG) represents a paradigm shift in how AI systems access and leverage knowledge. Unlike traditional language models constrained by static training data, RAG dynamically grounds responses in real-world information, combining the fluency of generative models with the precision of retrieval systems. This chapter unpacks the architectural pillars that make RAG uniquely powerful—from vector embeddings and semantic search to the seamless integration of external knowledge with large language models (LLMs).

You'll learn

- How vector search transforms unstructured text into retrievable, semantically rich data.

- The step-by-step workflow of RAG: indexing, retrieval, augmentation, and generation.

- Why hybrid approaches (like combining RAG with fine-tuning) unlock adaptability across domains.

- Key trade-offs between RAG and other LLM customization methods (e.g., pretraining, prompt engineering).

© Ranajoy Bose 2025
R. Bose, *Mastering Retrieval-Augmented Generation*,
https://doi.org/10.1007/979-8-8688-1808-0_2

- By breaking down these concepts, we'll illuminate how RAG addresses critical limitations of stand-alone LLMs—outdated knowledge, hallucinations, and lack of domain specificity—while preparing you to design and evaluate RAG systems in practice. The insights here form the bedrock for the hands-on implementation in Chapter 3 and the advanced patterns explored later in this book.

Retrieval-Augmented Generation (RAG) with Vector Search

Retrieval-augmented generation (RAG) is a transformative approach that combines the strengths of external information retrieval with the powerful generative capabilities of Large Language Models (LLMs), such as GPT or LLaMA. By enriching a user's prompt with relevant external data, RAG enables LLMs to generate responses that are more accurate, timely, and contextually appropriate.

While LLMs are exceptional at processing and generating human-like text, they are not inherently reliable sources of information. Their limitations include

- **Restricted Scope**: LLMs cannot access proprietary, domain-specific, or real-time data beyond their training set.

- **Hallucination Risks**: LLMs often fabricate answers rather than admit knowledge gaps.

- **Static Knowledge**: LLMs require retraining to incorporate new data, which can be both costly and time-intensive.

The Role of RAG in Addressing LLM Limitations

RAG addresses these challenges by dynamically supplying the LLM with external, context-specific information. This information is retrieved from various data sources and appended to the user's query, creating an expanded prompt. By doing so, RAG equips the LLM with the context to deliver precise and relevant responses.

For example, consider an LLM trained on publicly available data. If tasked with answering questions about internal company memos or project documents, it would likely hallucinate, generating plausible sounding but inaccurate responses. An RAG system can mitigate this by retrieving the required documents and embedding their content into the query, allowing the LLM to ground its response in factual and relevant information.

Versatility of RAG Data Sources

RAG systems are designed to interact with a wide array of data sources, including

- **Unstructured Text**: PDFs, articles, scraped websites, and codebases

- **Multimedia Content**: Podcasts, videos, and images

- **Dynamic Data**: Live search results and structured databases

This book focuses on RAG implementations that work with stored, unstructured data—such as documents and web pages—retrieved through **vector search**. These types of data are prevalent in enterprise applications and serve as a foundation for understanding the broader potential of RAG systems.

Vector Search: The Engine Behind RAG

At the heart of RAG lies **vector search**, a retrieval technique that uses dense embeddings to find semantically similar information in a vector database. Unlike traditional keyword-based searches, vector search identifies the meaning behind a query, enabling more contextually relevant results.

For example, when searching for "customer feedback strategies," a keyword-based system might only return documents containing those exact words. In contrast, vector search can retrieve information related to "customer experience improvement" or "client satisfaction frameworks," reflecting a deeper understanding of the query's intent.

This retrieval capability allows a RAG system to ground its generative responses in accurate, contextually aligned data. In the chapters, we will explore how vector search integrates seamlessly with LLMs and prompts, forming a cohesive framework for building sophisticated, real-world RAG solutions (see Figure 2-1).

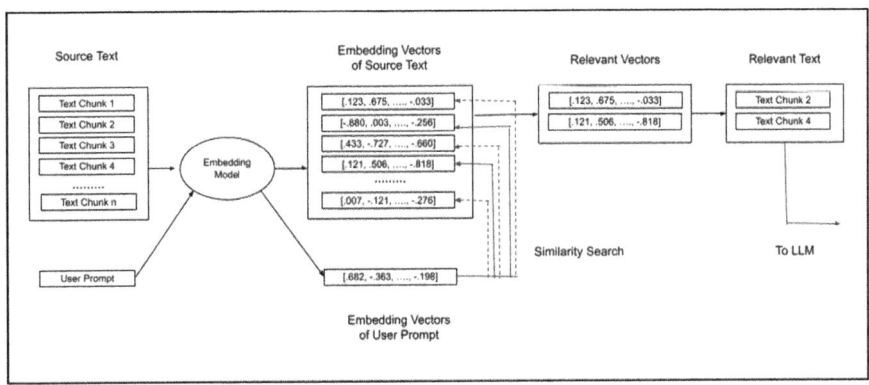

Figure 2-1. Vector search and embedding models

Basics of LLMs

Large Language Models (LLMs) represent a sophisticated class of generative AI designed to understand, process, and produce human-like text. These models are built on advanced deep learning architectures, primarily transformers, which enable them to capture complex relationships between words, phrases, and concepts across large datasets. LLMs have revolutionized natural language processing (NLP) by learning from massive amounts of text data, allowing them to generate fluent, coherent, and contextually appropriate text.

Key Concepts of LLMs

Transformer Architecture: The advent of the transformer architecture marked a critical breakthrough in NLP. Before transformers, models like recurrent neural networks (RNNs) were used for processing sequential data, but they struggled with long-range dependencies and context retention. Transformers, however, process entire sentences or documents simultaneously, enabling them to capture contextual relationships and dependencies in the whole text. This parallel processing capability allows transformers to handle large volumes of data more efficiently and accurately.

The core mechanism that powers transformers is self-attention, which enables the model to determine the importance of each word relative to others in each sentence. This self-attention mechanism helps the model disambiguate meanings, understand the context of words, and generate fluent text that mirrors human language.

Pre-training and Fine-Tuning: LLMs are typically pre-trained on vast datasets, including books, articles, websites, and other forms of textual data. During pre-training, the model learns the statistical properties of language, understanding word usage, grammar, and sentence structures. However, pre-training alone does not make an LLM effective for domain-specific tasks. To specialize, models undergo fine-tuning, training on more targeted datasets specific to a domain, such as medical, legal, or technical content. This enables LLMs to better respond to specialized queries, like medical diagnoses or legal interpretations.

Scale and Performance: The term "large" in large language models refer to the immense number of parameters—often billions or trillions—embedded within the model. These parameters are the internal variables that the model adjusts during training to learn language patterns. A more significant number of parameters allows an LLM to process more data, resulting in a deeper and more nuanced understanding of language.

However, the benefits of a larger model come with a cost: increased computational resources are required both during training and when deploying the model in production. The larger the model, the more computationally intensive its operations become, requiring substantial hardware and energy resources.

LLMs and Prompts in RAG

To use LLMs effectively in Retrieval-Augmented Generation (RAG) applications, it's essential to understand how prompts interact with these models. A prompt is the user input that guides the LLM in generating a response. Prompts vary greatly depending on the task, from completing a sentence to answering specific questions. In a typical LLM setup, a prompt is the starting point that dictates the model's output.

For example:

- **Text Completion**: If the prompt is "Jack and Jill went up the hill to…," the LLM continues the sentence with text that logically follows.

- **Question-Answering**: If the prompt is a question like "What happened to Jack after Jack and Jill went up the hill?" the LLM generates a response based on its internal understanding and learned context.

In RAG applications, the prompt provided by the user is often enhanced by incorporating additional context from external sources. In this approach, the LLM doesn't rely solely on its pre-trained knowledge. Instead, it combines the user's query with relevant, externally retrieved information, creating an "augmented" prompt. This enables the LLM to generate more accurate, contextually appropriate, and timely responses, even on topics it wasn't directly trained on.

LLMs can typically process prompts of several paragraphs, which makes them well-suited for tasks where the user needs detailed or complex answers. The prompt size in RAG can increase as external information is retrieved and appended to the original query. This augmentation is crucial for enhancing the performance of LLMs, ensuring they provide informed answers grounded in real-world data.

As we move forward, we'll examine where this external information is retrieved and how it works in synergy with LLMs to produce more reliable and accurate outputs.

Example of RAG in Action

To better understand the benefits of Retrieval-Augmented Generation (RAG), let's walk through two examples of how a question can be handled differently—one with an LLM without RAG, and the other with RAG, where the LLM has access to external documentation.

Example 1: Asking an LLM Without RAG

Imagine a user asks an LLM

"What is the process for filing a patent in the US?"

Without RAG, the LLM responds based on the knowledge it has learned from its training data. This means the LLM understands the patent filing process but cannot access real-time or specific data. Here's a typical response:

LLM Response (No RAG):
"Filing a patent in the US involves several steps. First, you must conduct a patent search to ensure your invention is unique. Then, a patent application will be prepared and filed with the United States Patent and Trademark Office (USPTO). After filing, the application undergoes an examination process, and if approved, the patent is granted."

This response gives a decent overview of the process. Still, it lacks crucial details, such as recent changes in patent law, specific filing procedures, or nuances regarding various types of patents (e.g., utility, design, etc.). It also doesn't address the user's specific context—whether they're filing as an individual inventor, a company, or an attorney. The model's limitations come from its inability to access specific, up-to-date information beyond what it was trained on.

Example 2: Using RAG to Give an LLM Access to Documentation

Now, let's consider the same question, but this time, the LLM is augmented with relevant external documentation using an RAG system. The system retrieves real-time information from sources like the USPTO guidelines, updated legal documents, or internal company procedures for filing patents. Here's how the response improves when RAG is involved:

Retrieval Step:
The RAG system first pulls in relevant documents from external sources, such as the most recent USPTO guidelines (Document ID #01234), legal updates related to patent law (e.g., changes effective January 2024), and company-specific internal processes regarding patent filing.

LLM Response (Augmented with RAG):
"To file a patent in the US, first ensure your invention meets the eligibility requirements under 35 U.S.C. § 101. Conduct a patent search using the USPTO's online database, or consider hiring a patent professional for a more comprehensive search. Prepare your patent application, including claims, drawings, and a description of the invention, adhering to the USPTO guidelines outlined in Document ID #01234. You can file the application electronically through the USPTO's EFS-Web system. The application will then be assigned to an examiner. You may receive an office action requiring amendments depending on the examiner's feedback. Once all requirements are satisfied, the patent is granted. It's important to note that recent changes in patent law (effective January 2024) have modified the examination procedures for AI-related inventions, which could impact the filing process for technology-based patents."

This augmented response is far richer and more specific. It references the latest legal changes, includes exact document IDs, and highlights the steps the user needs to take according to updated guidelines. The system has provided not only a general overview but also detailed, up-to-date information tailored to the user's needs.

By leveraging RAG, the LLM has transformed from a model that merely provides generic information to one that delivers accurate, actionable, and contextually aware responses, making it a more effective tool for users seeking precise information.

In this example, RAG enhances the LLM's ability to answer specialized questions by supplementing its training with real-time data and external documentation. It ensures that the LLM not only generates fluent responses but also provides answers grounded in the latest, most relevant information, reducing the chances of hallucinations and improving the quality of the response.

How RAG Leverages LLMs for Better Performance

The core innovation of Retrieval-Augmented Generation (RAG) lies in combining the generative strengths of Large Language Models (LLMs) with the precision of information retrieval systems. While LLMs alone are impressive in their fluency and coherence, they often face challenges when it comes to accuracy, especially when it comes to providing up-to-date or domain-specific information. RAG addresses these challenges by augmenting LLMs with external, real-time data retrieval, ensuring contextually accurate and factually grounded. Below, we explore how RAG enhances the capabilities of LLMs to deliver superior performance.

Augmenting Knowledge with Retrieval

Traditional LLMs are limited by the knowledge they have been trained on, which means they can only generate responses based on the patterns and information within their training data. As a result, they may provide outdated, incorrect, or shallow responses to specific queries, particularly when asked about rapidly changing fields or niche topics. For instance, an LLM trained in publicly available information may struggle with proprietary or domain-specific inquiries.

RAG improves on this limitation by introducing a retrieval mechanism. When a user query is received, RAG first uses a retrieval module to search through a database, document repository, or knowledge base to find the most relevant and up-to-date pieces of information. The retrieved data is then passed to the LLM, which uses this context to generate an informed, appropriate response.

Example: If a user asks a RAG system about the latest trends in renewable energy, the retrieval module might access a repository of recent industry reports, articles, or even real-time data feeds on energy markets. This ensures that the LLM's generated response is based on the most current knowledge available, as opposed to outdated or generalized training data.

Context-Aware Generation

While LLMs are skilled at generating fluent, coherent responses, they often "hallucinate" facts, generating answers that may sound plausible but are inaccurate. This is particularly problematic when dealing with specialized topics or recent events on which the model was not trained.

RAG minimizes this risk by supplying the LLM with additional, relevant context retrieved from reliable external sources. This grounding in external data allows the LLM to generate responses that are not only linguistically coherent but also factually accurate. Integrating external knowledge makes the LLM less likely to hallucinate or generate misleading responses.

Example: Consider the question, *"What is the process for filing a patent in the US?"* Without RAG, an LLM might provide a general overview of the patent filing process, which could be outdated or overly simplified. However, with RAG, the system retrieves up-to-date legal documentation, patent office guidelines, and recent regulatory changes, allowing the LLM to craft a more accurate and nuanced response.

Adaptability Across Domains

One limitation of traditional LLMs is their reliance on fine-tuning for specific domains. For instance, to perform well in a field like medicine, the LLM would need to be trained on a specialized medical corpus, which can be resource-intensive. RAG provides a more flexible solution. By keeping the LLM general purpose and integrating retrieval capabilities, RAG allows the system to adapt dynamically to various domains by pulling in relevant data when needed.

This adaptability is particularly valuable in enterprise settings where queries may span multiple domains, such as finance, healthcare, or law. Rather than fine-tuning an LLM for each specific area, a RAG system can seamlessly integrate specialized knowledge from diverse sources— whether a legal database, medical journal, or financial report—into a single query.

Example: A user asking a question about a patent filing might need different kinds of expertise depending on the context. If the question pertains to medical patents, the retrieval module might access a medical literature database, while a question about technological patents might prompt the retrieval of legal and regulatory documents. This flexibility enables RAG to handle complex, multidomain queries with ease.

Scalability and Efficiency

Training large-scale LLMs requires immense computational resources, which can be a barrier for many organizations. In contrast, RAG offers a more efficient approach by combining smaller, specialized LLMs with a powerful retrieval mechanism. Rather than relying on the LLM to store and process all knowledge internally, RAG systems retrieve relevant data as needed, reducing the computational load.

Furthermore, RAG systems can be more easily maintained and updated. Instead of retraining the entire model each time new information becomes available, organizations can update the retrieval database. This makes RAG systems more scalable, efficient, and adaptable to evolving data.

Example: A company maintaining an internal knowledge base might regularly update its policies, procedures, and product catalogs. With RAG, the system doesn't need to retrain the LLM every time a new product is introduced; instead, it can pull the latest data from the knowledge base when necessary, allowing for faster updates and a more efficient process.

Enhanced User Experience

The true power of LLMs lies in their ability to generate text that feels natural and intuitive, enabling more user-friendly interactions. RAG takes this a step further by ensuring that these interactions are conversational but also accurate and relevant. Users get the best of both worlds: the fluidity and conversational style of a generative model, combined with the precision and relevance of a retrieval system.

For example, whether a user is researching a complex scientific topic, asking for specific legal advice, or seeking up-to-date market trends, RAG allows the LLM to generate responses grounded in real-world, authoritative data. This makes the overall user experience more reliable and valuable, as users can confidently rely on the AI to deliver the most accurate and tailored information.

Conclusion

While Large Language Models (LLMs) have revolutionized AI by enabling more natural language understanding and generation, they have limitations, especially in terms of accuracy, timeliness, and domain-specific knowledge. Retrieval-augmented generation (RAG) builds upon these strengths by incorporating an external retrieval mechanism that allows LLMs to access real-time, specialized data. This combination results in more reliable, adaptable, and scalable AI systems, capable of delivering contextually aware, precise, and relevant responses across various domains.

By augmenting LLMs with retrieval technology, RAG systems offer an intelligent solution that mitigates the shortcomings of traditional models, making them more practical for real-world applications. The synergy between generative and retrieval capabilities marks a transformative leap in natural language processing, setting the stage for the next generation of AI systems that are both powerful and accurate.

RAG with Vector Search: A Step-by-Step Guide

Retrieval-Augmented Generation (RAG) with Vector Search integrates the capabilities of a large language model (LLM) with external data stored in a vector database. This approach ensures responses are enriched with relevant context retrieved from the database, making the system more accurate and practical. This section provides a detailed walkthrough of the foundational steps in implementing RAG with Vector Search.

Data Preparation: Importing External Data into a Vector Database

The first step in implementing RAG is importing and organizing the required data, typically unstructured text, into a vector database. This preparation ensures the data is ready for efficient retrieval. RAG systems benefit from the ability to update the vector database in real-time, allowing for the use of the latest information without retraining the LLM (see Figure 2-2). Below are the key steps involved:

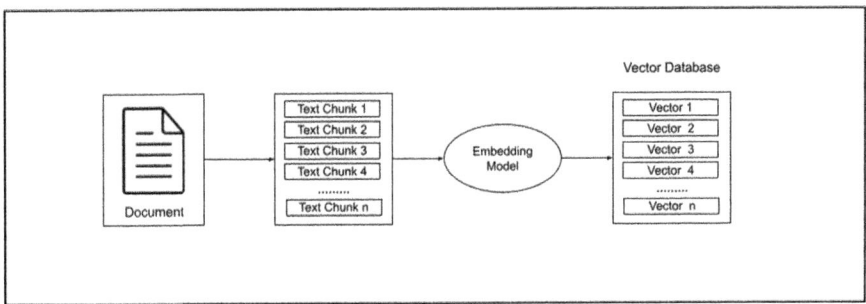

Figure 2-2. *Storing external data into a vector database*

Parsing Input Documents

Raw data, such as PDFs, web pages, or scanned documents, often requires preprocessing. Extraneous elements like headers, footnotes, or watermarks should be cleaned, and any non-text components, such as images or tables, must be converted into text where applicable. For example, text extracted from a technical manual might include irrelevant formatting that needs removal to ensure clarity and usability. The objective is to distill the content into meaningful text that can flow seamlessly into the next steps.

Splitting Documents into Chunks

Instead of working with entire documents, breaking them into smaller, manageable text chunks enables the system to retrieve precise and relevant information. For instance, a lengthy policy document could be split into sections, each addressing specific topics such as eligibility criteria or benefits. The ideal chunk size depends on the nature of the text and the LLM's context-processing capabilities. Small chunks may miss the necessary context, while overly large chunks could overwhelm the retrieval and generation processes. Experimentation is crucial to strike the right balance for a given use case.

Embedding the Text Chunks

Once split, each chunk is processed through an embedding model, which transforms the text into a high-dimensional vector—a numerical representation that encapsulates its semantic meaning. For example, the phrases "sunny weather" and "clear skies" would produce vectors with similar values due to their related meanings. These embeddings allow mathematical comparison, enabling the system to identify the most relevant text chunks in response to a user query.

Storing and Indexing the Embeddings

The generated embeddings are stored in a vector database optimized for rapid similarity searches. A vector index is typically created to organize the embeddings, ensuring quick and scalable retrieval even with extensive datasets. For example, if a customer support database contains thousands of FAQ entries, the index helps the system efficiently retrieve relevant chunks without combing through all entries manually.

Recording Metadata

Alongside embeddings, metadata such as document titles, publication dates, or source URLs can be stored. This information facilitates better filtering during retrieval and provides additional context for the results. For instance, metadata could allow users to prioritize recent articles or specify sources from trusted publications.

Example: Preprocessing and Embedding

Imagine a scenario where you're creating a knowledge base for an e-learning platform.

1. **Parsing the Input Documents**: Extract text from lecture notes and clean unnecessary formatting, such as headers or page numbers.

2. **Splitting the Documents**: Break the notes into smaller chunks, each covering a single topic, like "Introduction to Quantum Mechanics" or "Applications of Machine Learning."

3. **Embedding the Chunks**: Use an embedding model to convert each chunk into a numerical vector, capturing its meaning in mathematical form.

4. **Storing and Indexing**: Save the embeddings in a vector database, organizing them with a vector index for efficient search. Alongside each vector, metadata such as the lecture date and subject name are stored.

With the data prepared and indexed, the system is ready for the retrieval phase, where the most relevant information will be pulled based on a user query.

By completing these steps, your RAG system is equipped to handle user inputs effectively. The following section will explore how this data is queried and utilized to generate meaningful responses, bridging the gap between retrieval and generation.

Retrieval: Extracting Relevant Context

Once the vector database is populated with text chunks, embeddings, and metadata, the Retrieval phase forms the core of the Retrieval-Augmented Generation (RAG) workflow. This step ensures the system identifies and extracts the most relevant information from the database to enhance the user's input before sending it to the language model (LLM) (see Figure 2-3).

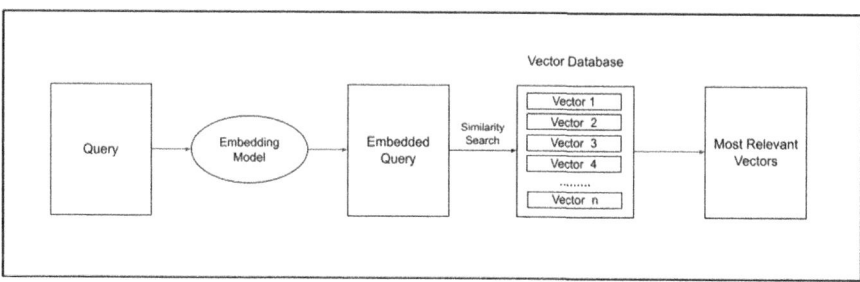

Figure 2-3. *Retrieving relevant context*

Querying the Vector Database

User queries, typically written in plain text, must first be transformed into the same embedding format as the stored data. This involves processing the query through the same embedding model used during data preparation and converting it into a high-dimensional vector representation.

With the embedded query ready, the system searches the vector database for embeddings most like the query vector. In smaller datasets, this process can involve direct comparisons where a similarity score is calculated for each embedding. Optimized methods like approximate nearest neighbor (ANN) algorithms enable faster retrieval while maintaining accuracy for larger datasets.

The database then returns the most relevant text chunks and associated metadata. These embeddings act as a bridge between the query and the stored data, eliminating the need to "translate" them back into text manually.

One crucial aspect of retrieval is determining how many results to return—retrieving too few chunks risks missing vital information, while too many can overwhelm the LLM, leading to less accurate responses. Fine-tuning this parameter for specific applications is critical to achieving a balance between relevance and focus.

Enhancing Retrieval Accuracy

Basic vector search is effective but can sometimes miss nuances or domain-specific information. To improve accuracy, several advanced techniques can be employed:

- **Hybrid Search**: Combines vector similarity with keyword-based search to handle cases where semantic and exact matches are relevant

- **Reranking**: Applies a secondary machine learning model to reorder the results, prioritizing the most contextually relevant chunks

- **Contextual Chunk Retrieval**: Includes adjacent chunks (e.g., preceding and following paragraphs) to provide a more comprehensive context

- **Prompt Refinement**: Automatically adjusts the user's query to better align with the database's contents, improving retrieval precision

- **Domain-Specific Embeddings**: Uses embedding models fine-tuned for specific industries or tasks, enhancing accuracy for niche applications

These enhancements ensure the retrieval phase captures the most relevant results, even for complex queries or specialized domains.

Example: Retrieving Relevant Information

Imagine a scenario where the vector database contains customer service documentation for an online store. If a user asks, *"How do I track my order?"* the retrieval process would follow these steps:

1. **Embed the Query**: The query is converted into a numerical vector using the same embedding model applied to the database.

2. **Search the Database**: The system identifies the top two most similar text chunks, such as sections about "tracking shipments" and "order status updates."

3. **Return Relevant Chunks**: The retrieved results might include specific details, like a step-by-step guide to using the order tracking tool, along with metadata such as the document's title and update date.

With this information retrieved, the system moves to the next phase: augmenting the user's prompt.

Augmentation: Enriching the User's Prompt

The retrieval phase provides the necessary context to help the LLM generate a well-informed response. The augmentation phase combines this retrieved data with the user's original query to form a structured prompt, maximizing the LLM's effectiveness.

Constructing the Augmented Prompt

At its simplest, augmentation involves appending the retrieved chunks to the user's query. However, this combination's structure significantly affects the generated response's quality. A typical augmented prompt includes:

1. **Retrieved Context**: The text chunks were retrieved during the previous phase.

2. **Instructions**: Guidance for the LLM, such as "answer only based on the provided context."

3. **User's Query**: The original question is often strategically placed to leverage the LLM's strengths.

For example, a prompt might instruct the LLM to focus strictly on the retrieved context, ensuring it doesn't rely on potentially incorrect prior training data (see Figure 2-4).

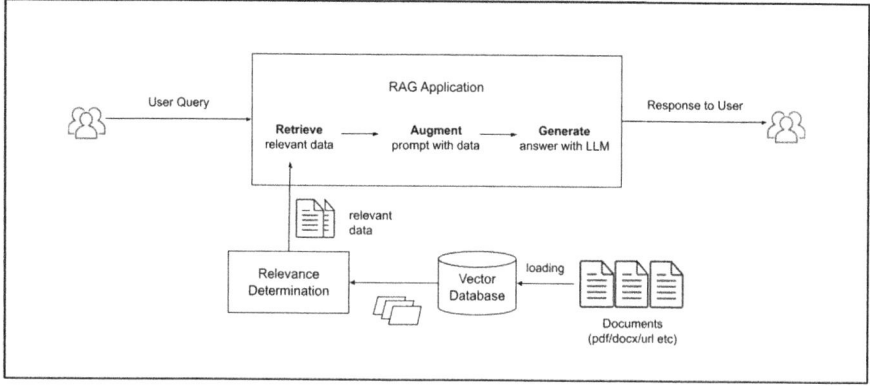

Figure 2-4. Enriching the user's prompt

Managing the Context Window

Large language models have a fixed "context window," defining the maximum text they can process in a single query. Designing the augmented prompt must account for this limitation. Including too much information risks overwhelming the model, leading to errors or incomplete outputs.

Even models with extended context windows may exhibit phenomena like the "lost in the middle" effect, where information in the middle of the context receives less weight. To address this, prioritize and organize retrieved chunks based on relevance and importance.

Example: Prompt Augmentation

Suppose a user asks the RAG system, *"How do I track my order?"* After retrieving two relevant chunks from the vector database, the augmented prompt might look like this:

- **Retrieved Context:**

  "Orders can be tracked via the 'My Account' section by navigating to 'Order History' and selecting the relevant order ID. Updates are displayed in real-time."

- **User's Question:**

  "How do I track my order?"

- **Instructions:**

  "Use the above context to explain how a customer can track their order."

This structured prompt ensures that the LLM has both the user's input and the necessary context to generate a precise response.

With the retrieval and augmentation steps complete, the system is ready to pass the enhanced prompt to the LLM, enabling it to produce an informed and contextually accurate output. These carefully designed processes ensure the RAG system delivers reliable, domain-specific results tailored to user queries.

Generation: Crafting Useful Outputs with an LLM

Once the retrieval and augmentation phases are complete, the system sends the enriched prompt containing relevant text and instructions to the language model (LLM). The generation phase is where the final output is produced, and it plays a pivotal role in shaping the user's experience. By applying various techniques, you can tailor the output to meet specific needs and ensure that it delivers value to the user (see Figure 2-5).

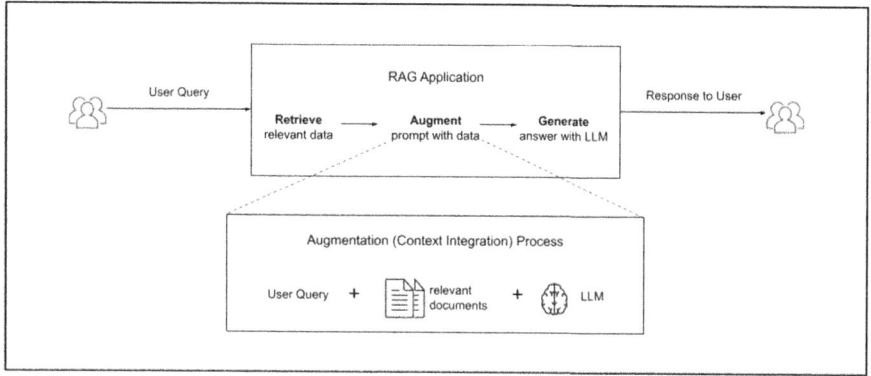

Figure 2-5. Generating outputs with LLM

Prompt Engineering

The structure of the augmented prompt is crucial in guiding the LLM's behavior. The way instructions are framed within the prompt can significantly impact the quality and relevance of the generated response. For example:

- **Context-Specific Guidance**: The LLM can be instructed to base its response solely on the provided context, minimizing the chances of introducing inaccurate or speculative information. For instance, the prompt might say, *"Answer the question using only the given context. Do not rely on external knowledge."*

- **Tone and Style**: The tone of the output can be adjusted depending on the intended audience. For example, if the LLM acts as a customer service assistant, the prompt might instruct it to respond politely and empathetically. In contrast, a technical assistant might be told to keep answers clear, concise, and precise.

- **Exclusion Criteria**: The LLM can also be instructed to avoid specific topics that do not align with business guidelines. For example, the prompt could specify, *"Do not discuss competitive products or services."*

By experimenting with different prompts, you can fine-tune the LLM's responses, ensuring that they meet your specific requirements.

Preprocessing and Postprocessing

While LLMs generate coherent and relevant outputs, they are not perfect. To ensure that the responses meet specific standards, preprocessing and postprocessing steps can be used:

- **Preprocessing**: Before the query reaches the LLM, the system can automatically enhance the prompt. For example, if the system detects ambiguity in the user's query, it could rephrase or clarify the request to help the LLM generate a more accurate response.

- **Postprocessing**: After the LLM generates a response, additional programmatic steps can refine the output. For example:

- **Appending References**: The system can add a list of source links to ensure transparency and provide users access to the original data.

- **Filtering Sensitive Information**: The system can automatically filter out sensitive or irrelevant content based on predefined rules.

- **Fallback Responses**: If the LLM encounters an unanswerable query, a fallback message (e.g., *"I am unable to provide an answer at this moment"*) can be generated automatically.

These processes help maintain the quality and reliability of the generated outputs.

Managing Conversation History

In advanced RAG systems, multiturn conversations are a key feature. This means the system can recall prior interactions, enabling users to ask follow-up questions or refine their queries based on previous responses. While this feature enhances user experience, it introduces challenges that must be carefully managed:

- **Context Management**: Storing conversation history takes up part of the LLM's context window, which can limit the space available for new queries and retrieved information. To manage this, the system may need to prioritize or summarize previous interactions to preserve space for new data.

- **Consistency**: It's essential to maintain coherence across multiple interactions. The system must reference past responses appropriately while remaining focused on the current query.

- **Framework-Specific Solutions**: Different systems handle conversation histories in various ways. Some may embed the entire history in every query, while others use external memory modules to retain only the relevant context, ensuring that the LLM's context window is not overloaded.

While multiturn conversations add complexity, they benefit applications such as virtual assistants, customer support bots, or personalized help systems.

Example: Customizing Outputs in Practice

Let's consider a scenario where a user asks for information about a product:

- The retrieved context contains information about the product's key features.

- The augmented prompt might include instructions like

  - *"Explain the features of this product in simple terms suitable for a general audience."*

  - *"Do not include proprietary or technical information that could compromise confidentiality."*

- Preprocessing steps could adjust an ambiguous query, such as rephrasing *"What does it do?"* to *"What are the key benefits of this product for end users?"*

- Postprocessing might add a disclaimer like: *"For more detailed information, visit the official product page."*

By applying prompt engineering, preprocessing, and postprocessing techniques, the system generates a response that is tailored to the audience, accurate, and compliant with business policies.

Through careful design and customization, the generation phase takes the retrieved data and transforms it into actionable, user-friendly responses, ensuring the success of the RAG system in meeting user needs.

Example: The RAG System in Action

After the retrieval and augmentation phases, the enriched prompt, now containing relevant context and clear instructions, is passed to the language model (LLM) for processing. Here's an example of how the system generates a response based on the augmented prompt:

User Query:

"What are the main features of this product?"

Retrieved Context (via RAG):

"This product simplifies the process of managing software deployments by bundling all necessary components into one package, including configurations, dependencies, and other required files.""These bundles help ensure consistency across different environments, making it easier to move from testing to production without configuration issues."

Augmented Prompt Sent to the LLM:

"Using the provided information, explain the main features of this product. Focus on how it simplifies deployment and the role it plays in ensuring consistency across environments. Do not speculate beyond the given context."

LLM Response:

"This product is designed to streamline the software deployment process by consolidating all dependencies, configurations, and related files into a single deployable package. This approach ensures that the setup remains consistent across various environments, such as development, staging, and production, thereby minimizing configuration mismatches and accelerating the deployment process."

This example demonstrates how the RAG system works: it retrieves the most relevant information, combines it with the user's query, and generates a focused, context-aware response that directly addresses the user's needs. By following this structured process, the RAG system is able to deliver accurate, relevant, and concise outputs, enhancing the overall user experience.

Evaluation: Measuring Performance of RAG

Measuring the performance of a Retrieval-Augmented Generation (RAG) system is essential to ensure that both the retrieval and generation components are functioning optimally. Given the complexity of RAG applications, with multiple adjustable variables affecting retrieval accuracy and response quality, evaluating the system effectively is crucial to achieving the desired results.

The Challenge of RAG Evaluation

RAG evaluation remains an evolving area of research and experimentation. Setting up a reliable evaluation process for a RAG system is challenging, as there is no one-size-fits-all solution. Since RAG involves a series of steps—retrieval, augmentation, and generation—it's beneficial to evaluate each phase separately. Poor retrieval can result in inaccurate or incomplete responses even with a high-performing language model (LLM)correct. Conversely, a strong retrieval system cannot compensate for a weak model when it comes to generating meaningful answers.

Key Elements of RAG Evaluation

The primary goal of RAG evaluation is to measure how well the system retrieves the most relevant information and generates useful responses. Here are the main elements to assess:

1. **Prompt Quality:** The effectiveness of the prompts used in the RAG system plays a crucial role in the quality of the output. Evaluating how well the system can generate meaningful prompts from user queries is the first step in the process.

2. **Retrieval Accuracy:** This involves assessing the system's ability to identify and retrieve the most relevant records for a given query. The evaluation here focuses on whether the retrieved information is pertinent to the user's needs.

3. **Response Quality:** After retrieval, the system generates a response. Evaluating this step focuses on how well the LLM uses the retrieved context to produce a coherent, accurate, and relevant answer.

To run a proper evaluation, each evaluation prompt is passed through the RAG application, and the actual results are compared to the desired or expected outcomes. This comparison highlights areas for improvement in the retrieval and generation stages.

Evaluation Methods: Leveraging Additional LLMs

One practical approach to evaluating RAG performance is using other LLMs to assess the quality of the responses. For example, after the RAG system generates a response, an additional LLM, known as the **judge LLM**, can be used to rate how closely the generated response aligns with the context retrieved.

A judge LLM typically performs a "faithfulness" evaluation, examining both the retrieved context and the response produced by the RAG system. The judge then provides a score or rating on how well the response adheres to the information provided in the context.

This method of evaluation can help in measuring not just the accuracy of the response but also how relevant and faithful the generated output is to the data retrieved from the database.

RAGAS Evaluation Framework

A widely discussed framework for evaluating RAG systems is **RAGAS** (Retrieval-Augmented Generation Assessment Suite). This framework is specifically designed to measure the performance of RAG systems across both retrieval and generation components. RAGAS provides a structured approach to assess the quality of retrieved data, how effectively it is used in the generation process, and the overall coherence of the response.

We will explore RAGAS and its application as an evaluation framework further on in this book. It provides a more comprehensive and systematic methodology for measuring the performance of RAG systems in real-world scenarios.

Utilizing RAG with Pretraining, Fine-Tuning, and Prompt Engineering

Retrieval-Augmented Generation (RAG) is one of several strategies for customizing and enhancing the performance of Large Language Models (LLMs). Each approach—whether **pretraining, fine-tuning, prompt engineering**, or RAG—has its unique benefits and trade-offs. These trade-offs revolve around three key factors:

- **Cost**: The financial resources needed to develop and operate the system

- **Complexity**: The technical challenges and infrastructure requirements

- **Expressiveness**: The system's ability to generate domain-specific, meaningful, and actionable outputs

While these methods can stand alone, combining them often yields the best results. Let's explore how RAG integrates with and complements these approaches.

Pretraining: Full Control at a High Cost

Pretraining involves training an LLM from scratch on a large dataset. This approach offers maximum control over the model's behavior and knowledge, as every data in the training corpus is handpicked.

- **Cost**: Extremely high, as pretraining requires substantial computational power and expertise

- **Complexity**: Very high, involving the management of massive datasets and training pipelines

- **Expressiveness**: Exceptionally high, as pretraining enables the creation of models tailored to niche requirements

For example, retraining may be necessary in industries like healthcare or finance, where regulatory or ethical constraints demand the exclusion of unreliable data sources. If an organization wants a model to generate legal advice but avoid specific jargon or untrusted sources, pretraining offers fine-grained control.

Fine-Tuning: Adapting Pretrained Models

Fine-tuning adjusts a pre-trained model's weights to specialize it for a specific domain or task. Compared to pretraining, it is a more targeted and cost-effective approach.

- **Cost**: Moderate to high, depending on the size of the model and the volume of fine-tuning data

- **Complexity**: Moderate, requiring a curated dataset and technical expertise to adapt the model

- **Expressiveness**: High, as fine-tuning can significantly improve performance for specific use cases

For instance, a general-purpose LLM could be fine-tuned using a dataset of customer support conversations to improve its ability to respond empathetically and accurately in support scenarios.

RAG: Dynamic Knowledge Integration

RAG enables models to retrieve external information dynamically and incorporate it into their outputs. It shines in scenarios requiring up-to-date or domain-specific information without retraining the model.

- **Cost**: Variable, depending on the scale of the retrieval system and the model used

- **Complexity**: Medium to high, which involves setting up and maintaining a retrieval system (e.g., a vector database)

- **Expressiveness**: High, as RAG allows models to access and integrate external knowledge on demand

For example, an RAG system could retrieve real-time market trends from a database to answer financial queries, ensuring responses reflect the latest information without retraining the model.

Prompt Engineering: A Low-Cost Starting Point

Prompt engineering involves designing effective instructions to guide an LLM's outputs. It is the most straightforward and most cost-efficient approach, but it is inherently limited by the base model's capabilities.

- **Cost**: Low, as it doesn't require retraining or additional infrastructure

- **Complexity**: Low, focusing primarily on crafting precise instructions

- **Expressiveness**: Limited to the model's inherent capabilities

For example, prompts like *"Explain this topic in layman's terms for a high school audience"* can improve the relevance of output for specific user groups without additional customization.

Combining Methods: Maximizing Potential

These approaches are not mutually exclusive and often complement each other effectively:

- **RAG + Prompt Engineering**: Merging context retrieved with user queries involves prompt engineering by design. Clear augmented prompts ensure RAG systems generate accurate and contextually relevant responses.

- **RAG + Fine-Tuning**: Fine-tuning the base model used in an RAG system enhances the domain-specific understanding, enabling the system to process and respond to retrieved data.

- **RAG + Pretraining**: Combining RAG with a custom-pretrained model provides complete control over the foundational knowledge while allowing dynamic updates through retrieval.

For instance, a healthcare organization might use a pre-trained model for medical terminology, fine-tune it to excel in diagnostics, and enhance it with RAG to retrieve the latest medical research.

Let's look at Table 2-1 below for a detailed comparison of various methods used to enhance and customize LLMs:

Table 2-1. *Comparison of Methods for Customizing and Enhancing LLMs*

Method	Description	Ideal use case	Data requirements	Key benefits	Key trade-offs
Prompt engineering	Designing tailored instructions to steer LLM responses	Quick adjustments or guidance for tasks	None required	Fast setup, low cost, no training needed	Limited control over behavior and output
Retrieval-Augmented Generation (RAG)	Augmenting LLMs with external knowledge for enhanced answers	Integrating external or frequently updated information	External data repository (e.g., vector database)	Access to dynamic, context-specific data	Increased prompt size and computational load
Fine-Tuning	Customizing a pretrained LLM for domain-specific needs	Enhancing performance on specialized tasks	Curated datasets with domain-specific examples	High accuracy for specific tasks, improved adaptability	Requires labeled data and computational effort
Pretraining	Building an LLM from scratch with a custom dataset	Developing unique or highly specialized models	Massive datasets (billions of tokens or more)	Complete control over model capabilities	Extremely expensive and time-intensive

Practical Recommendations

1. **Start Simple**: Begin with prompt engineering and basic RAG setups to address your use case.

2. **Evaluate Performance**: Assess the limitations of more straightforward methods before introducing more complex approaches.

3. **Iterate and Enhance**: Use fine-tuning or pretraining when higher expressiveness or precision is required.

In the following chapters, we will explore advanced topics like evaluation frameworks, including **RAGAS**, which provides tools to measure and refine the combined effectiveness of these methods.

Summary and Next Steps

Throughout this chapter on Core Concepts of Retrieval-Augmented Generation, we've explored the fundamental building blocks that make RAG systems uniquely powerful. Starting with the limitations of stand-alone LLMs, we've examined how RAG's hybrid architecture combines the strengths of information retrieval and generative AI to deliver accurate, context-aware responses.

Key Takeaways

Let's recap the key components we've covered:

- The role of vector search in enabling semantic understanding and efficient knowledge retrieval

- How RAG's three-phase workflow (indexing, retrieval, generation) creates a dynamic knowledge system

- The advantages of combining RAG with other LLM customization approaches like fine-tuning

- Practical comparisons between RAG and traditional generative models across accuracy, adaptability, and use cases

What makes these concepts particularly valuable is their universal applicability. Whether you're working with open-source tools or enterprise systems, these core principles remain constant, forming the foundation for all RAG implementations.

Next Steps

Looking Ahead

As you continue your RAG journey, consider

- Experimenting with different embedding models to see how they affect retrieval quality

- Exploring how these core concepts apply to your specific domain or use case

- Reflecting on which components (retrieval vs. generation) might need special attention for your needs

The understanding you've gained here will be crucial as we move forward. In Chapter 3, we'll put these concepts into practice by building a complete RAG application, while later chapters will explore advanced patterns and optimization techniques. Remember that mastering RAG is an iterative process—start with these core concepts, apply them in practice, and refine your approach as you go.

Building a Retrieval-Augmented Generation (RAG) Application

This chapter will guide you through the practical implementation of a Retrieval-Augmented Generation system, focusing on building a question-answering (Q&A) application that operates on unstructured textual data. We'll start with a foundational architecture and progressively explore each component, demonstrating how they work together to create an effective RAG system.

You'll learn how to

- Design and implement a basic RAG architecture

- Process and index unstructured text data for efficient retrieval

- Create an effective retrieval mechanism

- Integrate a language model for generation

- Evaluate and optimize your RAG system's performance

© Ranajoy Bose 2025
R. Bose, *Mastering Retrieval-Augmented Generation*,
https://doi.org/10.1007/979-8-8688-1808-0_3

While we'll begin with a straightforward Q&A implementation, the principles and techniques covered here form the basis for more sophisticated RAG applications. Throughout the chapter, we'll highlight decision points where you might want to consider alternative approaches for your specific use case, and we'll point you to advanced techniques that can enhance your system's capabilities.

By the end of this chapter, you'll have a working RAG application and a solid understanding of the practical considerations involved in building production-ready RAG systems.

Technology Stack Selection

For this implementation, we'll leverage OpenAI's language models and the LangChain framework, creating a powerful foundation for our RAG system. While this combination represents just one of many possible technology stacks, it provides an excellent starting point for understanding RAG concepts due to its

- Robust documentation and community support

- Clear API interfaces

- Straightforward implementation patterns

- Production-ready capabilities

Our technology choices for this chapter include

- **LLM Provider:** OpenAI

  - Offers reliable API access to powerful language models

  - Well-documented integration patterns

  - Consistent performance characteristics

- **Framework**: LangChain

  - Provides abstracted interfaces for RAG components

  - Enables easy swapping of underlying technologies

  - Offers built-in tools for common RAG operations

Understanding LangChain: LangChain is an open-source development framework designed to simplify the creation of applications using large language models. It provides several key advantages:

- **Component Abstraction**: LangChain breaks down complex LLM applications into modular components, such as document loaders, text splitters, embeddings, vector stores, and LLM interfaces, making it easier to build and maintain RAG systems.

- **Chain Construction**: The framework allows you to create "chains"—sequences of operations that combine different components to process input and generate output. This is particularly useful in RAG applications where you need to coordinate document retrieval, context injection, and LLM generation.

- **Flexibility and Integration**: LangChain supports multiple LLM providers, embedding models, and vector stores, allowing you to experiment with different technologies without significant code changes.

Technology Flexibility: It's crucial to understand that the concepts and techniques we'll explore are transferable across different technology stacks. Whether you ultimately choose to use

- Alternative LLMs (like Anthropic's Claude, Meta's Llama, or open-source models)

- Different frameworks (such as LlamaIndex, Haystack, or custom implementations)

- Various vector stores (like Pinecone, Weaviate, FAISS, pgvector, or Milvus)

The fundamental principles of RAG remain consistent. Focus on understanding

- How components interact within the RAG pipeline

- Key design decisions and their implications

- Performance optimization strategies

- Evaluation methodologies

As you progress in your RAG journey, you can easily adapt these concepts to different technology stacks based on your specific requirements, whether they're related to cost, scalability, privacy, or other considerations.

Setup and Development Environment

The examples in this chapter can be implemented in several development environments, with Jupyter notebooks being the most convenient option due to their interactive nature and excellent support for both code execution and markdown documentation. We'll present two primary setup approaches:

Option 1: Local Development Environment. To set up your local environment, you'll need Python 3.8 or higher installed. Create a new virtual environment and install the required packages:

```
# Create and activate a virtual environment
python -m venv rag-env
source rag-env/bin/activate  # On Windows, use: rag-env\
Scripts\activate
```

```
# Install required packages
pip install jupyter
pip install langchain chromadb openai sentence-transformers
```

Option 2: Google Colab. For those preferring a cloud-based solution, Google Colab provides a zero-setup environment with

- Free GPU access (with usage limits)

- Pre-installed common ML libraries

- Persistent storage via Google Drive integration

- Easy sharing and collaboration features

To get started with Colab

1. Visit colab.research.google.com.

2. Create a new notebook.

3. Install additional required packages using: !pip install.

```
!pip install langchain chromadb openai sentence-transformers
```

Dependencies Overview. Throughout this chapter, we'll be using these key libraries:

- `langchain`: For RAG pipeline orchestration

- `chromadb`: For vector storage and similarity search

- `sentence-transformers`: For text embeddings

- `openai`: For accessing language models

Note   Keep your environment consistent throughout the chapter to avoid version conflicts. If you're using a local setup, consider creating a `requirements.txt` file to maintain dependency versions.

Preview

In this guide, we'll build a RAG application that answers questions about web content. Using the LangChain ecosystem, we'll create a powerful question-answering system in roughly 50 lines of code. Our example will demonstrate how to transform web documents into an interactive, AI-powered knowledge base, showcasing the potential of retrieval-augmented generation technologies.

```python
# Install required packages
!pip install langchain-community langchain-openai chromadb

# Importing necessary libraries for RAG application
import os  # OS interactions and environment variables

# Document loading and vector store components
from langchain_community.document_loaders import
WebBaseLoader  # Load documents from web
from langchain_community.vectorstores import Chroma
# Vector storage and retrieval
from langchain_text_splitters import
RecursiveCharacterTextSplitter  # Splitting documents
into chunks

# Embedding and language model components
from langchain_openai import OpenAIEmbeddings
# Generate vector embeddings
from langchain_openai import ChatOpenAI
# OpenAI language model

# LangChain core components for RAG pipeline
from langchain_core.prompts import ChatPromptTemplate
# Create prompt templates
```

```python
from langchain_core.output_parsers import StrOutputParser
# Parse model outputs
from langchain_core.runnables import RunnablePassthrough
# Create runnable sequences

# Set OpenAI API Key (replace with your actual key)
os.environ["OPENAI_API_KEY"] = "your-openai-api-key"

# Load documents from a webpage
loader = WebBaseLoader("https://docs.smith.langchain.com")
docs = loader.load()

# Split documents into chunks
text_splitter = RecursiveCharacterTextSplitter(chunk_size=1000,
chunk_overlap=200)
splits = text_splitter.split_documents(docs)

# Create vector store
vectorstore = Chroma.from_documents(documents=splits,
embedding=OpenAIEmbeddings())

# Create retriever
retriever = vectorstore.as_retriever()

# Define RAG prompt
prompt = ChatPromptTemplate.from_template("""
Answer the question based only on the following context:
{context}

Question: {question}
""")

# Initialize language model
llm = ChatOpenAI(model_name="gpt-3.5-turbo", temperature=0)
```

```
# Create RAG chain
rag_chain = (
    {"context": retriever, "question": RunnablePassthrough()}
    | prompt
    | llm
    | StrOutputParser()
)

# Example usage
def ask_question(question):
    return rag_chain.invoke(question)

# Demonstrate the RAG application
print("Question: What is LangSmith?")
response = ask_question("What is LangSmith?")
print("\nResponse:", response)

# Add source retrieval (new lines)
print("\nSources:")
sources = retriever.invoke("What is LangSmith?")
for i, source in enumerate(sources, 1):
    print(f"\nSource {i}:")
    print("Content:", source.page_content[:500] + "..." if
    len(source.page_content) > 500 else source.page_content)
    print("Source URL:", source.metadata.get('source', 'Unknown
    source'))
```

Detailed walkthrough

Let's go through the above code step by step to really understand what's going on.

Package Installation

```
# Install required packages
!pip install langchain-community langchain-openai chromadb
```

Package Installation Explained

This command installs three critical libraries for our Retrieval-Augmented Generation (RAG) application:

1. **langchain-community**: Provides community-contributed tools and integrations for LangChain

2. **langchain-openai**: Enables seamless integration with OpenAI's language models and embeddings

3. **chromadb**: An open-source vector database for efficient semantic search and storage

These packages form the technological foundation of our RAG system, enabling web document loading, vector embedding, and intelligent retrieval.

Key Considerations

- Ensures all necessary dependencies are available

- Prepares the environment for building our RAG application

- Uses pip for straightforward package management

Note   The !pip install syntax is specific to Jupyter notebooks and Google Colab environments.

Import Statements: Building the RAG Toolkit

Each import in our RAG application represents a critical component of our technological stack, carefully selected to simplify complex AI workflows. Let's examine how these modules collaborate to form an intelligent document retrieval and generation system.

```python
# Importing necessary libraries for RAG application
import os  # OS interactions and environment variables

# Document loading and vector store components
from langchain_community.document_loaders import
WebBaseLoader  # Load documents from web
from langchain_community.vectorstores import Chroma
# Vector storage and retrieval
from langchain_text_splitters import
RecursiveCharacterTextSplitter  # Splitting documents
into chunks

# Embedding and language model components
from langchain_openai import OpenAIEmbeddings
# Generate vector embeddings
from langchain_openai import ChatOpenAI
# OpenAI language model

# LangChain core components for RAG pipeline
from langchain_core.prompts import ChatPromptTemplate
# Create prompt templates
from langchain_core.output_parsers import StrOutputParser
# Parse model outputs
from langchain_core.runnables import RunnablePassthrough
# Create runnable sequences
```

Breaking Down the Imports

1. **os**: The foundational module for interacting with the operating system, crucial for managing environment variables like API keys.

2. **WebBaseLoader**: Our digital librarian, capable of fetching and loading web content dynamically. It transforms web pages into processable document objects.

3. **Chroma**: A vector database that allows us to store and efficiently search document embeddings. Think of it as an intelligent indexing system for our documents.

4. **RecursiveCharacterTextSplitter**: An intelligent text segmentation tool that breaks large documents into meaningful, manageable chunks while preserving context.

5. **OpenAIEmbeddings**: Converts text into dense vector representations, capturing semantic meanings and relationships between words.

6. **ChatOpenAI**: Our language model interface, providing access to OpenAI's powerful text generation capabilities.

7. **ChatPromptTemplate**: A sophisticated prompt engineering tool that structures and guides our AI's response generation. Allows us to create structured prompts that guide our AI's response generation.

8. **StrOutputParser**: Transforms complex model outputs into simple, readable strings.

9. **RunnablePassthrough**: A flexible component that allows data to pass through our processing pipeline without modification.

Each module plays a unique role in our RAG architecture, working in concert to transform raw web content into an intelligent, queryable knowledge base.

API Key Configuration: Unlocking OpenAI's Capabilities

```
# Set OpenAI API Key (replace with your actual key)
```

Understanding API Authentication

In the world of AI-powered applications, authentication is the gateway to powerful language models. By setting the **OPENAI_API_KEY** environment variable, we're establishing a secure connection to OpenAI's advanced language generation services.

Key Considerations:

- Always use a personal, valid OpenAI API key.

- Keep your API key confidential.

- Environment variables provide a secure method of storing sensitive credentials.

- Replace "your-openai-api-key" with your actual OpenAI API key.

Best Practices:

- Use environment management tools for production environments.

- Never hard-code API keys directly in source code.

- Consider using secure secret management systems for sensitive credentials.

This single line of code bridges our application with OpenAI's sophisticated language model, enabling intelligent document processing and generation.

OBTAINING YOUR OPENAI API KEY

Getting an API key from OpenAI is straightforward. Visit platform.openai.com, create an account, and navigate to the API key section. Click "Create new secret key" to generate your unique access token.

Key considerations:

- Save the key immediately (it's shown only once).

- Store securely using environment variables.

- Never share or embed directly in code.

- Monitor usage in your OpenAI dashboard.

New accounts receive initial free credits, allowing you to explore OpenAI's capabilities before setting up billing.

Indexing

Loading Documents

```
# Load documents from a webpage
loader = WebBaseLoader("https://docs.smith.langchain.com")
docs = loader.load()
```

Document Loading Explained

The **WebBaseLoader** serves as our digital research assistant, dynamically extracting web content and converting it into a format ready for AI processing.

Mechanism of Web Document Retrieval

1. **URL-Based Content Extraction:** By specifying a web URL, we instruct the loader to fetch the entire content of the webpage. This goes beyond simple text scraping, capturing the structural and contextual nuances of the document.

2. **Comprehensive Document Parsing:** The loader intelligently processes HTML, extracting not just text, but maintaining underlying document structure. It handles various web page complexities, such as nested elements, scripts, and dynamic content.

Technical Significance

- **Dynamic Knowledge Acquisition:** Enables real-time information retrieval from web sources

- **Structured Data Transformation:** Converts unstructured web content into machine-readable document objects

- **Flexible Information Gathering:** Supports diverse web document types and structures

This initial loading stage is crucial, preparing web content for subsequent processing in our RetrievalAugmented Generation pipeline.

Indexing: Splitting Documents

```
# Split documents into chunks
text_splitter = RecursiveCharacterTextSplitter(chunk_size=1000,
chunk_overlap=200)
splits = text_splitter.split_documents(docs)
```

Document Segmentation: The Art of Intelligent Text Fragmentation

Imagine a massive book that's too complex to read in one sitting. Our `RecursiveCharacterTextSplitter` is like a masterful editor, breaking this book into digestible chapters while preserving the narrative's essence.

Key Segmentation Strategies:

- **Semantic Preservation:** By using a recursive approach, we ensure that text chunks maintain contextual integrity. Each fragment is a mini-narrative that makes sense on its own.

- **Intelligent Chunking:** The 1000-character chunk size strikes a balance between granularity and comprehensive information. It's large enough to capture meaningful context, yet small enough for efficient processing.

- **Contextual Continuity:** The 200-character overlap is our narrative bridge. It ensures that no critical information is lost between chunks, allowing our AI to understand nuanced connections across text segments.

Why This Matters:

- Enables more accurate vector embeddings

- Improves retrieval precision

- Allows the AI to understand complex, multipart documents

- Prepares data for semantic search and question-answering

This segmentation transforms our raw web document into a collection of intelligently parsed text segments, ready for the next stage of our RAG journey.

Storing Documents: Creating Our Semantic Knowledge Base

One of the most crucial steps in building a RAG system is transforming our processed documents into a format that enables intelligent retrieval. Let's examine how we create a vector store that serves as the foundation for semantic search:

```
# Create vector store
vectorstore = Chroma.from_documents(documents=splits,
embedding=OpenAIEmbeddings())
```

```
# Create vector store
vectorstore = Chroma.from_documents(documents=splits,
embedding=OpenAIEmbeddings())
```

Think of this step as building a highly organized library where each piece of text is cataloged not just by its content but by its meaning. Our vector store transforms raw text into mathematical representations that capture the subtle nuances and relationships between different pieces of information.

Understanding Vector Embeddings

When we create our vector store, each document chunk undergoes a fascinating transformation. The OpenAIEmbeddings model acts like a skilled translator, converting human language into high-dimensional vectors—sequences of numbers that represent meaning in a way computers can understand and compare.

For example, when we process a chunk of text about "machine learning algorithms," the embedding model might create a vector that places this content closer to other AI-related concepts and further from unrelated topics. This mathematical representation enables our system to comprehend not only the presence of words but also their contextual relationships and semantic meaning.

The Role of ChromaDB

ChromaDB serves as our sophisticated document management system. Unlike traditional databases, which store text in rows and columns, ChromaDB creates an intelligent structure optimized for semantic similarity searches. It's comparable to having a librarian who understands not just where every book is located but also comprehends the relationships between different topics and can instantly find related information.

The vector store performs several critical functions:

First, it processes each document chunk through the embedding model, creating dense vector representations that capture semantic meaning. These vectors typically have hundreds of dimensions, allowing them to encode complex relationships between concepts.

Next, it organizes these vectors in a way that enables efficient similarity searches. When we later query our knowledge base, ChromaDB can quickly identify the most relevant content by comparing vector similarities, rather than relying on simple keyword matching.

Finally, it maintains the connection between these mathematical representations and the original text, ensuring we can retrieve the actual content when needed.

Technical Implementation Details

Our implementation makes two important choices:

1. We use **OpenAIEmbeddings()** as our embedding model because it excels at capturing semantic relationships and has been trained on a vast corpus of text, enabling nuanced understanding of content.

2. We use ChromaDB's **from_documents** method, which handles the entire pipeline of processing documents, generating embeddings, and storing them in an optimized format. This method abstracts away much of the complexity while providing a robust foundation for our RAG system.

The Impact on System Performance

The quality of our vector store directly influences how well our RAG system can understand and respond to queries. Well-structured vector representations enable

More accurate content retrieval, as the system can better understand the semantic relationship between a query and stored documents

Faster search capabilities, since ChromaDB optimizes vector comparisons for efficient similarity matching

Better handling of nuanced queries, because the embedding model captures subtle differences in meaning that might be missed by traditional keyword-based searches

As we move forward in building our RAG system, this vector store will serve as the foundation for intelligent information retrieval, enabling our system to find and utilize relevant information effectively.

Retrieval and Generation

Building Our Knowledge Base: Transforming Text into Searchable Vectors

The creation of our retriever represents a crucial moment in our RAG application's architecture. While the code appears remarkably concise, it encapsulates sophisticated information retrieval mechanisms that are worth understanding in detail:

```
# Create vector store
vectorstore = Chroma.from_documents(documents=splits,
embedding=OpenAIEmbeddings())
```

This single line of code transforms our static vector store into an active, intelligent retrieval system. To truly appreciate what's happening here, let's explore the underlying mechanics and significance of this transformation.

Understanding the Retrieval Process

When we call **as_retriever()** on our vector store, we're essentially creating a specialized search engine that operates in the realm of semantic meaning rather than simple text matching. Think of it as the difference between a librarian who only looks at book titles vs. one who understands the content and themes of every page in every book.

The retriever works through a sophisticated process:

First, when it receives a query, it transforms that question into the same high-dimensional vector space where our document embeddings reside. This transformation ensures we're comparing apples with apples— both the query and our stored documents are represented in the same semantic space.

Next, the retriever performs what is known as a similarity search, measuring the semantic distance between the query vector and all the stored document vectors. This is analogous to our librarian understanding not just what books contain the exact words from your question but which passages best capture the meaning and intent behind your query.

Finally, it returns the most relevant document chunks, ranked by their semantic similarity to the query. This means even if a passage uses different words but captures the same meaning as the query, it can still be retrieved effectively.

Customizing Retrieval Behavior

While we're using the default configuration in our example, the retriever can be fine-tuned through several parameters:

```
# Example of a more customized retriever
retriever = vectorstore.as_retriever(
    search_type="similarity",
    search_kwargs={"k": 4}
)
```

Here, **search_type="similarity"** specifies that we want to use similarity-based search, while k=4 indicates we want to retrieve the top 4 most relevant chunks for each query. This level of control allows us to balance between providing enough context for accurate answers and maintaining computational efficiency.

The Role of the Retriever in the RAG Pipeline

The retriever acts as the crucial bridge between user questions and our knowledge base. When a user asks a question, the retriever's job is to

1. Understand the semantic meaning of the question

2. Search through our vector store efficiently

3. Return the most relevant context that will help the language model generate an accurate response

Think of the retriever as a skilled research assistant who knows exactly which passages from which documents will be most helpful in answering a specific question. This is fundamentally different from traditional keyword search, as it can understand conceptual relationships and thematic similarities that might not be apparent from the words alone.

Impact on Response Quality

The quality of our retriever directly impacts the final output of our RAG system. If the retriever fails to find relevant context, even the most sophisticated language model won't be able to generate an accurate response. Conversely, when the retriever successfully identifies the most pertinent passages, it sets the stage for highly accurate and contextually grounded answers.

Crafting the Intelligent Dialog: The Art of RAG Prompting

In our RAG application, the prompt template serves as the critical interface between raw information and intelligent responses. Let's examine the carefully crafted prompt that orchestrates this interaction:

```
# Define RAG prompt
prompt = ChatPromptTemplate.from_template("""
Answer the question based only on the following context:
{context}
Question: {question}
""")
```

This seemingly simple prompt template embodies sophisticated prompt engineering principles that are essential for reliable and accurate responses. Let's break down its design and understand why each element matters.

The Anatomy of Our RAG Prompt

The prompt's structure reflects a careful balance between constraint and clarity. Consider how each component serves a specific purpose:

The opening instruction, "Answer the question based only on the following context," serves as a crucial constraint. This directive acts like placing guardrails around the language model's vast knowledge, effectively telling it: "For this task, your world of knowledge should be limited to exactly what I'm about to show you." This restriction is fundamental to RAG's promise of grounded, verifiable responses.

The placement of **{context}** before the **{question}** is intentional. By presenting the context first, we prime the language model with relevant information before it encounters the question. Think of this like giving someone reference materials to review before asking them to solve a problem—it ensures they have the necessary information fresh in their mind.

The **Question: {question}** format provides a clear demarcation between context and query. This structural clarity helps the model distinguish between its reference material (the context) and the task it needs to perform (answering the specific question).

The Power of Template Variables

Our prompt uses two template variables:

- **{context}:** This will be filled with the relevant document chunks identified by our retriever.

- **{question}:** This will contain the user's actual query.

When the prompt template is rendered, these placeholders will be replaced with actual content. For example:

```
# Example of how the prompt might be rendered
rendered_prompt = """
Answer the question based only on the following context:
The first computer mouse was invented by Douglas Engelbart in
the 1960s.
It was made of wood and had two wheels.

Question: When was the first computer mouse invented?
"""
```

The Psychology Behind the Prompt

Our prompt's design leverages several psychological principles that influence the language model's behavior:

91

The directive to use "only" the provided context creates a clear boundary for the model. This is like telling a student: "Base your essay only on the passages provided in the exam." It helps prevent the model from incorporating external knowledge that might be incorrect or irrelevant.

The simplicity and directness of the prompt reduce cognitive load and potential confusion. Rather than providing complex, multistep instructions, we've created a clear, single-purpose directive that the model can follow consistently.

Prompt As a Control Mechanism

This prompt template acts as a control mechanism in our RAG system, ensuring that

1. Responses are grounded in the provided context.

2. The model stays focused on answering the specific question.

3. Generated answers maintain consistency with our source material.

4. The system provides verifiable responses rather than hallucinated information.

The Impact of Prompt Design on Response Quality

The effectiveness of our prompt design becomes evident in the quality of responses it elicits. A well-crafted prompt like this one helps ensure that

- Answers are relevant and focused

- Responses stick to the provided context

- The model doesn't introduce unverified information

- Generated content maintains a consistent style and format

Configuring the Language Model: Selecting Our AI Brain

Now that we have our knowledge base and retrieval system in place, let's configure the component that will generate our responses. For this example implementation, we'll use GPT-3.5 Turbo, though keep in mind, this is just one of many possible choices:

```
# Initialize language model
llm = ChatOpenAI(model_name="gpt-3.5-turbo", temperature=0)
```

While this configuration may look simple, it represents an important architectural decision in our RAG system. Let's examine the key components and considerations.

Language Model Selection

For this initial implementation, we're using OpenAI's GPT-3.5 Turbo as our language model. However, it's important to note that RAG systems are model-agnostic—you can use any language model that fits your specific needs. In later chapters, we'll explore implementing RAG with various alternatives, including

- Anthropic's Claude

- Meta's Llama 2

- Open-source models like Mistral

- Enterprise-focused models like Azure OpenAI

- Local models for privacy-sensitive applications

Each model comes with its trade-offs in terms of performance, cost, and deployment requirements, which we'll discuss in detail in Chapter 7.

Understanding Temperature

Regardless of which language model you choose, the temperature parameter plays a crucial role in RAG applications:

```
temperature=0  # Maximizing precision and consistency
```

Think of temperature as a precision control for your language model. At zero, we're instructing the model to be as deterministic as possible, prioritizing accuracy and consistency over creativity. This is particularly important in RAG systems where we want the model to

1. Stick closely to the retrieved context

2. Provide consistent responses for identical queries

3. Minimize the risk of hallucination

Best Practices for Model Configuration

When configuring your language model for RAG applications, consider these key principles:

1. **Consistency**: Use a temperature of 0 (or very close to it) to ensure reliable responses.

2. **Context Window**: Choose a model with sufficient context window to handle your typical document chunks plus overhead for prompts.

3. **Cost-Performance Balance**: Start with a more economical model like GPT-3.5 Turbo, and upgrade only if needed.

Implementation Note   The code examples in this book use OpenAI's interface, but the concepts apply to any language model. The LangChain framework makes it easy to swap models by changing just a few lines of code.

Looking Ahead

In the next section, we'll see how our configured language model works together with the retrieval system and prompt template to create a complete RAG pipeline. The principles we'll discuss apply regardless of which specific language model you ultimately choose for your application.

Building the RAG Pipeline: Assembling Our Components

We've now reached a crucial stage in our implementation where we connect all our carefully crafted components into a cohesive RAG pipeline. Let's examine the code that brings everything together:

```
# Create RAG chain
rag_chain = (
    {"context": retriever, "question": RunnablePassthrough()}
    | prompt
    | llm
    | StrOutputParser()
)
```

This elegant chain construction represents the full workflow of our RAG system. Let's break down each component and understand how information flows through this pipeline.

Understanding the Chain Structure

Our RAG chain is built using LangChain's modern pipe syntax (|), which makes the data flow clear and easy to follow. Here's what happens at each stage:

1. Input Processing Stage

    ```
    {"context": retriever, "question":
    RunnablePassthrough()}
    ```

    - Creates a dictionary with two key components:

        - **Context**: Uses our retriever to fetch relevant documents

        - **Question**: Passes the original question through unchanged

    - The **RunnablePassthrough()** acts like a direct connection, passing the user's question forward without modification.

2. Prompt Integration

    ```
    | prompt
    ```

    - Takes the retrieved context and question

    - Formats them into our carefully designed prompt template

    - Prepares the input for our language model

3. Language Model Processing

    ```
    | llm
    ```

- Sends the formatted prompt to our configured language model

- Generates a response based on the provided context and question

4. Output Formatting

   | StrOutputParser()

   - Ensures the model's response is converted to a clean string format

   - Removes any unnecessary formatting or metadata

The Flow of Information

When a user asks a question, here's what happens behind the scenes:

1. The question enters the pipeline.

2. The retriever finds relevant context from our vector store.

3. The prompt template combines this context with the question.

4. The language model generates a response.

5. The output parser formats the response for the user.

Design Note   This pipeline architecture makes it easy to modify individual components without affecting the rest of the system. Want to swap in a different retriever or language model? Just change that component while keeping the overall structure intact.

Why This Design Works

The pipe-based syntax (|) not only makes our code more readable but also provides several advantages:

1. **Clarity:** The data flow is immediately apparent from the code structure.

2. **Modularity:** Each component has a single, clear responsibility.

3. **Maintainability:** Components can be tested and modified independently.

4. **Flexibility:** Easy to add or remove processing steps.

In the next section, we'll see how to use this chain to answer questions and examine some example interactions. We'll also explore how to add error handling and implement best practices for production environments.

Making Our RAG System Interactive: Creating the Query Interface

Now that we've built our sophisticated RAG pipeline, we need a clean, simple way for users to interact with it. Let's look at how we create a straightforward interface for our system:

```
# Example usage
def ask_question(question):
    return rag_chain.invoke(question)
```

This concise function encapsulates all the complexity we've built so far into a simple, user-friendly interface. While it may look basic at first glance, this function serves as the primary gateway to our RAG system's capabilities.

Understanding the Interface Design

Let's break down why this implementation is particularly effective:

1. **Simplicity**

   - **Single Parameter**: Just pass in your question.

   - **Direct Return**: Get your answer without any extra steps.

   - **No Exposed Complexity**: Users don't need to understand the underlying RAG mechanics.

2. **Functionality**

   - `rag_chain.invoke()`: Activates our entire RAG pipeline

   - Handles the complete process from question to answer

   - Manages all intermediate steps automatically

Using the Function

Here's how you might use this interface in practice:

```
# Example queries
response = ask_question("What is LangSmith?")
print(response)

response = ask_question("How can I improve my RAG system's
performance?")
print(response)
```

> **Implementation Tip**   While this simple interface works well for basic use cases, you might want to add error handling and input validation for production environments. We'll cover these enhancements in Chapter 6.

What Happens Behind the Scenes

When you call **ask_question()**, here's the sequence of events:

1. Your question is passed to the RAG chain.

2. The retriever finds relevant context from your documents.

3. The prompt template formats everything.

4. The language model generates a response.

5. The formatted answer is returned to you.

In the next section, we'll explore some practical examples of using this interface and examine the kinds of responses it generates. We'll also look at how to troubleshoot common issues and optimize response quality.

Testing Our RAG System: Running Our First Query

Now comes the exciting part—seeing our RAG system in action! Let's test our implementation with a real question:

```
# Demonstrate the RAG application
print("Question: What is LangSmith?")
response = ask_question("What is LangSmith?")
print("\nResponse:", response)
```

When this code runs, it initiates a sophisticated sequence of operations that showcase the power of our RAG architecture. The question "What is LangSmith?" triggers a chain reaction through our system, demonstrating how each component contributes to generating an informed response.

The Journey of a Query

When we execute this code, our question travels through several crucial stages. First, the retriever transforms our question into a vector embedding and searches through our vector store, identifying the most semantically relevant passages about LangSmith. These passages are then woven together with our question using our carefully designed prompt template, creating the perfect context for our language model to generate a response.

Understanding the Output

The response format we've chosen is intentionally clean and easy to read. The question appears first, clearly labeled to establish context, followed by a newline separator for visual clarity. The system's response follows, presenting synthesized information drawn directly from our source documents. This structure makes it easy to follow the question-answer flow and verify the accuracy of responses.

Best Practices for Initial Testing

When first testing your RAG system, it's crucial to start with questions you know have clear answers in your document base. This allows you to verify that all components are working correctly and helps you understand how the system handles different types of queries. Our choice of "What is LangSmith?" serves as an excellent initial test because it's specific enough to have clear information in our documentation while still requiring the system to synthesize information effectively.

The real power of this implementation becomes apparent when you observe how it handles both specific technical queries and broader conceptual questions. Unlike traditional keyword-based search systems, our RAG architecture understands the semantic meaning behind questions, allowing it to pull relevant information even when exact keyword matches aren't present.

Summary and Next Steps

Throughout this chapter on building your RAG journey, we've walked through the practical implementation of a Retrieval-Augmented Generation system, transforming abstract concepts into working code. Starting from a basic setup and progressing through each component, we've built a functional Q&A application that can intelligently process and respond to queries about our document base.

Key Takeaways

Let's recap the key components we've explored:

- Document processing and vectorization using LangChain and ChromaDB

- Creation of an efficient retrieval system for finding relevant context

- Integration with OpenAI's language models for response generation

- Implementation of a transparent system that shows its sources

- Building a simple but powerful interface for user interaction

What makes this implementation particularly valuable is its flexibility. While we used specific technologies like OpenAI's GPT-3.5 Turbo and ChromaDB, the architecture we've built can easily adapt to different language models, embedding systems, or vector stores. This modularity ensures that your RAG system can evolve alongside your needs and the rapidly advancing field of AI.

Next Steps

Looking ahead as you continue your RAG journey, consider exploring:

- Fine-tuning retrieval parameters for better context selection

- Experimenting with different language models and their impact on response quality

- Implementing more sophisticated error handling for production environments

- Adding features like streaming responses or multidocument querying

The RAG system we've built provides a solid foundation, but it's just the beginning. In the following chapters, we'll delve deeper into advanced topics, such as optimization techniques, scaling strategies, and specialized use cases, that can elevate your implementation to the next level.

Remember, building an effective RAG system is an iterative process. Begin with this basic implementation, thoroughly test it with your specific use case, and gradually refine it based on your needs and observations. The code we've explored here gives you the building blocks—how you assemble and refine them will depend on your unique requirements and goals.

PART II

Core Components

## CHAPTER 4

Document Loaders—The Gateway to Knowledge

In the previous chapter, we built a functional RAG application, providing you with hands-on experience on how these systems work in practice. Now, it's time to take a closer look under the hood and understand each component in detail. Think of this chapter as opening up the engine of a car after learning to drive it—we'll examine each part, understand its purpose, and learn how to optimize it for peak performance.

Understanding the RAG Ecosystem

Before we delve into document loaders, let's take a moment to understand the comprehensive landscape of RAG components. Picture a RAG system as a sophisticated assembly line, where each component performs a specialized task:

1. **Document Loaders**: The initial gatekeepers that bring your data into the system

2. **Text Splitters**: The precision cutters that segment your documents into optimal chunks

R. Bose, *Mastering Retrieval-Augmented Generation*,
https://doi.org/10.1007/979-8-8688-1808-0_4

3. **Embedding Models**: The translators that convert text into vector representations

4. **Vector Stores**: The intelligent librarians that organize and retrieve your knowledge

5. **Retrievers**: The research assistants that find the most relevant information

6. **Prompt Templates**: The communication experts who structure interactions with the LLM

While each of these components deserves its own detailed exploration (which we'll cover in subsequent chapters), this chapter focuses on document loaders—the crucial first step in any RAG system. As we'll see, the quality of your document loading directly impacts the performance of every component that follows.

Document Loaders: Your Gateway to Knowledge

Think of document loaders as skilled librarians who can read and understand information in any format. Just as a librarian needs different techniques to handle ancient manuscripts vs. modern digital texts, our document loaders need specialized approaches for various types of information.

In the vast ecosystem of document types, we'll focus on mastering the most commonly used and strategically essential formats. Our journey will take us through

1. **Plain Text Files**: We'll start with the fundamentals, understanding how to process simple text documents. This will establish core principles that apply across all document types.

2. **PDF Documents**: We'll tackle one of the most common yet challenging formats, learning how to extract and preserve complex document structures.

3. **Web Content**: We'll explore how to load and process

   - Web pages with their dynamic content

   - HTML documents with their structured markup

   - JSON data, increasingly common in modern web applications

4. **Structured Data**: We'll learn to handle

   - CSV files, the workhorses of data analysis

   - Directory structures, managing multiple documents efficiently

   - JSON and XML: handling hierarchical data

5. **Rich Documents**:

   - Microsoft Office files (Word, Excel, PowerPoint)

   - Markdown documents, popular in technical documentation

6. **Custom Documents**: Finally, we'll learn how to create custom loaders for specialized document types unique to your needs.

Let's begin our exploration with plain text files, the simplest yet most fundamental document type. Understanding how we handle these basic documents will give us a strong foundation for tackling more complex formats.

Loading Plain Text Documents: Starting with the Basics

When you open a book, you're interacting with text in its most fundamental form—words on a page. Plain text files represent this same pure form of information in the digital world. While they might seem simple at first glance, processing text files effectively requires careful attention to several important details.

To help you follow along with the concepts in this chapter, I've prepared a companion Jupyter notebook that you can run in Google Colab. This notebook contains all the code examples we'll discuss, along with additional testing scenarios and error handling examples. You can find it as **"RAG_Document_Loading_Examples.ipynb"** in the repository for the book. Let's explore these concepts together, using both theory and hands-on practice.

Setting Up Your Environment

First, let's set up our development environment. Open the companion notebook in Google Colab and run the first cell to install the required packages:

```
# Install required packages
!pip install chardet pandas numpy

# Import necessary libraries
import os
import chardet
from typing import List, Optional
import pandas as pd
import numpy as np
from google.colab import files
import io
```

```
# Test imports
print("Setup complete! All required packages installed.")
```

This setup provides us with the essential tools we'll need for document loading: chardet for encoding detection, pandas for data handling, and numpy for numerical operations.

Creating a Sample Document Collection

Before we dive into implementation, it's valuable to have a diverse set of documents that represent real-world scenarios. The companion notebook includes a function to create these sample documents. Let's understand what each sample demonstrates:

1. **Business Memo**: This document represents structured business content with headers, bullet points, and hierarchical information. It helps us understand how to preserve document structure during loading.

2. **Technical Documentation**: This sample includes code snippets, API endpoints, and JSON data. It demonstrates how our loader handles formatted technical content and special characters.

3. **Multilingual Content**: This document contains text in multiple languages and scripts, including Latin, Chinese, Japanese, Hindi, and Bengali characters. It tests our encoding detection and handling capabilities.

Let's create these sample documents using the following code from our notebook:

```python
def create_sample_files():
    """Creates sample files demonstrating different document
    loading scenarios"""

    # Create directory for our samples
    !mkdir -p rag_sample_data

    # Create business memo
    business_content = """
QUARTERLY BUSINESS REVIEW
Date: February 9, 2025
Department: Engineering

PERFORMANCE METRICS:
• Project completion rate: 95%
• Code quality score: 9.2/10
• Customer satisfaction: 4.8/5

ACTION ITEMS:
1. Review Q2 objectives
2. Update team KPIs
3. Schedule stakeholder meeting
"""

    with open('rag_sample_data/business_memo.txt', 'w') as f:
        f.write(business_content)

    # Create technical documentation
    technical_content = """
API DOCUMENTATION

================
```

```
Endpoint: /api/v1/documents
Method: POST

Request Format:
{
    "document_id": "string",
    "content": "string",
    "metadata": {
        "author": "string",
        "date": "ISO-8601 timestamp"
    }
}
"""

with open('rag_sample_data/technical_doc.txt', 'w') as f:
    f.write(technical_content)

# Create multilingual content
multilingual_content = """
Global Documentation Guidelines
==============================
English: Please follow the style guide
中文: 请遵守文体指南
日本語: スタイルガイドに従ってください
हिदी: कृपया स्टाइल गाइड का पालन करें
"""

with open('rag_sample_data/multilingual.txt', 'w',
    encoding='utf-8') as f:
        f.write(multilingual_content)
# Create sample files
create_sample_files()

# Verify creation
!ls -l rag_sample_data/
```

Building a Robust Text Loader

Now that we have our sample documents, let's build a text loader that can handle real-world challenges. Our implementation in the companion notebook is called **EnhancedTextLoader**. Let's examine its key components:

1. **Initialization and Configuration:** The loader accepts a file path and optional encoding specification:

```
class EnhancedTextLoader:
    """

    A comprehensive text loader designed for RAG systems.
    Handles encoding detection, metadata extraction, and
    content cleaning.
    """

    def __init__(self, file_path: str, encoding: Optional[str]
    = None):
        """Initialize the loader with a file path and optional
        encoding."""
        self.file_path = file_path
        self.encoding = encoding
        self.metadata = {}
```

2. **Encoding Detection:** Our loader automatically detects file encoding using the chardet library:

```
def detect_encoding(self) -> str:
    """

    Automatically detect file encoding.
    Returns the detected encoding string.
    """
```

```
with open(self.file_path, 'rb') as file:
    raw_data = file.read()
    result = chardet.detect(raw_data)
    return result['encoding']
```

3. **Metadata Extraction:** We extract valuable metadata that can help in document organization and retrieval:

```
def extract_metadata(self, content: str) -> dict:
    """Extract useful metadata about the document."""
    lines = content.split('\n')
    metadata = {
        'filename': os.path.basename(self.file_path),
        'file_size': os.path.getsize(self.file_path),
        'line_count': len(lines),
        'word_count': len(content.split()),
        'char_count': len(content),
        'avg_line_length': sum(len(line) for line in lines)
        / len(lines) if lines else 0,
        'has_unicode': any(ord(c) > 127 for c in content)
    }
    return metadata
```

4. **Content Cleaning:** The loader includes sophisticated cleaning routines that preserve document structure:

```
def clean_text(self, content: str) -> str:
    """Clean text while preserving document structure."""
    # Remove null bytes and other control characters
    content = content.replace('\x00', '')
```

```
# Normalize line endings
content = content.replace('\r\n', '\n').
replace('\r', '\n')

# Split into lines for processing
lines = content.splitlines()
cleaned_lines = []

for line in lines:
    # Preserve indentation
    indent = len(line) - len(line.lstrip())
    cleaned_line = line.strip()

    # Skip empty lines
    if cleaned_line:
        # Restore indentation with spaces
        cleaned_lines.append(' ' * indent +
        cleaned_line)
    else:
        cleaned_lines.append('')

return '\n'.join(cleaned_lines)
```

5. **Main Loading Method:** Here's the primary method that brings everything together:

```
def load(self) -> tuple[str, dict]:
    """
    Load and process the text file.
    Returns tuple of (cleaned_content, metadata).
    """
    try:
        # Detect encoding if not specified
        if not self.encoding:
            self.encoding = self.detect_encoding()
```

```
# Read the file
with open(self.file_path, 'r', encoding=self.
encoding) as file:
    content = file.read()

# Clean the content
cleaned_content = self.clean_text(content)

# Extract metadata
self.metadata = self.extract_metadata(cleaned_
content)

return cleaned_content, self.metadata

except Exception as e:
    print(f"Error loading file {self.file_path}:
    {str(e)}")
    return None, None
```

Testing the Implementation

Let's test our implementation with the sample documents we created:

```
def test_loader():
    """Test the EnhancedTextLoader with our sample
    documents."""
    sample_files = [
        "rag_sample_data/business_memo.txt",
        "rag_sample_data/technical_doc.txt",
        "rag_sample_data/multilingual.txt"
    ]
```

```
for file_path in sample_files:
    print(f"\nProcessing: {file_path}")
    loader = EnhancedTextLoader(file_path)
    content, metadata = loader.load()

    print("Metadata:")
    for key, value in metadata.items():
        print(f"{key}: {value}")

    print("\nFirst 100 characters of content:")
    print(content[:100])
    print("-" * 50)
# Run the test
test_loader()
```

Best Practices for Document Loading

Through our implementation and testing, we've discovered several best practices for document loading in RAG systems. Let's explore each one with concrete examples from our companion notebook:

1. **Never Assume Encoding:** Always use robust encoding detection or allow manual specification. Here's how our loader implements this principle:

```
def handle_different_encodings():
    """Demonstrate handling of different text encodings"""

    # Create test files with different encodings
    test_content = "This is a test file with special
    characters: é, ñ, 漢"
```

```
# UTF-8 encoded file
with open('rag_sample_data/utf8_test.txt', 'w',
encoding='utf-8') as f:
    f.write(test_content)

# Try loading without specifying encoding
loader = EnhancedTextLoader('rag_sample_data/utf8_test.txt')
content, metadata = loader.load()
print("Auto-detected encoding:", metadata.get('encoding',
'unknown'))

# Try loading with explicit encoding
loader_explicit = EnhancedTextLoader('rag_sample_data/utf8_
test.txt', encoding='utf-8')
content_explicit, metadata_explicit = loader_explicit.load()
print("Explicit encoding results match:", content ==
content_explicit)
```

```
# Test encoding handling
handle_different_encodings()
```

2. **Preserve Document Structure:** When cleaning text, maintain meaningful formatting and structure. Our implementation demonstrates this through careful handling of whitespace and special characters:

```
def test_structure_preservation():
    """Demonstrate how document structure is preserved"""

    # Create a test file with specific formatting
    structured_content = """
SECTION 1:
    • First bullet point
    • Second bullet point
        - Sub bullet
```

```
    SECTION 2:
        1. Numbered item
        2. Another item
            More details here
    """

    with open('rag_sample_data/structured_test.txt', 'w') as f:
        f.write(structured_content)

    # Load and verify structure preservation
    loader = EnhancedTextLoader('rag_sample_data/structured_
    test.txt')
    content, _ = loader.load()
    print("Original structure maintained:")
    print(content)

# Test structure preservation
test_structure_preservation()
```

3. **Extract Rich Metadata:** Collect as much useful information about documents as possible. Here's how we can extend our metadata extraction:

```
def demonstrate_metadata_extraction():
    """Show comprehensive metadata extraction"""

    test_files = [
        'rag_sample_data/business_memo.txt',
        'rag_sample_data/technical_doc.txt',
        'rag_sample_data/multilingual.txt'
    ]

    for file_path in test_files:
        loader = EnhancedTextLoader(file_path)
        _, metadata = loader.load()
```

```
    print(f"\nMetadata for {os.path.basename(file_path)}:")
    for key, value in metadata.items():
        print(f"{key}: {value}")

# Test metadata extraction
demonstrate_metadata_extraction()
```

4. **Handle Errors Gracefully:** In production systems,
   document loading should never crash your
   application. Let's test our error handling:

```
def test_error_handling():
    """Demonstrate robust error handling"""

    # Test with non-existent file
    loader = EnhancedTextLoader('non_existent.txt')
    content, metadata = loader.load()
    print("Non-existent file handled gracefully:", content
    is None)

    # Test with corrupt file
    with open('rag_sample_data/corrupt.txt', 'wb') as f:
        f.write(b'\x80\x81\x82\x83')  # Invalid bytes

    loader = EnhancedTextLoader('rag_sample_data/corrupt.txt')
    content, metadata = loader.load()
    print("Corrupt file handled gracefully:", content is None)

# Test error handling
test_error_handling()
```

5. **Test with Diverse Documents:** The sample
   documents in our implementation help ensure the
   loader can handle various real-world scenarios:

```python
def comprehensive_testing():
    """Run comprehensive tests with various document types"""

    # Test results container
    test_results = {
        'total_tests': 0,
        'passed': 0,
        'failed': 0,
        'issues': []
    }

    # Test all sample files
    for file_path in os.listdir('rag_sample_data'):
        test_results['total_tests'] += 1
        try:
            loader = EnhancedTextLoader(f'rag_sample_data/
            {file_path}')
            content, metadata = loader.load()
            if content and metadata:
                test_results['passed'] += 1
            else:
                test_results['failed'] += 1
                test_results['issues'].append(f"Failed to load
                {file_path}")
        except Exception as e:
            test_results['failed'] += 1
            test_results['issues'].append(f"Error processing
            {file_path}: {str(e)}")

    # Print test results
    print("\nTest Results:")
    print(f"Total Tests: {test_results['total_tests']}")
    print(f"Passed: {test_results['passed']}")
```

```
print(f"Failed: {test_results['failed']}")
if test_results['issues']:
    print("\nIssues Found:")
    for issue in test_results['issues']:
        print(f"- {issue}")

# Run comprehensive tests
comprehensive_testing()
```

Practical Exercises

To get hands-on experience with document loading, let's work through several exercises in our companion notebook. Each exercise builds upon our **EnhancedTextLoader** implementation and introduces new challenges you might encounter in real-world scenarios.

1. **Working with Document Hierarchies:** Extend the loader to handle hierarchical document structures.

2. **Enhanced Metadata Analysis:** Create additional metadata analyzers for deeper document insights.

3. **Multilingual Support:** Add support for multilingual text.

4. **Combined Challenge:** Comprehensive Document Loader.

Moving Forward

The concepts and implementation we've covered provide a solid foundation for document loading in RAG systems. In the following sections, we'll build upon these basics to handle more complex document types like PDFs and web content. The companion notebook will continue to be your practical guide for implementing and testing these advanced features.

Remember, you can always refer back to the companion notebook ("RAG_Document_Loading_Examples.ipynb") to experiment with the code and see these concepts in action. The notebook includes additional advanced usage examples and error-handling scenarios that you can use to deepen your understanding of document loading in RAG systems.

Understanding and Loading PDF Documents

After mastering text files, we're ready to tackle one of the most common yet complex document formats: PDF (Portable Document Format). In this section, we'll build a deep understanding of PDFs and learn how to effectively process them in our RAG systems.

To help you follow along with practical examples, I've prepared a companion notebook, "PDF_Processing_in_RAG_Systems.ipynb," which you can find in the book's repository. I encourage you to open this notebook in Google Colab as we explore these concepts together.

What Makes PDFs Special?

Before we dive into code, let's understand what makes PDFs different from the text files we worked with earlier. Imagine you're creating a digital version of a beautiful magazine. You want every reader to see exactly the same layout, fonts, and images, regardless of what device or software they're using. This is exactly what PDF was designed to do.

A PDF is like a digital snapshot of a document, preserving every aspect of its appearance. Think of it as a collection of layers:

1. **The Visual Layer**: This is what you see when you open a PDF—the exact arrangement of text, images, and graphics on each page.

2. **The Content Layer**: Behind the scenes, this layer contains the actual text, images, and other elements that make up the document.

3. **The Metadata Layer**: This contains information about the document itself, like who created it, when it was made, and how it should be displayed.

Let's set up our environment to work with PDFs. Open the companion notebook and run the first cell:

```
# First, install our PDF processing toolkit
!pip install pypdf pdfplumber pdf2image pytesseract pillow
reportlab

# Import necessary libraries
import os
import pypdf
import json
import pdfplumber
from reportlab.pdfgen import canvas
from reportlab.lib.pagesizes import letter

print("PDF processing environment setup complete!")
```

Types of PDFs You'll Encounter

In the real world, you'll encounter several types of PDFs, each requiring different handling approaches:

1. **Digital-Native PDFs**: Created directly from digital sources (like Microsoft Word or LaTeX)

2. **Scanned Documents**: Like digital photographs of paper documents

3. **Hybrid PDFs**: Combining both digital text and scanned images

4. **Form PDFs**: Containing interactive elements

Let's create some example PDFs to understand these different types:

```python
def create_sample_pdfs():
    """Creates sample PDFs demonstrating different document
    types"""

    # Create directory for our samples
    os.makedirs("rag_sample_data", exist_ok=True)

    # Create a simple digital PDF
    c = canvas.Canvas("rag_sample_data/digital_pdf.pdf",
    pagesize=letter)
    c.drawString(72, 800, "Digital PDF Example")
    c.drawString(72, 780, "This is a digital-native PDF with
    text content.")
    c.drawString(72, 760, "Created: February 9, 2025")

    # Add some structured content
    y_position = 700
    for i in range(1, 4):
        c.drawString(72, y_position, f"• Point {i}: Sample
        content for demonstration")
        y_position -= 20

    c.save()

    # Create a PDF with tables and structured data
    c = canvas.Canvas("rag_sample_data/structured_pdf.pdf",
    pagesize=letter)
    c.drawString(72, 800, "Structured PDF Example")

    # Add table headers
    headers = ["Product", "Q1 Sales", "Q2 Sales", "Growth"]
    y_position = 750
```

```
for idx, header in enumerate(headers):
    c.drawString(72 + idx*100, y_position, header)

# Add table data
data = [
    ["Product A", "$10,000", "$12,500", "+25%"],
    ["Product B", "$8,000", "$9,600", "+20%"],
    ["Product C", "$15,000", "$16,500", "+10%"]
]

for row in data:
    y_position -= 20
    for idx, cell in enumerate(row):
        c.drawString(72 + idx*100, y_position, cell)

    c.save()

# Create our sample PDFs
create_sample_pdfs()

# Verify creation
print("Created PDFs:")
!ls -l rag_sample_data/*.pdf
```

Building Our PDF Processing Toolkit

To work with PDFs effectively, we need several specialized tools. You'll
find these tools installed and imported in Section 1 of the companion
notebook. Let's understand what each tool does and why we need it:

1. **pypdf**: This is our Swiss Army knife for basic PDF
   operations. It can

   - Read and write PDFs

   - Extract text and metadata

- Manipulate pages
- Handle encrypted documents

2. **pdfplumber**: Think of this as a more sophisticated text extractor. It understands

   - Text positioning and layout
   - Tables and structured data
   - Character formatting

3. **pdf2image**: This tool converts PDF pages to images, which is crucial when we need to

   - Handle scanned documents
   - Process PDFs that resist normal text extraction
   - Prepare content for OCR

4. **pytesseract**: This is our OCR (Optical Character Recognition) engine. It can

   - Read text from images
   - Handle multiple languages
   - Recognize different character sets

Let's implement our PDF loader step by step, understanding each component:

```
class EnhancedPDFLoader:
    """

    A comprehensive PDF loader that can handle various types of
    PDF documents.
    It combines multiple approaches to ensure reliable text
    extraction.
    """
```

```python
def __init__(self, file_path: str, use_ocr: bool = False):
    """

    Initialize the PDF loader with configuration options.
    Args:
        file_path: Path to the PDF file
        use_ocr: Whether to use OCR for text extraction
        (helpful for scanned docs)
    """

    self.file_path = file_path
    self.use_ocr = use_ocr
    self.metadata = {}

def extract_metadata(self) -> dict:
    """

    Extract useful information about the PDF document.
    This helps us understand the document's properties
    and origin.
    """

    with open(self.file_path, 'rb') as file:
        reader = pypdf.PdfReader(file)
        info = reader.metadata

        # Collect comprehensive metadata
        metadata = {
            'title': info.get('/Title', '') if info else '',
            'author': info.get('/Author', '') if info else '',
            'creation_date': info.get('/CreationDate', '')
            if info else '',
            'page_count': len(reader.pages),
            'file_size': os.path.getsize(self.file_path),
            'is_encrypted': reader.is_encrypted
        }
        return metadata
```

Let's pause here and know what we've built so far. Our loader class begins with metadata extraction, which is like reading the "label" on our document. This information helps us understand what we're dealing with and how to process it effectively.

Next, let's add the ability to extract text while preserving the document's layout:

```python
def extract_text_with_layout(self, page: int) -> dict:
    """
    Extract text while preserving its position and
    formatting on the page.
    Args:
        page: Page number to process (0-based index)
    Returns:
        Dictionary containing text elements with their
        positions
    """
    with pdfplumber.open(self.file_path) as pdf:
        pdf_page = pdf.pages[page]

        # Extract words with their positions
        words_with_positions = pdf_page.extract_words(
            keep_blank_chars=True,
            x_tolerance=3,   # How far apart words can be
            horizontally
            y_tolerance=3    # How far apart words can be
            vertically
        )

        # Extract tables if present
        tables = pdf_page.extract_tables()
```

```
return {
    'words': words_with_positions,
    'tables': tables,
    'page_height': pdf_page.height,
    'page_width': pdf_page.width
}
```

The **extract_text_with_layout** method does more than just extract text. It maintains information about where each piece of text appears on the page. This is like creating a map of the document's content, which can be crucial for understanding its structure.

Now, let's add OCR capabilities for handling scanned documents:

```
def perform_ocr(self, page_number: int) -> str:
    """

    Use Optical Character Recognition to extract text from
    scanned pages.
    This is our fallback method when regular text
    extraction fails.
    Args:
        page_number: The page to process (0-based index)
    Returns:
        Extracted text from the page image
    """

    # Convert PDF page to image
    images = pdf2image.convert_from_path(
        self.file_path,
        first_page=page_number + 1,
        last_page=page_number + 1
    )
```

```
    # Extract text from the image using OCR
    text = pytesseract.image_to_string(
        images[0],
        lang='eng+fra+deu+spa+ita',  # Support multiple
        languages
        config='--psm 1'  # Automatic page segmentation
        with OSD
    )

    return text
```

The OCR component is like having a skilled assistant who can read and transcribe text from images. This is essential for handling scanned documents or when normal text extraction fails.

Finally, let's put everything together in our main loading method:

```
def load(self) -> tuple[list, dict]:
    """

    Process the PDF document, combining all our extraction
    methods.
    This is our main method that orchestrates the entire
    extraction process.
    Returns:
        Tuple containing:
        - List of dictionaries with page content
        - Document metadata
    """
    self.metadata = self.extract_metadata()
    pages_content = []

    with open(self.file_path, 'rb') as file:
        reader = pypdf.PdfReader(file)

        for page_num in range(len(reader.pages)):
```

```python
        # Initialize storage for this page's content
        page_content = {'page_number': page_num + 1}

        # Try normal text extraction first
        text = reader.pages[page_num].extract_text()

        # If text extraction fails or OCR is
        requested, use OCR
        if not text.strip() or self.use_ocr:
            text = self.perform_ocr(page_num)

        # Get layout information
        layout_info = self.extract_text_with_
        layout(page_num)

        # Combine all information for this page
        page_content.update({
            'text': text,
            'layout': layout_info,
            'has_images': bool(reader.pages[page_
            num].images)
        })

        pages_content.append(page_content)

    return pages_content, self.metadata
```

Let's test our complete implementation:

```python
def test_pdf_processing():
    """Test the complete PDF processing implementation"""
    # Process the digital PDF
    loader = EnhancedPDFLoader("rag_sample_data/digital_
    pdf.pdf")
    content, metadata = loader.load()
```

```python
print("Processing digital PDF:")
print("\nMetadata:")
print(json.dumps(metadata, indent=2))

print("\nFirst page content preview:")
if content:
    page = content[0]
    print(f"Text extract (first 200 chars): {page['text']}
    [:200]}")
    print(f"Page has images: {page['has_images']}")
    print(f"Layout elements: {len(page['layout']['words'])}
    words found")

# Run the test
test_pdf_processing()
```

Best Practices for PDF Processing

Through our implementation and testing, we've discovered several important practices for handling PDFs effectively:

1. **Always Start with Metadata:** Always examine document properties before processing. Our loader demonstrates this through its initial metadata extraction:

```python
def demonstrate_metadata_first():
    """Show why examining metadata first is important"""

    loader = EnhancedPDFLoader("rag_sample_data/digital_
    pdf.pdf")
    metadata = loader.extract_metadata()

    # Check important properties before processing
    print("Document Properties:")
```

```python
    print(f"- Number of pages: {metadata['page_count']}")
    print(f"- File size: {metadata['file_size']} bytes")
    print(f"- Is encrypted: {metadata['is_encrypted']}")

    # This helps determine processing approach
    if metadata['is_encrypted']:
        print("Document is encrypted - need password handling")
    if metadata['file_size'] > 10_000_000:  # 10MB
        print("Large document - consider batch processing")

# Test metadata-first approach
demonstrate_metadata_first()
```

2. **Use Multiple Extraction Methods:** Different PDFs require different approaches. Let's test various extraction methods:

```python
def test_extraction_methods():
    """Demonstrate when to use different extraction
    approaches"""

    loader = EnhancedPDFLoader("rag_sample_data/digital_
    pdf.pdf")

    # Try simple extraction first
    with open(loader.file_path, 'rb') as file:
        reader = pypdf.PdfReader(file)
        basic_text = reader.pages[0].extract_text()

    # Try layout-aware extraction
    layout_info = loader.extract_text_with_layout(0)

    print("Extraction Results Comparison:")
    print("\nBasic Extraction:")
    print(basic_text[:200])
```

```python
print("\nLayout-Aware Extraction:")
print(f"Found {len(layout_info['words'])} positioned words")
if layout_info['tables']:
    print(f"Found {len(layout_info['tables'])} tables")

# Test different extraction methods

test_extraction_methods()
```

3. **Preserve Document Structure:** When extracting text, maintain information about its position and formatting:

```python
def demonstrate_structure_preservation():
    """Show how to preserve document structure"""

    loader = EnhancedPDFLoader("rag_sample_data/structured_
    pdf.pdf")
    content, _ = loader.load()

    if content:
        page = content[0]
        layout = page['layout']

        print("Document Structure Analysis:")
        print(f"Page dimensions: {layout['page_width']}
        x{layout['page_height']}")
        print("\nText Positions:")

        # Show how words are positioned on the page
        for word in layout['words'][:5]:  # First 5 words
            print(f"Word: {word['text']}, Position:
            ({word['x0']}, {word['top']})")

# Test structure preservation
demonstrate_structure_preservation()
```

4. **Handle Errors Gracefully:** PDFs can be complex and sometimes problematic. Always include error handling:

```python
def demonstrate_error_handling():
    """Show robust error handling practices"""

    test_files = [
        "rag_sample_data/digital_pdf.pdf",   # Should work
        "non_existent.pdf",                  # Missing file
        "rag_sample_data/empty.pdf"          # Empty file
    ]

    for file_path in test_files:
        try:
            loader = EnhancedPDFLoader(file_path)
            content, metadata = loader.load()
            print(f"\nSuccessfully processed: {file_path}")
            print(f"Pages: {metadata['page_count']}")
        except FileNotFoundError:
            print(f"\nFile not found: {file_path}")
        except Exception as e:
            print(f"\nError processing {file_path}: {str(e)}")

# Test error handling
demonstrate_error_handling()
```

5. **Monitor Processing Performance:** Keep track of processing times and success rates:

```python
def monitor_performance():
    """Demonstrate performance monitoring"""
    import time

    loader = EnhancedPDFLoader("rag_sample_data/digital_
pdf.pdf")
```

```
    start_time = time.time()
    content, metadata = loader.load()
    processing_time = time.time() - start_time

    print("Processing Performance:")
    print(f"Total time: {processing_time:.2f} seconds")
    print(f"Pages processed: {metadata['page_count']}")
    print(f"Average time per page: {processing_time/
metadata['page_count']:.2f} seconds")

# Monitor processing performance
monitor_performance()
```

Practical Exercises

To get hands-on experience with PDF processing, try these exercises in the companion notebook:

1. Modify the sample PDF generation to include different fonts and text styles.

2. Add table extraction capabilities to handle structured data.

3. Implement password protection handling for encrypted PDFs.

4. Create a method to extract and process images from PDFs.

Moving Forward

The concepts and implementation we've covered provide a solid foundation for handling PDF documents in RAG systems. In the following sections, we'll explore web content loading, where we'll deal with dynamic content and different HTML structures.

Remember, you can always refer back to the companion notebook ("PDF_Processing_in_RAG_Systems.ipynb") to experiment with these concepts. The notebook includes additional examples and edge cases that will help you build confidence in handling PDF documents in your RAG applications.

Working with Web Content: The Dynamic Knowledge Layer

After mastering static documents like PDFs, we're ready to tackle the dynamic world of web content. The web presents unique challenges because it combines multiple types of content that work together to create rich, interactive experiences. To help you follow along with hands-on examples, I've prepared a companion notebook called "Web_Content_Loading_in_RAG_Systems.ipynb" that you can find in the book's repository.

Understanding Web Content Types

Before we dive into implementation, let's understand the three main types of web content we'll encounter and how they interact with each other. Imagine you're viewing a modern e-commerce website. The product listings you see are HTML documents providing structure, the interactive shopping cart uses dynamic JavaScript content, and the product data itself might come from a JSON API.

Let's explore how to effectively handle each of these content types within our RAG system. Open the companion notebook and start by setting up our web content processing environment:

```python
# First, install required packages
!pip install fastapi uvicorn beautifulsoup4 requests playwright

# Install Playwright browsers
!playwright install

# Import necessary libraries
from fastapi import FastAPI
import uvicorn
import json
from threading import Thread
import requests
from bs4 import BeautifulSoup
from playwright.async_api import async_playwright
import asyncio
import nest_asyncio
# Enable nested asyncio for Colab
nest_asyncio.apply()

print("Web processing environment setup complete!")
```

Web Pages with Dynamic Content

Dynamic web pages represent one of the biggest challenges in web content loading. Unlike static documents, these pages can change their content after they load, often using JavaScript to fetch and display information. Let's implement a dynamic content loader and test it with a simple server:

```python
# Create a simple FastAPI server with dynamic content
app = FastAPI()

@app.get("/")
async def read_root():
```

```
    return {
        "html": """
        <!DOCTYPE html>
        <html>
        <body>
            <div id="content">Loading...</div>
            <script>
                setTimeout(() => {
                    document.getElementById('content').
                    innerText = 'Loaded Content';
                }, 2000);
            </script>
        </body>
        </html>
        """
    }

# Start server in a separate thread
def run_server():
    uvicorn.run(app, host="127.0.0.1", port=8000)

server_thread = Thread(target=run_server, daemon=True)
server_thread.start()

class DynamicWebLoader:
    """
    Handles web pages that load content dynamically through
    JavaScript.
    """

    def __init__(self, url: str, wait_time: int = 5):
        self.url = url
        self.wait_time = wait_time
```

```python
async def load_dynamic_content(self) -> str:
    """
    Load a webpage and wait for dynamic content to render.
    Uses Playwright to handle JavaScript execution.
    """
    async with async_playwright() as p:
        browser = await p.chromium.launch()
        page = await browser.new_page()

        # Load the page and wait for dynamic content
        await page.goto(self.url)
        await page.wait_for_timeout(self.wait_time * 1000)

        # Extract the rendered content
        content = await page.content()
        await browser.close()

        return content

# Test function using asyncio
async def test_dynamic_loader():
    """Test the dynamic content loader"""
    loader = DynamicWebLoader("http://127.0.0.1:8000")
    content = await loader.load_dynamic_content()
    print("Dynamic content loaded:", 'Loaded Content' in
content)

# Run the test
await test_dynamic_loader()
```

When working with dynamic content, we need to consider

1. **The context of dynamic loading**

   - Our **DynamicWebLoader** class accounts for JavaScript
     execution time.

- We wait for content to be rendered before extraction.

- The companion notebook includes examples of handling different loading patterns.

2. **JavaScript-driven updates**

- Our implementation watches for dynamic DOM changes.

- We handle asynchronous content loading.

- This ensures we capture content that appears after initial page load.

3. **Real-time updates and dynamic loading**

- The loader can handle infinite scrolling and lazy loading.

- We capture content that updates periodically.

- Examples in the notebook show how to handle streaming content.

4. **Cross-origin data restrictions**

- Our implementation respects same-origin policies.

- We handle CORS and other security restrictions.

- The companion notebook includes examples of proper security handling.

The companion notebook includes examples of handling common dynamic content scenarios like infinite scrolling, lazy loading, and real-time updates.

HTML Documents with Structured Markup

HTML documents form the backbone of web content, providing structure and meaning through their markup. While they might look simple at first glance, HTML documents can contain rich semantic information that we need to preserve for our RAG system. Let's build a structured HTML processor:

```python
class StructuredHTMLProcessor:
    """

    Processes HTML documents while preserving their semantic
    structure.
    """
    def __init__(self, html_content: str):
        self.soup = BeautifulSoup(html_content, 'html.parser')

    def extract_main_article(self) -> dict:
        """Extract the main content from article tags or main
        content area."""
        article = self.soup.find('article') or self.soup.
        find('main')
        if article:
            return {
                'content': article.get_text(strip=True),
                'has_article_tag': bool(self.soup.
                find('article')),
                'word_count': len(article.get_text().split())
            }
        return {}

    def extract_heading_hierarchy(self) -> list:
        """Extract headings while preserving their hierarchical
        structure."""
```

```python
    headings = []
    for level in range(1, 7):
        for heading in self.soup.find_all(f'h{level}'):
            headings.append({
                'level': level,
                'text': heading.get_text(strip=True),
                'id': heading.get('id', ''),
                'has_links': bool(heading.find_all('a'))
            })
    return headings

def extract_lists(self) -> dict:
    """Extract ordered and unordered lists."""
    lists = {
        'ordered': [],
        'unordered': [],
        'definition': []
    }

    # Process ordered lists
    for ol in self.soup.find_all('ol'):
        lists['ordered'].append([
            item.get_text(strip=True) for item in ol.find_
            all('li')
        ])

    # Process unordered lists
    for ul in self.soup.find_all('ul'):
        lists['unordered'].append([
            item.get_text(strip=True) for item in ul.find_
            all('li')
        ])
```

```python
        # Process definition lists
        for dl in self.soup.find_all('dl'):
            defs = []
            for dt, dd in zip(dl.find_all('dt'), dl.find_
            all('dd')):
                defs.append({
                    'term': dt.get_text(strip=True),
                    'definition': dd.get_text(strip=True)
                })
            lists['definition'].append(defs)

        return lists

    def extract_metadata(self) -> dict:
        """Extract metadata from meta tags and other
        sources."""
        metadata = {
            'title': self.soup.title.string if self.soup.title
            else '',
            'meta': {},
            'links': []
        }

        # Extract meta tags
        for meta in self.soup.find_all('meta'):
            name = meta.get('name', meta.get('property', ''))
            if name:
                metadata['meta'][name] = meta.get('content', '')

        # Extract important links
        for link in self.soup.find_all('a'):
            metadata['links'].append({
                'text': link.get_text(strip=True),
```

```python
        'href': link.get('href', ''),
        'title': link.get('title', '')
    })

    return metadata

# Let's test our HTML processor with a sample document
def test_html_processor():
    """Test the structured HTML processor with a sample
    document"""

    sample_html = """
<!DOCTYPE html>
<html>
<head>
    <title>Sample Document</title>
    <meta name="description" content="A test document">
    <meta name="keywords" content="test, sample, document">
</head>
<body>
    <article>
        <h1>Main Title</h1>
        <p>This is the introduction.</p>

        <h2>First Section</h2>
        <ul>
            <li>First point</li>
            <li>Second point</li>
        </ul>

        <h2>Second Section</h2>
        <ol>
            <li>Step one</li>
            <li>Step two</li>
        </ol>
```

```
            <dl>
                <dt>Term 1</dt>
                <dd>Definition 1</dd>
                <dt>Term 2</dt>
                <dd>Definition 2</dd>
            </dl>
        </article>
    </body>
</html>
"""

# Process the HTML
processor = StructuredHTMLProcessor(sample_html)

# Extract and display different components
print("Article Content:")
print(json.dumps(processor.extract_main_article(),
indent=2))

print("\nHeading Hierarchy:")
print(json.dumps(processor.extract_heading_hierarchy(),
indent=2))

print("\nLists:")
print(json.dumps(processor.extract_lists(), indent=2))

print("\nMetadata:")
print(json.dumps(processor.extract_metadata(), indent=2))
# Run the test
test_html_processor()
```

When processing HTML documents, we need to pay special attention to

1. **Semantic markup** (article, section, nav, etc.)

   - Our **`extract_main_article`** method specifically looks for semantic tags like `<article>` and `<main>`.

   - This helps us understand the document's logical structure beyond its visual layout.

   - The companion notebook includes examples of handling different semantic elements.

2. **Heading hierarchy and document structure**

   - Our **`extract_heading_hierarchy`** method preserves the hierarchical relationship between headings.

   - We track heading levels (h1 through h6) to maintain document outline.

   - This structure is crucial for understanding the relationship between different sections of content.

3. **Lists, tables, and other organized content**

   - The **`extract_lists`** method handles different types of lists (ordered, unordered, definition).

   - We preserve the original structure and relationships between list items.

   - Similar approaches are used for other structured elements like tables and forms.

4. **Metadata and SEO elements**

   - Our **`extract_metadata`** method captures important meta tags and SEO information.

- We extract metadata like title, description, and keywords.

- This information helps in understanding the document's context and purpose.

The companion notebook includes examples of processing different HTML structures and preserving semantic relationships. These examples demonstrate how to handle various real-world scenarios you'll encounter when processing web content.

Web-Embedded JSON in Modern Web Applications

Modern web applications frequently embed JSON data within their pages or load it dynamically. This is different from stand-alone JSON files that we'll explore in our next section on structured data formats. Think of web-embedded JSON as ingredients mixed into a recipe (the web page), while stand-alone JSON files are like complete recipe books on their own.

Let's implement a processor that can handle different types of web-embedded JSON:

```python
from typing import List, Dict, Any
import json
from bs4 import BeautifulSoup

class WebEmbeddedJSONProcessor:
    """
    Handles JSON data embedded in or loaded by web pages.
    """
    def __init__(self, html_content: str):
        self.soup = BeautifulSoup(html_content, 'html.parser')
```

1. **JSON-LD (Linked Data)**: This is structured metadata embedded in HTML pages that helps search engines understand your content. Our processor extracts it like this:

```python
def extract_json_ld(self) -> List[dict]:
        """Extract JSON-LD metadata from HTML content."""
        json_ld_tags = self.soup.find_all('script',
        type='application/ld+json')
        results = []

        for tag in json_ld_tags:
            try:
                data = json.loads(tag.string)
                results.append(data)
            except json.JSONDecodeError:
                continue

        return results
```

2. **API Responses**: When a web page makes AJAX calls, it receives JSON data that updates the page content. We process these responses like this:

```python
def process_api_response(self, response_text: str) -> dict:
        """
        Process JSON from API responses.
        Handles JSON that's dynamically loaded into the page.
        """
        try:
            data = json.loads(response_text)
            return {
                'data': self.normalize_api_data(data),
                'metadata': self.extract_api_metadata(data)
            }
```

```python
        except json.JSONDecodeError:
            return {'error': 'Invalid JSON in API response'}

    def normalize_api_data(self, data: Any) -> Any:
        """Normalize API response data for consistent
        processing."""
        if isinstance(data, dict):
            return {
                key: self.normalize_api_data(value)
                for key, value in data.items()
            }
        elif isinstance(data, list):
            return [self.normalize_api_data(item) for item
            in data]
        return data

    def extract_api_metadata(self, data: dict) -> dict:
        """Extract metadata from API response."""
        metadata = {}
        if isinstance(data, dict):
            # Extract common metadata fields
            metadata = {
                'total_items': data.get('total'),
                'page': data.get('page'),
                'has_more': data.get('has_more', False),
                'timestamp': data.get('timestamp')
            }
        return metadata
```

Let's test both types of JSON processing:

```python
def test_json_processor():
    """Test processing different types of web-embedded JSON"""

    # Create a test page with embedded JSON-LD
    test_html = """
<html>
<head>
    <script type="application/ld+json">
    {
        "@context": "https://schema.org",
        "@type": "Article",
        "headline": "Understanding RAG Systems",
        "author": {
            "@type": "Person",
            "name": "John Doe"
        }
    }
    </script>
</head>
<body>
    <div id="content">Main content here</div>
</body>
</html>
    """

    # Test JSON-LD extraction
    processor = WebEmbeddedJSONProcessor(test_html)
    json_ld = processor.extract_json_ld()

    print("Extracted JSON-LD:")
    print(json.dumps(json_ld, indent=2))
```

```python
# Test API response processing
api_response = {
    "products": [
        {"id": 1, "name": "Product A", "price": 29.99},
        {"id": 2, "name": "Product B", "price": 39.99}
    ],
    "metadata": {
        "total": 2,
        "page": 1
    }
}

normalized_data = processor.process_api_response(json.
dumps(api_response))
print("\nProcessed API Response:")
print(json.dumps(normalized_data, indent=2))

# Run the test
test_json_processor()
```

When working with web-embedded JSON, we need to consider

1. **The context of the JSON within the web page**

    - How the JSON relates to the visible content

    - The purpose of the embedded data (metadata, dynamic content, etc.)

    - The relationship between different JSON blocks

2. **Different types of JSON embeddings**

    - JSON-LD for structured metadata

    - Dynamic API responses

    - Configuration and state data

3. **Data validation and normalization**

    - Ensuring JSON is well-formed

    - Handling missing or malformed data

    - Normalizing data structures for consistent processing

4. **Security considerations**

    - Validating JSON content

    - Handling cross-origin data

    - Protecting against JSON injection

The companion notebook includes examples of handling these different scenarios and best practices for processing web-embedded JSON safely and effectively.

This is different from processing stand-alone JSON files, which we'll explore in the next section on structured data formats.

Bringing It All Together

In real-world applications, you'll often need to handle all three types of web content together. Let's implement a comprehensive web content processor that combines our previous components:

```
class ComprehensiveWebProcessor:
    """

    A unified processor that handles dynamic content, HTML
    structure, and embedded JSON in web pages.
    """

    def __init__(self, url: str, wait_time: int = 5):
        self.url = url
        self.wait_time = wait_time
```

```python
        self.raw_html = None
        self.processed_content = {}

    async def process_page(self) -> dict:
        """
        Process a webpage combining all our processing
        capabilities.
        Returns a complete analysis of the page content.
        """
        # 1. Load dynamic content
        dynamic_loader = DynamicWebLoader(self.url)
        self.raw_html = await dynamic_loader.load_dynamic_
        content()

        # 2. Process HTML structure
        html_processor = StructuredHTMLProcessor(self.raw_html)
        html_structure = {
            'main_content': html_processor.extract_main_
            article(),
            'headings': html_processor.extract_heading_
            hierarchy(),
            'lists': html_processor.extract_lists(),
            'metadata': html_processor.extract_metadata()
        }

        # 3. Handle embedded JSON
        json_processor = WebEmbeddedJSONProcessor(self.
        raw_html)
        json_ld = json_processor.extract_json_ld()

        # 4. Combine all information
        self.processed_content = {
            'structural_content': html_structure,
            'embedded_json': json_ld,
```

```
        'metadata': {
            'url': self.url,
            'processing_time': self.wait_time,
            'has_dynamic_content': bool(html_structure.
            get('main_content'))
        }
    }

    return self.processed_content
```

Let's test this implementation with a page that contains all types of content:

```
async def test_comprehensive_processor():
    """Test the comprehensive web content processor"""

    # Start our test server with a complex page
    app = FastAPI()

    @app.get("/")
    async def read_root():
        return {
            "html": """
            <!DOCTYPE html>
            <html>
            <head>
                <title>Test Complex Page</title>
                <script type="application/ld+json">
                {
                    "@context": "https://schema.org",
                    "@type": "Article",
                    "headline": "Test Article"
                }
                </script>
            </head>
```

```
        <body>
            <article>
                <h1>Main Content</h1>
                <p>Static content here</p>
                <div id="dynamic-content">Loading...</div>
                <script>
                    setTimeout(() => {
                        document.getElementById('dynamic-
                        content').innerText =
                            'Dynamically Loaded Content';
                    }, 1000);
                </script>
                <ul>
                    <li>First item</li>
                    <li>Second item</li>
                </ul>
            </article>
        </body>
        </html>
        """

    }

# Start server in thread
server_thread = Thread(target=lambda: uvicorn.run(app,
host="127.0.0.1", port=8000))
server_thread.daemon = True
server_thread.start()

# Allow server to start
await asyncio.sleep(1)

# Test the processor
processor = ComprehensiveWebProcessor("http://127.0.0.1:8000")
content = await processor.process_page()
```

```
print("Comprehensive Processing Results:")
print("\n1. Structural Content:")
print(json.dumps(content['structural_content'], indent=2))

print("\n2. Embedded JSON:")
print(json.dumps(content['embedded_json'], indent=2))

print("\n3. Metadata:")
print(json.dumps(content['metadata'], indent=2))

# Run the comprehensive test
await test_comprehensive_processor()
```

Our comprehensive processor demonstrates how to

1. Load and process dynamic content

2. Extract structured HTML content

3. Handle embedded JSON data

4. Combine everything into a unified format

The companion notebook includes additional examples and test cases to help you understand how these components work together in different scenarios.

Best Practices and Common Challenges

Through our implementation and testing, we've discovered several important practices for handling web content effectively in RAG systems. The companion notebook includes practical implementations for each of these best practices:

1. **Dynamic Content Timing**

- Implement smart waiting strategies for dynamic content.

- Use configurable timeouts based on content type.

- Handle different loading patterns effectively.

- Monitor and adapt to page loading behaviors.

2. **Error Recovery**

- Implement automatic retries with exponential backoff.

- Handle network timeouts and connection issues.

- Recover from partial content loads.

- Log and track error patterns for debugging.

3. **Content Validation**

- Verify the completeness of extracted content.

- Validate structural integrity of processed data.

- Check for required metadata presence.

- Monitor content quality metrics.

4. **Performance Monitoring**

- Track processing time for different components.

- Monitor memory usage during processing.

- Identify and address performance bottlenecks.

- Collect metrics for system optimization.

These practices help ensure your web content processing is

- Reliable and robust

- Properly validated

- Performance optimized
- Easy to maintain and debug

The companion notebook includes complete implementations and examples for each of these best practices, allowing you to test and adapt them for your specific RAG system needs.

Moving Forward

As we conclude our exploration of web content processing, let's recall the key components we've developed for handling web content in RAG systems:

1. Dynamic content loading with Playwright
2. Structured HTML processing with BeautifulSoup
3. JSON data extraction and normalization
4. Comprehensive content validation and error handling

In the next section, we'll explore structured data formats like CSV, JSON, and XML files. While we've dealt with JSON in web contexts, stand-alone data files present different challenges and opportunities. You'll learn how these stand-alone data files require different processing strategies compared to their web-embedded counterparts.

Working with Structured Data: Organizing Knowledge for RAG Systems

After mastering web content, we're ready to tackle structured data formats—the organized filing cabinets of the digital world. While web content can be dynamic and complex, structured data follows specific patterns and rules that make it both powerful and predictable.

To help you follow along with practical examples, I've prepared a companion notebook called "Structured_Data_Loading_in_RAG.ipynb" that you can find in the book's repository. Let's explore how to effectively process these structured formats in your RAG system.

Understanding Structured Data Formats

Think of structured data like different types of filing systems, each designed for specific purposes:

- CSV files are like spreadsheets, organizing data in rows and columns.

- JSON files are like nested folders, containing hierarchical information.

- XML files are like annotated documents, with tags that describe their content.

- Directory structures are like file cabinets, organizing related documents together.

Let's explore how to handle each format effectively in our RAG system. Open the companion notebook and start by setting up our environment:

```
# Install required packages
!pip install pandas numpy chardet xmltodict
# Import necessary libraries
import os
import json
import pandas as pd
import numpy as np
import chardet
import xmltodict
from typing import List, Dict, Any
```

```python
from pathlib import Path
from datetime import datetime

print("Structured data processing environment setup complete!")
```

CSV Files: The Workhorses of Data Analysis

CSV (Comma-Separated Values) files might seem simple, but they're incredibly powerful for handling tabular data. Think of a CSV file like a well-organized spreadsheet where each row represents a record, and each column represents a specific attribute. Let's implement a robust CSV loader:

```python
class EnhancedCSVLoader:
    """

    A comprehensive CSV loader for RAG systems.
    Handles encoding detection, delimiter inference, and data
    validation.
    """

    def __init__(self, file_path: str):
        self.file_path = file_path
        self.metadata = {}
        self.encoding = None
        self.delimiter = None
```

This initializes our CSV loader with basic properties. The encoding and delimiter are initially None because we'll detect them automatically.

```python
def detect_file_properties(self) -> dict:
    """

    Automatically detect CSV file properties including
    encoding and delimiter.
    """
```

```python
with open(self.file_path, 'rb') as file:
    # Read a sample of the file - we only need a small
    portion to detect properties
    raw_data = file.read(10000)  # Read first 10KB

    # Use chardet to automatically detect the file's
    encoding
    result = chardet.detect(raw_data)
    self.encoding = result['encoding']

    # Find the most commonly used delimiter in the file
    sample_text = raw_data.decode(self.encoding)
    delimiters = [',', ';', '\t', '|']
    delimiter_counts = {d: sample_text.count(d) for d
    in delimiters}
    self.delimiter = max(delimiter_counts.items(),
    key=lambda x: x[1])[0]

    return {
        'encoding': self.encoding,
        'delimiter': self.delimiter,
        'confidence': result['confidence']
    }
```

This method examines the file contents to determine its properties. We read only the first 10KB for efficiency, detect the encoding using chardet, and find the most common delimiter character.

```python
def validate_data(self, df: pd.DataFrame) -> dict:
    """
    Perform validation checks on the loaded data.
    Returns a dictionary of validation results.
    """
```

```python
# Basic statistics about the dataset
validation = {
    'total_rows': len(df),
    'total_columns': len(df.columns),
    'missing_values': df.isnull().sum().to_dict(),
    'column_types': df.dtypes.astype(str).to_dict(),
    'duplicate_rows': df.duplicated().sum()
}

# Check for data quality issues
validation['warnings'] = []

# Alert if any column has more than 20% missing values
missing_percentages = (df.isnull().sum() /
len(df)) * 100
for column, pct in missing_percentages.items():
    if pct > 20:
        validation['warnings'].append(
            f"Column '{column}' has {pct:.1f}%
            missing values"
        )

# Check for columns that mix numeric and non-
numeric values
for column in df.columns:
    if df[column].dtype == 'object':
        try:
            pd.to_numeric(df[column], errors='raise')
            validation['warnings'].append(
                f"Column '{column}' contains mixed
                numeric and non-numeric values"
            )
```

```
        except:
            pass

    return validation
```

```
print(df.head())
print("\nMetadata:")
print(json.dumps(metadata, indent=2))
```

This method performs comprehensive data validation, checking for common issues like missing values and mixed data types. It helps identify potential data quality problems early.

```
def load(self) -> tuple[pd.DataFrame, dict]:
    """
    Load and process the CSV file.
    Returns both the data and metadata about the file and
    its contents.
    """
    try:
        # First detect the file's properties to ensure
        correct loading
        properties = self.detect_file_properties()

        # Load the CSV using detected encoding and delimiter
        df = pd.read_csv(
            self.file_path,
            encoding=self.encoding,
            delimiter=self.delimiter,
            on_bad_lines='warn'  # Warn instead of failing
            on problematic lines
        )

        # Run validation checks on the loaded data
        validation_results = self.validate_data(df)
```

```
        # Collect comprehensive metadata about the file
        self.metadata = {
            'file_properties': properties,
            'validation': validation_results,
            'file_size': os.path.getsize(self.file_path),
            'last_modified': os.path.getmtime(self.
            file_path)
        }

        return df, self.metadata

    except Exception as e:
        print(f"Error loading CSV: {str(e)}")
        return None, None
```

This is our main loading method that brings everything together. It first detects file properties, loads the data using pandas, validates the content, and collects metadata. The method is designed to fail gracefully if any issues occur.

```
def test_csv_loader():
    """Test the CSV loader with sample data"""

    # Create a test CSV file with various data scenarios
    sample_data = """
name,age,city,salary
John Doe,30,New York,75000
Jane Smith,25,Los Angeles,82000
Bob Johnson,,Chicago,68000
Alice Brown,35,Houston,91000.5
"""

    # Ensure we have a directory for our samples
    os.makedirs('rag_sample_data', exist_ok=True)
```

```python
# Write the sample data to a file
with open('rag_sample_data/sample.csv', 'w') as f:
    f.write(sample_data.strip())

# Test our loader with the sample file
loader = EnhancedCSVLoader('rag_sample_data/sample.csv')
df, metadata = loader.load()

# Convert numpy types to Python native types for JSON
serialization
metadata['validation']['total_rows'] =
int(metadata['validation']['total_rows'])
metadata['validation']['total_columns'] =
int(metadata['validation']['total_columns'])
metadata['validation']['duplicate_rows'] =
int(metadata['validation']['duplicate_rows'])

# Display the results
print("Loaded Data Preview:")
print(df.head())

print("\nMetadata and Validation Results:")
print(json.dumps(metadata, indent=2))

# Create sample directory and run the test
os.makedirs('rag_sample_data', exist_ok=True)
test_csv_loader()
```

This test function creates a sample CSV file with various scenarios (missing values, different data types) and demonstrates how our loader handles them. The sample data includes common cases you might encounter in real-world CSV files.

The test includes

- A missing value (Bob Johnson's age)

- Mixed numeric types (integer and decimal salaries)

- Different string formats in the city column

- Multiple columns with different data types

Directory Structures: Managing Document Collections

When working with multiple documents, organizing them effectively becomes crucial. Let's create a directory manager that can handle collections of documents:

```python
from pathlib import Path
from typing import Dict, List, Any
import os
import json
from datetime import datetime

class DocumentCollectionManager:
    """

    Manages collections of documents organized in directories.
    Supports various file types and maintains a
    searchable index.
    """

    def __init__(self, root_dir: str):
        self.root_dir = root_dir
        self.file_index = {}
        self.metadata = {}
```

This initializes our directory manager with a root directory and storage for file indexing.

```python
def _count_file_types(self) -> dict:
    """Count the number of files of each type."""
    type_counts = {}
    for file_info in self.file_index.values():
        ext = file_info['extension']
        type_counts[ext] = type_counts.get(ext, 0) + 1
    return type_counts

    def _calculate_max_depth(self) -> int:
        """Calculate the maximum directory depth."""
        max_depth = 0
        for file_path in self.file_index:
            relative_path = os.path.relpath(file_path, self.
            root_dir)
            depth = len(relative_path.split(os.sep))
            max_depth = max(max_depth, depth)
        return max_depth

    def scan_directory(self) -> dict:
        """
        Scan the directory structure and catalog all documents.
        Returns information about the document collection.
        """
        for root, dirs, files in os.walk(self.root_dir):
            for file in files:
                file_path = os.path.join(root, file)
                file_ext = os.path.splitext(file)[1].lower()

                # Get file metadata
                stats = os.stat(file_path)
```

```python
            # Store file information
            self.file_index[file_path] = {
                'extension': file_ext,
                'size': stats.st_size,
                'modified': datetime.fromtimestamp(stats.
                st_mtime).isoformat(),
                'relative_path': os.path.relpath(file_path,
                self.root_dir)
            }

        # Collect collection statistics
        self.metadata = {
            'total_files': len(self.file_index),
            'file_types': self._count_file_types(),
            'total_size': sum(f['size'] for f in self.file_
            index.values()),
            'directory_depth': self._calculate_max_depth()
        }

        return self.metadata
```

This method scans the directory structure and builds an index of all files with their metadata.

Let's test our directory manager:

```python
def test_document_manager():
    """Test the document collection manager"""

    # Create a test directory structure
    base_dir = 'rag_sample_data/documents'
    os.makedirs(f'{base_dir}/texts', exist_ok=True)
    os.makedirs(f'{base_dir}/pdfs', exist_ok=True)
```

```python
# Create some sample files
with open(f'{base_dir}/texts/doc1.txt', 'w') as f:
    f.write("Sample text document")
with open(f'{base_dir}/pdfs/doc2.pdf', 'w') as f:
    f.write("Sample PDF content")

# Test the manager
manager = DocumentCollectionManager(base_dir)
metadata = manager.scan_directory()

print("Directory Analysis:")
print(json.dumps(metadata, indent=2))

print("\nFile Index:")
for path, info in manager.file_index.items():
    print(f"\nFile: {path}")
    print(json.dumps(info, indent=2))

# Run the test
test_document_manager()
```

This implementation demonstrates how to

- Organize and track documents in a directory structure

- Extract and maintain file metadata

- Create a searchable index of documents

- Monitor directory statistics

The companion notebook includes additional examples of handling different file organizations and advanced directory operations.

JSON and XML: Handling Hierarchical Data

Now let's add support for JSON and XML files, which often contain rich, hierarchical data structures. These formats are commonly used for configuration files, API responses, and data exchange. While JSON is typically used for API responses and configuration, XML is often found in legacy systems and document-centric applications.

Let's implement a processor that can handle both formats:

```python
import json
import xmltodict
from typing import Union, Any

class HierarchicalDataLoader:
    """

    Handles both JSON and XML files with support for schema
    validation
    and data transformation.
    """

    def __init__(self, file_path: str):
        self.file_path = file_path
        self.file_type = os.path.splitext(file_path)[1].lower()
        self.metadata = {}
```

For JSON files, we need to handle nested structures and maintain the hierarchical relationships:

```python
def _load_json(self) -> tuple[dict, dict]:
    """Load and parse JSON files with error handling."""
    try:
        with open(self.file_path, 'r',
        encoding='utf-8') as f:
            data = json.load(f)
```

```python
        # Extract metadata about the JSON structure
        self.metadata = {
            'depth': self._calculate_json_depth(data),
            'keys': self._extract_json_keys(data),
            'size': os.path.getsize(self.file_path)
        }

        return data, self.metadata

    except json.JSONDecodeError as e:
        print(f"Error parsing JSON: {str(e)}")
        return None, None

def _calculate_json_depth(self, obj: Any, current_depth:
int = 1) -> int:
    """Calculate the maximum depth of a JSON object."""
    if isinstance(obj, dict):
        if not obj:
            return current_depth
        return max(
            self._calculate_json_depth(value, current_
            depth + 1)
            for value in obj.values()
        )
    elif isinstance(obj, list):
        if not obj:
            return current_depth
        return max(
            self._calculate_json_depth(item, current_depth + 1)
            for item in obj
        )
    return current_depth
```

For XML files, we need to handle elements, attributes, and maintain the document structure:

```python
def _load_xml(self) -> tuple[dict, dict]:
    """Load and parse XML files with error handling."""
    try:
        with open(self.file_path, 'r', encoding='utf-8') as f:
            data = xmltodict.parse(f.read())

            # Extract metadata about the XML structure
            self.metadata = {
                'root_tag': list(data.keys())[0],
                'size': os.path.getsize(self.file_path),
                'depth': self._calculate_xml_depth(data)
            }

            return data, self.metadata

    except Exception as e:
        print(f"Error parsing XML: {str(e)}")
        return None, None

def _extract_json_keys(self, obj: Any, prefix: str = '') -> list:
    """Extract all keys from a JSON object with their full
    paths."""
    keys = []
    if isinstance(obj, dict):
        for key, value in obj.items():
            full_key = f"{prefix}.{key}" if prefix else key
            keys.append(full_key)
            if isinstance(value, (dict, list)):
                keys.extend(self._extract_json_keys(value,
                full_key))
```

```
        elif isinstance(obj, list):
            for i, item in enumerate(obj):
                full_key = f"{prefix}[{i}]"
                if isinstance(item, (dict, list)):
                    keys.extend(self._extract_json_keys(item,
                    full_key))
        return keys
```

Let's test our implementation with both JSON and XML files:

```
def test_hierarchical_loader():
    """Test the hierarchical data loader"""

    # Create sample JSON file
    json_data = {
        "users": [
            {
                "id": 1,
                "name": "John Doe",
                "address": {
                    "city": "New York",
                    "country": "USA"
                }
            }
        ]
    }

    with open('rag_sample_data/sample.json', 'w') as f:
        json.dump(json_data, f)

    # Create sample XML file
    xml_data = """<?xml version="1.0" encoding="UTF-8"?>
<root>
    <users>
```

```
        <user>
            <id>1</id>
            <name>John Doe</name>
            <address>
                <city>New York</city>
                <country>USA</country>
            </address>
        </user>
    </users>
</root>
"""

with open('rag_sample_data/sample.xml', 'w') as f:
    f.write(xml_data)

# Test both formats
json_loader = HierarchicalDataLoader('rag_sample_data/
sample.json')
json_content, json_metadata = json_loader._load_json()

print("JSON Processing Results:")
print("\nContent:")
print(json.dumps(json_content, indent=2))
print("\nMetadata:")
print(json.dumps(json_metadata, indent=2))

xml_loader = HierarchicalDataLoader('rag_sample_data/
sample.xml')
xml_content, xml_metadata = xml_loader._load_xml()

print("\nXML Processing Results:")
print("\nContent:")
print(json.dumps(xml_content, indent=2))
```

```
print("\nMetadata:")
print(json.dumps(xml_metadata, indent=2))

# Run the test
test_hierarchical_loader()
```

Bringing It All Together

In real-world applications, you'll often need to work with multiple data formats together. Let's create a unified data processor that can handle all the formats we've discussed:

```
class UnifiedDataProcessor:
    """

    A comprehensive processor that handles CSV, JSON, XML,
    and directory structures in a unified way.
    """

    def __init__(self, base_dir: str):
        self.base_dir = base_dir
        self.collection_manager =
        DocumentCollectionManager(base_dir)
        self.data_cache = {}

    def process_document(self, file_path: str) ->
    tuple[Any, dict]:
        """

        Process any supported document type and return
        its content
        and metadata.
        """

        file_ext = os.path.splitext(file_path)[1].lower()
```

```python
    if file_ext == '.csv':
        loader = EnhancedCSVLoader(file_path)
        return loader.load()

    elif file_ext in ['.json', '.xml']:
        loader = HierarchicalDataLoader(file_path)
        if file_ext == '.json':
            return loader._load_json()
        else:
            return loader._load_xml()
    else:
        raise ValueError(f"Unsupported file type:
{file_ext}")

def build_knowledge_base(self) -> dict:
    """

    Scan the directory and process all supported documents
    into a unified knowledge base.
    """

    # Scan directory structure
    collection_metadata = self.collection_manager.scan_
    directory()

    # Process each file
    for file_path in self.collection_manager.file_index:
        try:
            content, metadata = self.process_
            document(file_path)
            if content is not None:
                self.data_cache[file_path] = {
                    'content': content,
                    'metadata': metadata
                }
```

```
        except Exception as e:
            print(f"Error processing {file_path}:
            {str(e)}")

    return {
        'collection_metadata': collection_metadata,
        'processed_files': len(self.data_cache),
        'total_files': len(self.collection_manager.
        file_index)
    }
```

Let's test our unified processor with different types of files:

```python
import numpy as np
from json import JSONEncoder

class CustomJSONEncoder(JSONEncoder):
    """Custom JSON encoder to handle numpy and other special
    types."""
    def default(self, obj):
        if isinstance(obj, np.integer):
            return int(obj)
        elif isinstance(obj, np.floating):
            return float(obj)
        elif isinstance(obj, np.ndarray):
            return obj.tolist()
        return super().default(obj)

def test_unified_processor():
    """Test the unified data processor with multiple file types"""

    # Create test files of different types
    os.makedirs('rag_sample_data/unified_test', exist_ok=True)
```

```python
# Create a CSV file
with open('rag_sample_data/unified_test/data.csv',
'w') as f:
    f.write("name,age\nJohn,30\nJane,25")

# Create a JSON file
with open('rag_sample_data/unified_test/config.json',
'w') as f:
    json.dump({"setting1": "value1", "setting2": 42}, f)

# Create an XML file
with open('rag_sample_data/unified_test/data.xml',
'w') as f:
    f.write('<root><item>Test</item></root>')

# Test the unified processor
processor = UnifiedDataProcessor('rag_sample_data/unified_test')
results = processor.build_knowledge_base()

print("Unified Processing Results:")
print(json.dumps(results, indent=2, cls=CustomJSONEncoder))

print("\nProcessed Documents:")
for path, data in processor.data_cache.items():
    print(f"\nFile: {path}")
    print("Metadata:", json.dumps(data['metadata'],
    indent=2, cls=CustomJSONEncoder))

# Run the test
test_unified_processor()
```

Best Practices and Common Challenges

Through our implementation and testing, we've discovered several important practices for handling structured data in RAG systems. The companion notebook provides complete implementations and examples for each of these practices.

1. **Data Validation and Quality Control:** Our first best practice focuses on comprehensive data validation. The companion notebook includes implementations for

   - Missing value detection and handling

   - Data type consistency checks

   - Structural validation for hierarchical data

   - Quality metrics and recommendations

2. **Memory Management for Large Datasets:** When working with large datasets, memory management becomes crucial. The notebook demonstrates

   - Chunk-based processing for large files

   - Streaming data handlers

   - Memory-efficient data transformations

   - Progress monitoring for long-running processes

3. **Error Recovery and Logging:** Robust error handling and logging are essential for production systems. In the notebook, you'll find

   - Comprehensive error tracking

   - Detailed logging mechanisms

- Error frequency analysis
- Recovery strategies for common failures

These practices help ensure your structured data processing is

- Reliable and robust
- Properly validated
- Performance optimized
- Easy to maintain and debug

The companion notebook includes practical implementations of all these concepts, along with examples using real-world data scenarios. You can experiment with different approaches and adapt them to your specific needs.

Rich Documents: Beyond Basic Text

In the digital world, not all documents are created equal. While we've mastered handling plain text and PDFs, organizations often work with more sophisticated document formats—from polished PowerPoint presentations to intricately formatted Word documents and complex Excel spreadsheets. In this section, we'll explore how to unlock these rich document formats for your RAG system.

Our companion notebook "Rich_Document_Loading.ipynb" demonstrates how to handle these complex formats effectively. Let's dive into each type:

Microsoft Office Documents

Think of Office documents as multilayered content containers. A Word document isn't just text—it's a combination of styled paragraphs, tables, images, and formatting. Similarly, an Excel spreadsheet contains not just data, but formulas, relationships, and multiple worksheets.

Let's look at our implementation:

```python
# Install required packages
!pip install python-docx openpyxl python-pptx pandas

from typing import List, Dict, Any
import docx
import openpyxl
from pptx import Presentation
import os
import pandas as pd

class OfficeDocumentLoader:
    """Handle various Microsoft Office document formats"""
    def __init__(self):
        self.supported_formats = {
            '.docx': self._load_word,
            '.xlsx': self._load_excel,
            '.pptx': self._load_powerpoint
        }
```

The companion notebook demonstrates several key scenarios:

1. **Word Documents**

   - Preserving document structure and formatting

   - Handling tables and embedded objects

   - Maintaining style hierarchies

2. **Excel Workbooks**

   - Processing multiple worksheets

   - Preserving formulas and relationships

   - Handling different data types

3. **PowerPoint Presentations**

- Extracting slide content and notes
- Processing embedded media
- Maintaining presentation structure

Markdown Documents

Markdown has become the lingua franca of technical documentation. Its simple yet powerful syntax makes it perfect for everything from API documentation to technical blogs. Here's how we approach it:

```python
import mistune
import frontmatter
from typing import Dict, Any

class MarkdownDocumentLoader:
    """Process Markdown documents while preserving structure"""
    def __init__(self):
        self.markdown_parser = mistune.create_markdown()

    def load(self, file_path: str) -> Dict[str, Any]:
        """Load and process a Markdown document"""
        with open(file_path, 'r', encoding='utf-8') as f:
            post = frontmatter.load(f)

        return {
            'content': self._parse_content(post.content),
            'metadata': dict(post.metadata) if post.metadata
            else {},
            'structure': self._analyze_structure(post.content)
        }
```

```python
def _parse_content(self, content: str) -> List[Dict]:
    """Parse markdown content into structured sections"""
    sections = []
    # Parsing logic (see companion notebook)
    return sections

def _analyze_structure(self, content: str) -> Dict:
    """Analyze document structure for headers,
    lists, etc."""
    structure = {
        'headers': [],
        'code_blocks': [],
        'lists': []
    }
    # Analysis logic (see companion notebook)
    return structure
```

The companion notebook provides comprehensive examples for handling

- Complex document hierarchies

- Code blocks with syntax highlighting

- Lists and tables

- Links and references

Each implementation is designed to preserve the rich structure and formatting that makes these document types valuable for your RAG system. The complete code, along with detailed examples and test cases, can be found in the companion notebook.

Custom Documents: Tailoring Your RAG System

Imagine walking into a specialized library where every book follows its own unique format—some written in ancient scripts, others using industry-specific notations, and still others combining multiple formats in novel ways. This is the challenge many organizations face with their custom document formats. Whether it's proprietary XML schemas, legacy file formats, or domain-specific documentation, handling these specialized formats is crucial for a comprehensive RAG system.

Think of custom document loaders as skilled translators who understand these unique "languages" and can convert them into a format your RAG system understands. Just as a translator must understand both the source language's nuances and the target language's requirements, your custom loaders need to handle both the specific format's intricacies and your RAG system's needs.

Our companion notebook "Custom_Document_Loading.ipynb" demonstrates how to build these translators. Here's the foundation we'll build upon:

```python
from abc import ABC, abstractmethod
from typing import Dict, Any

class CustomDocumentLoader(ABC):
    """Base class for custom document loaders"""

    def __init__(self, file_path: str):
        self.file_path = file_path
        self.metadata = {}

    @abstractmethod
    def _parse_content(self, content: str) -> Dict[str, Any]:
        """Parse the document content based on its format"""
        pass
```

```python
def load(self) -> Dict[str, Any]:
    """Load and process the document"""
    with open(self.file_path, 'r', encoding='utf-8') as f:
        content = f.read()

    return {
        'content': self._parse_content(content),
        'metadata': self.metadata
    }
```

The companion notebook shows you how to extend this foundation to handle

- Industry-specific XML formats that might contain technical specifications or compliance documents

- Legacy formats from older systems that still hold valuable information

- Domain-specific formats used in specialized fields like healthcare or finance

- Hybrid documents that combine multiple formats in a single file

Think of each custom loader as adding a new "language" to your RAG system's repertoire, expanding its ability to understand and process your organization's unique knowledge base.

Summary and Next Steps

Throughout this chapter, we've explored the diverse landscape of document processing in RAG systems. Like a skilled librarian handling everything from ancient manuscripts to digital publications, we've built a toolkit to process

- Plain text documents (straightforward yet vital)
- PDFs (complex layouts and structures)
- Web content (dynamic and interactive elements)
- Structured data (CSV, JSON, XML)
- Rich documents (Office files, Markdown)
- Custom formats (domain-specific or legacy systems)

Key Takeaways

1. **Flexibility is crucial**: Real-world documents come in countless formats; your RAG system must adapt while preserving content integrity.

2. **Structure preservation**: Maintain hierarchies (e.g., markdown headers) and layouts (e.g., PDFs) for context retention.

3. **Error handling**: Robust validation ensures graceful handling of imperfect documents.

Next Steps

In the next chapter, we'll dive into text splitters—components that break processed documents into optimal chunks for embedding and retrieval. You'll learn to balance chunk size with semantic meaning, ensuring your RAG system efficiently finds and uses information.

The companion notebooks for this chapter provide hands-on implementations to experiment with these concepts.

CHAPTER 5

Text Splitters in RAG Systems

Imagine you're handed a massive book and asked to find every mention of a specific concept quickly. Would you read through the entire book page by page? Or would you prefer the book to be divided into smaller, manageable sections that you can quickly scan and reference? This is exactly the challenge that RAG systems face when working with large documents.

In this chapter, we'll explore text splitting—a crucial component that transforms lengthy documents into digestible chunks that your RAG system can effectively process and retrieve. Let's visualize this concept (see Figure 5-1):

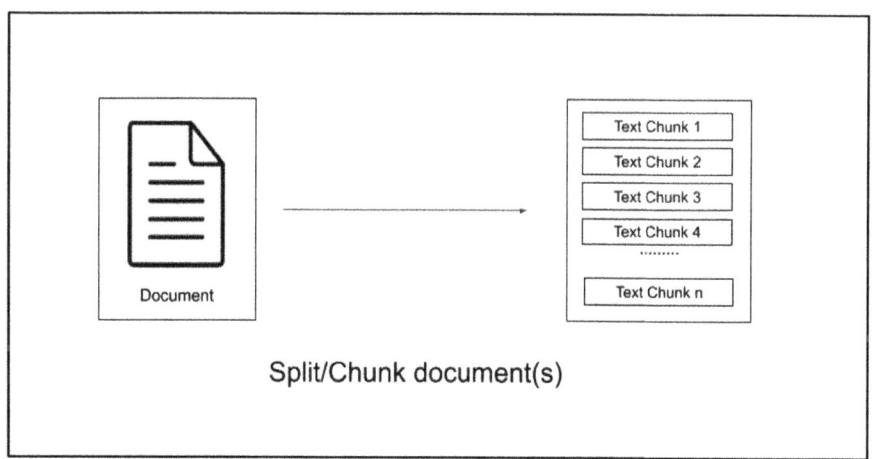

Figure 5-1. *Text splitters—splitting documents into smaller chunks*

© Ranajoy Bose 2025
R. Bose, *Mastering Retrieval-Augmented Generation*,
https://doi.org/10.1007/979-8-8688-1808-0_5

Understanding Text Splitting

Think of text splitting as the art of breaking down a book while preserving its meaning. Just as a book is thoughtfully divided into chapters, sections, and paragraphs, your RAG system needs documents to be split in a way that maintains context and relevance. But why is this so important?

Imagine trying to find information about "climate change impacts on coral reefs" in a 300-page environmental report. If your RAG system treats the document as one massive chunk, it might struggle to

- Pinpoint the exact sections about coral reefs

- Distinguish between different types of environmental impacts

- Return concise, relevant information to user queries

This is where intelligent text splitting comes into play.

The Benefits of Smart Text Splitting

1. **Enhanced Retrieval Precision**

   - Instead of matching against entire documents, your system can pinpoint relevant chunks.

   - More granular content allows for more precise semantic matching.

   - Reduces noise in search results by focusing on specific sections.

2. **Optimized Processing**

- Language models have token limits—typically 4,096 or 8,192 tokens.

- Splitting ensures your content fits within these limits.

- Enables parallel processing of different chunks.

3. **Improved Context Management**

- Well-designed chunks maintain semantic coherence

- Overlapping chunks preserve context across boundaries

- Natural language flow is maintained within each section

4. **Better Resource Utilization**

- Smaller chunks mean more efficient memory usage

- Faster processing times for embeddings and retrieval

- More cost-effective API usage with language models

To help you implement these concepts, we've prepared a series of companion notebooks, which can be found in the book's repository. Each notebook focuses on specific splitting strategies and includes practical examples you can run and modify.

The Art of Chunk Size

One of the most critical decisions in text splitting is determining the optimal chunk size. Too large, and you lose precision; too small, and you fragment context. Consider these factors:

- **Model Context Window**: Your language model's token limit

- **Query Nature**: The typical length of information needed to answer queries

- **Content Structure**: Natural boundaries in your documents (paragraphs, sections)

- **Retrieval Goals**: Balance between precision and context preservation

Types of Text Splitters We'll Explore

In this chapter, we'll cover several text splitting strategies, each designed for specific use cases and document types. Here's what we'll explore:

1. **Recursive Character Text Splitting**: The foundation for handling general text documents

2. **HTML Document Splitting**: Specialized for web content and HTML files

3. **Code Text Splitting**: Designed for source code and technical documentation

4. **Markdown Header Splitting**: Perfect for markdown-based documentation

5. **JSON/Dictionary Splitting**: Handles structured data formats

6. **Semantic Text Splitting**: Uses meaning-based approaches for intelligent splitting

For each strategy, we've prepared companion notebooks, which can be found in the book's repository. These notebooks contain practical examples and exercises to help you implement these concepts in your own RAG systems.

Let's start with our first and most versatile approach—recursive character splitting. We'll work through examples in our companion notebook.

Recursive Character Text Splitting: The Foundation

When you're building a RAG system, recursive character splitting is often your first and most versatile tool. Think of it as a smart document slicer that tries to cut at natural boundaries—like paragraphs and sentences—before resorting to more aggressive splits.

The core idea is simple but powerful: start with the most natural text separators (like paragraph breaks) and progressively try finer splits (like sentence endings, then word spaces) until you get chunks of the desired size. This preserves as much context and readability as possible.

Our companion notebook "Text_Splitting_Strategies.ipynb" will help you implement this approach. Let's start with a basic setup:

```
# Install required packages
!pip install langchain-text-splitters tiktoken

# Import necessary libraries
from langchain_text_splitters import
RecursiveCharacterTextSplitter
import tiktoken
from typing import List, Dict

print("Environment setup complete!")
```

Now, let's create a smart text splitter that understands document structure:

```
# Create and activate a virtual environment
python -m venv rag-env
source rag-env/bin/activate  # On Windows, use: rag-env\
Scripts\activate
```

```
# Install required packages
pip install jupyter
pip install langchain chromadb openai sentence-transformers
```

Let's see how this works with a real-world example:

```
def demonstrate_recursive_splitting():
    """
    Demonstrate how recursive splitting handles different text
    structures.
    """
    # Sample research abstract
    research_text = """
    Machine Learning in Healthcare: A Comprehensive Review

    This review examines the impact of machine learning in
    healthcare settings.
    Recent advances have shown promising results in diagnosis
    and treatment.

    Key Applications:
    • Medical imaging analysis has achieved expert-level
      accuracy
    • Patient risk prediction models are improving care
      outcomes
    • Drug discovery is being accelerated through ML techniques

    Methods and Results:
    Our analysis covers 500 research papers from 2020-2025. The
    findings
    indicate a 45% improvement in early diagnosis rates when ML
    is applied.
    """
```

```
# Create splitter
splitter = create_smart_splitter(chunk_size=200, chunk_
overlap=50)

# Split text
chunks = splitter.split_text(research_text)

# Analyze and display results
print(f"Original Length: {len(research_text)} characters")
print(f"Number of Chunks: {len(chunks)}\n")

for i, chunk in enumerate(chunks, 1):
    print(f"Chunk {i} ({len(chunk)} chars):")
    print("-" * 50)
    print(chunk)
    print("\n")

# Run demonstration
demonstrate_recursive_splitting()
```

The companion notebook includes more examples and exercises. Three key features make this approach particularly effective:

1. **Natural Boundaries First**

   • Tries to split at paragraph breaks before sentences

   • Preserves natural text structure where possible

   • Only resorts to word-level splits when necessary

2. **Smart Overlap**

   • Each chunk shares some content with its neighbors.

   • Helps maintain context across chunk boundaries.

   • Overlap size is configurable for your needs.

3. **Flexible Sizing**

- Targets consistent chunk sizes while respecting content
- Adapts to document structure when possible
- Prevents awkward mid-sentence splits

Advanced Configurations and Special Cases

While the basic recursive splitter works well for most scenarios, real-world documents often present unique challenges. Let's explore how to handle these special cases using advanced configurations. The companion notebook "Text_Splitting_Strategies.ipynb" includes complete implementations of all these scenarios.

Handling Multilingual Content: One of the most common challenges is processing documents that contain multiple languages, especially those without clear word boundaries, such as Chinese, Japanese, or Thai. Our companion notebook demonstrates how to configure the recursive splitter with language-specific separators and handling rules to ensure proper splitting across different writing systems.

Custom Document Formats: Different document types often require specialized splitting approaches:

- Scientific papers with sections, citations, and references
- Legal documents with articles, clauses, and numbered sections
- Technical documentation with code blocks and figures
- Business reports with tables and structured data

The notebook includes examples of how to customize separator patterns and chunk sizes for each of these formats.

Preserving Metadata: When splitting documents that contain important metadata (like author information, dates, or categories), you'll want to ensure this information is preserved across chunks. The companion notebook shows how to

- Carry forward document-level metadata to all chunks

- Add chunk-specific metadata (like chunk numbers and positions)

- Handle documents with nested metadata structures

Advanced Configurations: For more complex use cases, you can fine-tune the splitter's behavior by

- Adjusting chunk overlap based on document structure

- Implementing custom length functions for special content types

- Adding document-specific preprocessing steps

- Handling edge cases like very long paragraphs or unusual formatting

The companion notebook includes practical examples and exercises for each scenario, enabling you to tailor the recursive splitter to your specific needs.

HTML Document Splitting: Preserving Web Content Structure

When working with web content, simple text splitting isn't enough. HTML documents contain structural elements like headers, paragraphs, lists, and tables that carry important semantic meaning. Breaking these structures incorrectly can lead to loss of context and reduced retrieval quality.

Our companion notebook "HTML_Splitting_Strategies.ipynb" demonstrates how to handle these challenges effectively. Let's explore the implementation:

```python
from langchain_text_splitters import HTMLHeaderTextSplitter
from bs4 import BeautifulSoup
from typing import List, Dict

class HTMLSplitter:
    """Specialized splitter for HTML content"""
    def __init__(self,
                    chunk_size: int = 1000,
                    chunk_overlap: int = 200):
        self.chunk_size = chunk_size
        self.chunk_overlap = chunk_overlap
        # Headers to split on with their metadata keys
        self.headers_to_split_on = [
            ("h1", "Header 1"),
            ("h2", "Header 2"),
            ("h3", "Header 3"),
            ("div", "Division")
        ]

    def split_html_content(self, html_content: str) ->
List[str]:
        """Split HTML while preserving structure"""
        # Create splitter with headers
        splitter = HTMLHeaderTextSplitter(headers_to_split_
on=self.headers_to_split_on)

        # Split the text
        splits = splitter.split_text(html_content)
        return splits
```

Let's handle common HTML elements:

```python
def split_html_content(self, html_content: str) -> List[str]:
    """Split HTML while preserving structure"""

    splitter = HTMLHeaderTextSplitter(
        tags_to_split_on=self.tags_to_split_on
    )

    splits = splitter.split_text(html_content)
    return splits
```

We can test this with a typical HTML document:

```python
def test_html_splitting():
    """Test HTML splitter with a sample document"""
    sample_html = """
    <html>
        <body>
            <h1>AI Applications</h1>
            <div class="section">
                <h2>Machine Learning</h2>
                <p>Machine learning is transforming
                industries.</p>
                <ul>
                    <li>Predictive Analytics</li>
                    <li>Pattern Recognition</li>
                </ul>
            </div>
            <div class="section">
                <h2>Deep Learning</h2>
                <p>Neural networks are advancing rapidly.</p>
            </div>
        </body>
```

```
</html>
"""

splitter = HTMLSplitter()
chunks = splitter.split_html_content(sample_html)

for i, chunk in enumerate(chunks, 1):
    print(f"\nChunk {i}:")
    print(chunk)
```

The companion notebook includes additional examples for

- Complex HTML layouts

- Nested structures

- Dynamic content

- Forms and interactive elements

Code Text Splitting: Maintaining Programming Logic

When splitting code documents, we face unique challenges compared to regular text. Code has strict structural rules, dependencies, and logical flows that must be preserved. A poorly split code block could break function definitions, class relationships, or make the code completely unreadable.

Our companion notebook "Code_Splitting_Strategies.ipynb" demonstrates effective approaches to handling code documents. Let's explore the implementation:

```
from langchain_text_splitters import
RecursiveCharacterTextSplitter, Language
from typing import List, Dict
```

```python
class CodeSplitter:
    """Splitter specialized for source code documents"""
    def __init__(self, language: str = "python"):
        self.language = language
        self.splitter = self._create_language_splitter()

    def _create_language_splitter(self):
        """Create a splitter for specific programming
        language"""
        return RecursiveCharacterTextSplitter.from_language(
            language=self.language,
            chunk_size=1000,
            chunk_overlap=200
        )

    def split_code(self, code: str) -> List[str]:
        """Split code while preserving structure"""
        return self.splitter.split_text(code)
```

Let's test this with different programming languages and scenarios:

```python
def test_code_splitting():
    """Test different code splitting scenarios"""

    # Python code example
    python_code = '''
class DataProcessor:
    """Handle data processing operations."""

    def __init__(self, data_path: str):
        self.data_path = data_path
        self.processed = False

    def process(self):
        """Process the data."""
```

```
        print(f"Processing data from {self.data_path}")
        self.processed = True
'''

python_splitter = CodeSplitter(language="python")
chunks = python_splitter.split_code(python_code)

for i, chunk in enumerate(chunks, 1):
    print(f"\nChunk {i}:")
    print(chunk)
```

Multilanguage Support: The splitter handles various programming languages, each with its unique syntax and structure:

- Python: Preserves indentation and decorators

- JavaScript: Maintains function scope and closures

- Java: Respects class and method boundaries

- SQL: Keeps queries and transactions intact

- Shell scripts: Preserves command sequences

Complex Code Structures: The notebook includes examples of handling

- Class definitions with multiple methods

- Nested functions and callbacks

- Decorator patterns

- Module-level imports and configurations

- Mixed code and documentation

Documentation Handling: Special attention is paid to preserving

- Docstrings and comments
- Function and class documentation
- API documentation blocks
- Code examples within documentation
- Type hints and annotations

Special Cases: The implementation also covers

- Template code with placeholders
- Code with embedded DSLs (Domain Specific Languages)
- Multiline strings and comments
- Preprocessor directives
- Compile-time annotations

Each of these scenarios is demonstrated with practical examples in the companion notebook, showing how to maintain code integrity while creating meaningful chunks for your RAG system.

Markdown Header Splitting: Preserving Document Hierarchy

When dealing with technical documentation, particularly markdown files, preserving the document's hierarchical structure is crucial. Markdown headers (using #, ##, ###, etc.) naturally organize content into logical sections and subsections.

Our companion notebook "Markdown_Splitting_Strategies.ipynb" demonstrates how to effectively handle markdown documents. Let's explore the key aspects:

```python
from langchain_text_splitters import MarkdownHeaderTextSplitter
from typing import List, Dict

class MarkdownSplitter:
    """Splitter specialized for Markdown documents"""
    def __init__(self):
        self.headers_to_split_on = [
            ("#", "Header 1"),
            ("##", "Header 2"),
            ("###", "Header 3")
        ]

    def split_markdown(self, markdown_text: str) -> List[str]:
        """Split markdown while preserving header hierarchy"""
        splitter = MarkdownHeaderTextSplitter(
            headers_to_split_on=self.headers_to_split_on
        )
        splits = splitter.split_text(markdown_text)
        return splits
```

The companion notebook includes examples that demonstrate

- Handling nested header hierarchies

- Processing code blocks within markdown

- Managing inline formatting

- Preserving links and references

- Handling markdown tables

Each of these scenarios is crucial for maintaining the document's readability and structure while creating meaningful chunks for your RAG system.

JSON and Dictionary Splitting: Handling Structured Data

When working with JSON or dictionary data, maintaining the structural relationships between data elements is crucial. Unlike plain text, JSON documents have nested hierarchies, arrays, and key-value relationships that need to be preserved during splitting.

Our companion notebook "JSON_Splitting_Strategies.ipynb" demonstrates effective approaches for handling structured data. Let's start with the basic implementation:

```python
from langchain_text_splitters import
RecursiveCharacterTextSplitter
import json
from typing import Dict, List, Any

class JSONSplitter:
    """Splitter specialized for JSON/Dictionary data"""
    def __init__(self,
                chunk_size: int = 1000,
                chunk_overlap: int = 200):
        self.chunk_size = chunk_size
        self.chunk_overlap = chunk_overlap
        self.splitter = RecursiveCharacterTextSplitter(
            chunk_size=chunk_size,
            chunk_overlap=chunk_overlap
        )
```

The companion notebook includes examples that demonstrate

- Handling deeply nested JSON structures

- Processing arrays and lists

- Maintaining key-value relationships

- Managing complex data types

- Preserving JSON schema integrity

Each of these scenarios shows how to split JSON data while ensuring that

1. The resulting chunks are valid JSON.

2. Related data stays together.

3. Context is maintained across splits.

4. References and relationships are preserved.

Semantic Text Splitting: Context-Aware Chunking

Unlike previous approaches that rely on structural markers or character counts, semantic text splitting focuses on maintaining the meaning and context of the content. This approach utilizes embeddings to identify natural breakpoints where the semantic flow changes.

Our companion notebook "Semantic_Splitting_Strategies.ipynb" demonstrates this advanced approach. Let's look at the implementation:

```
# Install required packages
!pip install langchain
!pip install langchain-community
```

```
!pip install langchain-text-splitters
!pip install transformers
!pip install torch

# Now import the required modules
from langchain_text_splitters import
RecursiveCharacterTextSplitter
from langchain_community.embeddings import
HuggingFaceEmbeddings
import numpy as np
from typing import List, Dict

class SemanticTextSplitter:
    """Split text based on semantic meaning using embeddings"""
    def __init__(self,
                    chunk_size: int = 1000,
                    chunk_overlap: int = 200):
        self.chunk_size = chunk_size
        self.chunk_overlap = chunk_overlap
        self.embeddings = HuggingFaceEmbeddings()
```

The companion notebook includes examples that demonstrate

- Using embeddings to find semantic boundaries

- Measuring semantic similarity between chunks

- Adjusting chunk sizes based on content meaning

- Handling topic transitions

- Preserving contextual relationships

Each example shows how semantic splitting can produce more meaningful chunks for your RAG system by

1. Keeping related concepts together

2. Identifying natural topic boundaries

3. Maintaining semantic coherence

4. Improving retrieval relevance

Summary and Next Steps

As we conclude our journey through text splitting strategies, let's take a moment to reflect on what we've learned. Think of text splitting as the art of breaking down a complex puzzle while ensuring each piece maintains its meaning and can be reassembled perfectly when needed.

Throughout this chapter, we've explored six distinct approaches to text splitting, each with its own strengths and ideal use cases. Like a master chef choosing the right knife for different ingredients, a skilled RAG system developer must select the appropriate splitting strategy for different types of content.

The Text Splitting Arsenal

1. **Recursive Character Splitting:** Like a skilled editor who knows exactly where to break paragraphs and sentences, this approach respects natural language boundaries while maintaining readability. We saw how it intelligently works its way from paragraph breaks down to individual characters, always trying to preserve the most natural text divisions possible.

2. **HTML Document Splitting:** Imagine trying to disassemble a beautiful piece of furniture without damaging any of its intricate decorations. That's what HTML splitting does—it preserves not just the content but the structure, styling, and interactive elements that make web content rich and functional. Through our examples, we discovered how to maintain everything from navigation menus to complex forms while creating manageable chunks.

3. **Code Text Splitting:** Perhaps our most precise instrument, like a surgeon's scalpel. When working with code, we can't afford to break function definitions or class structures haphazardly. We learned how to respect the syntax and logic of different programming languages while creating meaningful segments that preserve the code's functionality and readability.

4. **Markdown Header Splitting:** The organizer's dream tool. By respecting header hierarchies, this approach maintains the logical flow of documentation while preserving formatting, links, and other markdown elements. It's particularly valuable for technical documentation and wikis where structure is as important as content.

5. **JSON/Dictionary Splitting:** Our specialist for structured data. Like a librarian who understands how different books relate to each other, this

approach maintains the relationships between data elements while creating manageable chunks. We explored how to handle nested structures, arrays, and complex data types without losing their contextual connections.

6. **Semantic Text Splitting:** Our approach. Rather than relying on structural markers, it understands the meaning of content and creates chunks based on semantic relationships. Think of it as a thoughtful reader who naturally understands where one topic ends and another begins.

Key Takeaways
Making the Right Choice

Choosing the right splitting strategy is like selecting the right tool for a job—it requires understanding both your materials and your goals. Consider

- What type of content dominates your document collection?

- How vital is structural preservation?

- Do you need to maintain special formatting or relationships?

- What are your retrieval requirements?

The companion notebooks we've worked through provide hands-on experience with each strategy, allowing you to experiment and understand their practical implications. Try them with your own content, mix and match approaches, and observe how different strategies affect your retrieval results.

Next Steps

Looking Ahead

As we move into the next chapter, we'll discover how to transform these carefully crafted chunks into vector representations using embedding models. Think of it as converting our puzzle pieces into a format that computers can easily understand and compare, enabling quick and accurate retrieval in our RAG systems.

Remember, the way we split our documents lays the foundation for everything that follows. A well-chosen splitting strategy can mean the difference between a RAG system that merely works and one that truly excels at understanding and retrieving information.

## CHAPTER 6

Embedding Models: Converting Text to Vectors

In the architecture of Retrieval-Augmented Generation (RAG) systems, embedding models serve as the critical translation layer between human language and machine understanding. These sophisticated neural networks perform a remarkable alchemy—transforming the fluid, ambiguous nature of text into precise mathematical representations while preserving semantic relationships. Without this translation, LLMs would lack the contextual precision that makes RAG systems so powerful.

In this chapter, we'll explore

- The core principles of how embeddings convert language to vectors

- Key similarity metrics (cosine, Euclidean, dot product) and when to use each

- Industry-standard models (OpenAI, Sentence Transformers, Hugging Face) and their trade-offs

© Ranajoy Bose 2025
R. Bose, *Mastering Retrieval-Augmented Generation*,
https://doi.org/10.1007/979-8-8688-1808-0_6

215

- Implementation patterns for efficient embedding generation and caching

- Advanced techniques like custom embeddings and dimension optimization

Let's begin by understanding the two-stage process that makes semantic search possible. (See Figure 6-1 to understand how embeddings map language into a semantic space, where meaning translates to proximity.)

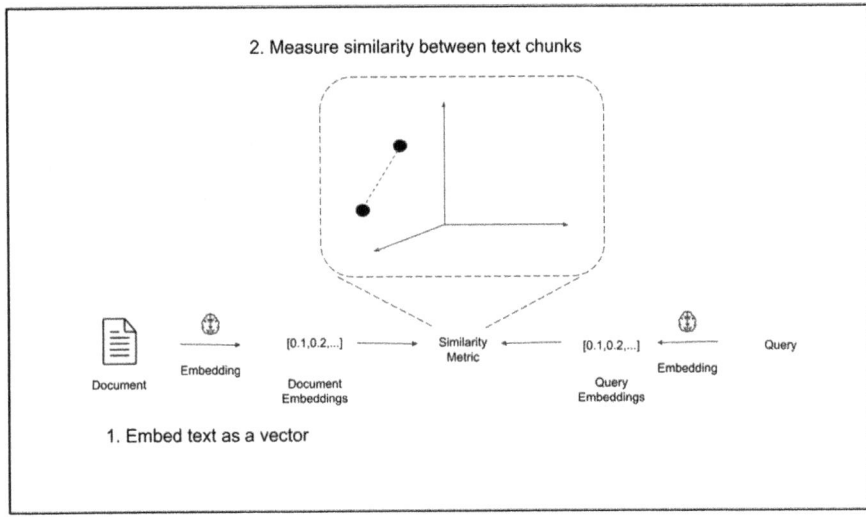

Figure 6-1. *Embedding process: key concepts*

The Two-Step Process

Step 1: From Words to Vectors

At its core, an embedding model performs a remarkable transformation: takes words, sentences, or entire documents and converts them into vectors, essentially long lists of numbers. But these aren't arbitrary numbers. Each dimension in this vector space captures some aspect of meaning, creating a rich mathematical representation of the text's semantics.

Consider a simple sentence: "The cat sat on the mat." When processed through an embedding model, this becomes a vector with hundreds or thousands of dimensions. Each dimension represents various aspects of meaning—from the basic subject–verb relationship to subtle connotations about indoor domestic scenes.

Step 2: Measuring Semantic Relationships

Once we have our text converted to vectors, we can perform precise mathematical operations to understand relationships between different pieces of text. This is where three key mathematical concepts come into play:

Cosine Similarity

Imagine two vectors as arrows pointing in a multidimensional space. Cosine similarity measures the angle between these arrows, providing a value between -1 and 1. When two pieces of text are semantically similar, their vectors point in nearly the same direction, resulting in a cosine similarity close to 1. This measure is compelling because it focuses on the direction of vectors rather than their magnitude.

Euclidean Distance

While cosine similarity looks at angles, Euclidean distance measures the straight-line distance between the tips of our vector arrows. This measure can be beneficial when the magnitude of vectors carries meaningful information. However, it's more sensitive to variations in text length and style.

Dot Product

The dot product combines aspects of both direction and magnitude. It can be thought of as measuring how much two vectors "agree" with each other, considering their alignment and lengths. This measurement becomes particularly useful when we want to consider both semantic similarity and the "strength" of the expressed meanings.

Choosing the Right Similarity Metric

The choice of similarity metric should be based on the model being used. For example, OpenAI recommends cosine similarity for their embeddings due to its effectiveness with their vector space characteristics. This can be implemented efficiently using standard numerical libraries:

```
def cosine_similarity(vec1, vec2):
    dot_product = np.dot(vec1, vec2)
    norm1 = np.linalg.norm(vec1)
    norm2 = np.linalg.norm(vec2)
    return dot_product / (norm1 * norm2)
```

Working with Embedding Models in Practice

When implementing embeddings in RAG applications, you'll typically follow these key steps:

1. Choosing an embedding model

2. Converting text to embeddings

3. Storing and retrieving embeddings

Let's explore these aspects with practical examples using popular embedding models.

Setting Up Your Environment

Before diving into embeddings, you must set up your environment with the necessary dependencies. We'll primarily use LangChain, which provides a unified interface for working with various embedding models.

Working with OpenAI Embeddings

OpenAI's embedding models are among the most popular choices for RAG applications due to their robust performance and ease of use. Here's how you can implement text embeddings using OpenAI's models:

```python
from langchain_openai import OpenAIEmbeddings

# Initialize the embedding model
embeddings_model = OpenAIEmbeddings(
    model="text-embedding-3-large",  # Specify the
                                     model version
    openai_api_key="your-api-key"    # Your API key
)

# Embed a single piece of text (query)
query = "What is machine learning?"
query_embedding = embeddings_model.embed_query(query)

# Embed multiple documents
documents = [
    "Machine learning is a subset of artificial intelligence.",
    "Deep learning is a type of machine learning.",
    "Neural networks are used in deep learning."
]
doc_embeddings = embeddings_model.embed_documents(documents)
```

Understanding Embedding Outputs

The embedding process generates vectors with specific characteristics:

1. **Dimensionality**: Each embedding vector has a fixed number of dimensions (e.g., 1536 for OpenAI's text-embedding-3-large model).

2. **Consistency**: The exact input text will always produce the same embedding vector.

3. **Normalization**: Embedding vectors are typically normalized to have a magnitude of 1.

Working with Multiple Embedding Models

LangChain's abstraction allows you to switch between different embedding models easily. Here's an example using Hugging Face's sentence transformers:

```
from langchain_community.embeddings import
HuggingFaceEmbeddings

# Initialize the HuggingFace embedding model
hf_embeddings = HuggingFaceEmbeddings(
    model_name="sentence-transformers/all-mpnet-base-v2"
)

# Use it the same way as any other embedding model
hf_query_embedding = hf_embeddings.embed_query(query)
hf_doc_embeddings = hf_embeddings.embed_documents(documents)
```

Best Practices for Embedding Generation

When working with embeddings, keep these practices in mind:

1. **Batch Processing**: When embedding many documents, process them in batches to optimize performance.

2. **Error Handling**: Always properly handle API calls and model loading.

3. **Text Preprocessing**: Clean and preprocess your text before generating embeddings.

4. **Model Selection**: Choose an embedding model based on your specific needs (speed, accuracy, cost).

Note   A comprehensive companion notebook for this section is available in the book's repository (**RAG_Chapter6_01_Embeddings.ipynb**). The notebook contains detailed, executable examples of all concepts covered here, including working with different embedding models, computing similarities, implementing best practices, and handling embeddings efficiently. You can run these examples directly in Google Colab to get hands-on experience with embedding models.

Caching Embedding Results: Optimizing Performance and Cost

When building production RAG systems, embedding computation management is one of the most critical yet often overlooked aspects. Every time your system needs to convert text into vectors, it either makes an

API call to services like OpenAI or performs complex local calculations. This process can quickly become both a performance bottleneck and a significant cost center. This section'll explore how to effectively implement embedding caching to optimize your RAG system.

Why Cache Embeddings?

At its core, embedding caching is about smart resource management. When your system encounters the same text multiple times, recomputing its embedding wastes computational resources and incurs unnecessary costs in the case of API-based services. Moreover, different embedding computations of the same text might produce slightly varying results, potentially affecting the consistency of your system's responses.

Consider a real-world scenario: a document search system that processes thousands of queries daily. Without caching, each identical search query requires a new embedding computation. With proper caching, the system can instantly retrieve pre-computed embeddings, dramatically improving response times and reducing operational costs. This optimization becomes particularly crucial when dealing with rate limits of embedding APIs or managing high-concurrency applications.

Caching Strategies
1. In-Memory Caching

In-memory caching represents the simplest form of embedding storage, where vectors are kept in RAM during runtime. While this approach offers the fastest possible access times, it comes with important considerations. When your application restarts, all cached embeddings are lost, and you're limited by available memory. This strategy works well for small-scale applications or development environments but may not be suitable for production systems.

Here's an implementation of a basic in-memory cache:

```
class SimpleEmbeddingCache:
    def __init__(self, embedding_model):
        self.cache = {}
        self.model = embedding_model
        self.hits = 0
        self.misses = 0

    def get_embedding(self, text: str) -> List[float]:
        """Get embedding from cache or compute it."""
        if text in self.cache:
            self.hits += 1
            return self.cache[text]

        self.misses += 1
        embedding = self.model.embed_query(text)
        self.cache[text] = embedding
        return embedding
```

This implementation acts like a smart memory bank. When you request an embedding for a piece of text, it first checks if it has seen this text before. If it has (a cache hit), it returns the stored embedding instantly. If not (a cache miss), it generates a new embedding and stores it for future use.

2. File-based Persistent Storage Caching

While in-memory caching is fast, it has limitations—primarily that all cached embeddings are lost when your application restarts. For more robust applications, we need embeddings to persist. This is where file-based caching comes in:

```
class FileCacheEmbeddings:
    def __init__(self, embedding_model, cache_dir="./embedding_
    cache"):
```

```python
        self.model = embedding_model
        self.cache_dir = cache_dir
        self.stats = {'hits': 0, 'misses': 0}
        os.makedirs(cache_dir, exist_ok=True)

    def _get_cache_path(self, text: str) -> str:
        """Generate cache file path for the text."""
        text_hash = hashlib.sha256(text.encode()).hexdigest()
        return os.path.join(self.cache_dir, f"{text_hash}.pkl")
```

This implementation creates a unique file for each embedding, using a hash of the input text as the filename. Think of it as creating a persistent library of embeddings, where each text has its own "book" that stores its numerical representation. When you need an embedding, the system first checks if this "book" exists before computing a new one.

3. Production-Ready Caching

In production environments, we need more sophisticated features. Imagine running a system where embedding models get updated periodically, or where different parts of your application need different types of embeddings. Our production-ready implementation handles these scenarios:

```python
class ProductionEmbeddingCache:
    def __init__(self,
                 embedding_model,
                 ttl_seconds: int = 86400,  # 24 hours
                 namespace: str = "default"):
        self.model = embedding_model
        self.ttl_seconds = ttl_seconds
        self.namespace = namespace
        self.cache: Dict[str, dict] = {}
        self.stats = {'hits': 0, 'misses': 0}
```

This implementation includes several critical features for production use:

1. **Time-to-Live (TTL):** Embeddings don't live forever. The system automatically invalidates old embeddings after a specified time period:

```python
def _is_cache_valid(self, cache_entry: dict) -> bool:
    """Check if cache entry is still valid based on TTL."""
    return time.time() - cache_entry['timestamp'] < self.ttl_seconds
```

2. **Namespacing:** Different parts of your application can use different cache spaces, preventing conflicts:

```python
def _generate_key(self, text: str) -> str:
    """Generate a consistent hash key for the text."""
    normalized_text = text.lower().strip()
    return hashlib.sha256(
        f"{self.namespace}:{normalized_text}".encode()
    ).hexdigest()
```

3. **Error Handling and Monitoring:** The system gracefully handles failures and provides insights into cache performance:

```python
def get_stats(self):
    """Return cache statistics."""
    total = self.stats['hits'] + self.stats['misses']
    hit_rate = (self.stats['hits'] / total * 100) if total > 0 else 0
    return {
        'hits': self.stats['hits'],
        'misses': self.stats['misses'],
```

```
    'hit_rate': f"{hit_rate:.2f}%",
    'cache_size': len(self.cache)
}
```

Making the Right Choice

Choosing the right caching strategy depends on your specific needs:

1. **Development Environment**: Use SimpleEmbeddingCache for quick iterations and testing.

2. **Small Production Systems**: FileCacheEmbeddings works well for applications with moderate traffic.

3. **Large-Scale Production**: ProductionEmbeddingCache provides the features needed for complex applications.

Note   A detailed companion notebook for this section is available in the book's repository (RAG_Chapter6_02_Caching.ipynb). The notebook provides hands-on examples of implementing these caching strategies, performance comparisons, and real-world usage patterns. You can experiment with different cache configurations and see their impact on performance and resource usage.

Creating Custom Embedding Models

When working with RAG systems, you might encounter pre-existing embedding models that don't quite meet your specific needs. Creating custom embedding models allows you to tailor the embedding process to your unique requirements. Let's explore how to build and integrate custom embedding solutions.

Why Create Custom Embeddings?

Before diving into implementation, it's essential to understand when and why you might need custom embeddings:

1. **Domain-Specific Requirements**: Standard embedding models might not capture the nuances of specialized domains like medical terminology or legal documents effectively.

2. **Performance Optimization**: You might need embeddings optimized for specific tasks, like matching technical documentation or finding similar product descriptions.

3. **Resource Constraints**: Some environments might require lightweight embedding solutions that can run with limited computational resources.

4. **Custom Tokenization**: Your application might need special handling of domain-specific tokens, abbreviations, or formatting.

Implementing Custom Embeddings

LangChain provides a flexible interface for creating custom embedding models. Here's how to implement one:

```
from typing import List
from langchain_core.embeddings import Embeddings

class CustomEmbeddings(Embeddings):
    """Custom embedding model implementation."""

    def __init__(self, dimension: int = 512):
        """Initialize the custom embedding model.
```

```
    Args:
        dimension: The dimension of the embedding vectors
    """
    self.dimension = dimension

def embed_documents(self, texts: List[str]) ->
List[List[float]]:
    """Generate embeddings for a list of documents.

    Args:
        texts: List of texts to embed

    Returns:
        List of embedding vectors
    """
    # Add your custom embedding logic here
    embeddings = []
    for text in texts:
        embedding = self._compute_embedding(text)
        embeddings.append(embedding)
    return embeddings

def embed_query(self, text: str) -> List[float]:
    """Generate embedding for a single query text.

    Args:
        text: Text to embed

    Returns:
        Embedding vector
    """
    return self._compute_embedding(text)
```

Types of Custom Embeddings

Let's explore different approaches to implementing custom embeddings:

Transformer-Based Custom Embeddings

This approach uses pre-trained transformer models with custom modifications:

```python
class TransformerCustomEmbeddings(Embeddings):
    def __init__(self, model_name: str = "bert-base-uncased"):
        from transformers import AutoTokenizer, AutoModel
        self.tokenizer = AutoTokenizer.from_
        pretrained(model_name)
        self.model = AutoModel.from_pretrained(model_name)

    def _compute_embedding(self, text: str) -> List[float]:
        inputs = self.tokenizer(text, return_tensors="pt",
                                padding=True, truncation=True)
        outputs = self.model(**inputs)
        # Use mean pooling of the last hidden state
        embeddings = outputs.last_hidden_state.mean(dim=1)
        return embeddings[0].tolist()
```

Domain-Specific Embeddings

For specialized domains, you might want to combine multiple embedding sources:

```python
class DomainSpecificEmbeddings(Embeddings):
    def __init__(self, base_embedder, domain_processor):
        self.base_embedder = base_embedder
        self.domain_processor = domain_processor
```

```python
def embed_documents(self, texts: List[str]) ->
List[List[float]]:
    """Generate domain-aware embeddings."""
    processed_texts = [
        self.domain_processor.preprocess(text)
        for text in texts
    ]
    base_embeddings = self.base_embedder.embed_
documents(processed_texts)

    # Enhance embeddings with domain-specific features
    enhanced_embeddings = [
        self._enhance_embedding(emb, text)
        for emb, text in zip(base_embeddings, texts)
    ]
    return enhanced_embeddings
```

Best Practices for Custom Embeddings

When implementing custom embeddings, follow these best practices:

1. **Input Validation:**

```python
def _validate_input(self, text: str) -> str:
    """Validate and clean input text."""
    if not isinstance(text, str):
        raise ValueError("Input must be a string")
    cleaned_text = text.strip()
    if not cleaned_text:
        raise ValueError("Input text cannot be empty")
    return cleaned_text
```

2. **Error Handling:**

```python
def embed_query(self, text: str) -> List[float]:
    try:
        validated_text = self._validate_input(text)
        embedding = self._compute_embedding(validated_text)
        return embedding
    except Exception as e:
        raise ValueError(f"Error generating embedding: {str(e)}")
```

3. **Performance Monitoring:**

```python
def _monitor_performance(self, text: str, start_time: float):
    """Monitor embedding generation performance."""
    end_time = time.time()
    processing_time = end_time - start_time
    self.metrics['processing_times'].append(processing_time)
    self.metrics['text_lengths'].append(len(text))
```

Note   A comprehensive companion notebook for this section is available in the book's repository (RAG_Chapter6_03_ Custom_Embedding.ipynb). The notebook demonstrates these custom embedding implementations with practical examples and performance comparisons.

Understanding the Embedding Model Landscape

As we conclude our exploration of embedding models, it's worth taking a broader look at the diverse ecosystem of available options. The field of text embeddings has evolved rapidly, with different models emerging to serve various needs and use cases.

Popular Embedding Models

OpenAI Embeddings

OpenAI's embedding models, particularly their text-embedding-3 series, represent some of the most advanced options available today. These models excel at capturing subtle semantic relationships and handling complex linguistic nuances. The larger variant, text-embedding-3-large, offers state-of-the-art performance across a wide range of tasks, while the smaller variant provides an excellent balance of efficiency and accuracy. Their particular strength lies in maintaining consistent quality across different languages and domains, making them a robust choice for multilingual applications.

Sentence Transformers

Sentence transformers provide a compelling alternative for those preferring open-source solutions. These models, such as "all-mpnet-base-v2," are specifically designed for generating sentence-level embeddings. What makes them particularly interesting is their architecture, which is optimized for capturing sentence-level semantics rather than just word-level meanings. This makes them especially effective for tasks like semantic search and document similarity comparison. Their ability to run locally also offers greater flexibility in deployment and customization.

Hugging Face Embeddings

Hugging Face's suite of embedding models presents perhaps the most diverse set of options. From BERT and RoBERTa to more specialized variants, these models can be fine-tuned for specific domains while maintaining strong baseline performance. This adaptability makes them particularly valuable in specialized fields like legal or medical text analysis, where domain-specific terminology and context are crucial.

Generic Text Embeddings (GTE)

The emergence of Generic Text Embeddings represents an interesting development in the field. These models are designed with a focus on versatility, aiming to perform well across different types of text without requiring domain-specific tuning. Their efficiency in compute requirements, combined with strong general-purpose performance, makes them an attractive option for applications where resource optimization is important.

BAAI General Embeddings (BGE)

BAAI's General Embeddings have made significant strides in multilingual capabilities. These models demonstrate remarkable consistency in performance across different languages, making them particularly valuable for global applications. Their architecture is optimized for both semantic search and retrieval tasks, providing flexibility in how they can be applied in RAG systems.

Making the Right Choice

When selecting an embedding model for your RAG system, several key factors come into play:

Use Case Requirements

The choice of an embedding model should be driven primarily by your specific use case. Consider whether your application needs real-time processing or can benefit from batch operations. Think about language support requirements—will your system need to handle multiple languages? Does your domain have specific terminology that requires special handling? These requirements will help narrow down the suitable options.

Resource Considerations

Different models come with varying resource demands. While larger models like text-embedding-3-large offer superior performance, they also require more robust infrastructure. Consider your available computational resources, latency requirements, and budget constraints. Sometimes, a lighter model might offer the perfect balance of performance and efficiency for your needs.

Integration Needs

Think about how the embedding model will fit into your broader system architecture. Consider factors like API availability and reliability, especially if you're using cloud-based solutions. How will you handle batch processing? What are your caching requirements? These practical considerations can significantly impact the success of your implementation.

Summary and Next Steps

In this chapter, we explored the critical role of embedding models in RAG systems, covering

- **Core concepts**: How embeddings transform text into semantic vector representations

- **Practical implementation**: Working with popular models like OpenAI, Sentence Transformers, and Hugging Face

- **Optimization techniques**: Caching strategies to reduce costs and latency

- **Custom embeddings**: Building domain-specific solutions for specialized use cases

Key Takeaways

1. Embeddings bridge human language and machine understanding, enabling semantic search.

2. Model selection depends on your requirements (speed, accuracy, cost, and domain specificity).

3. Caching and quantization significantly improve performance in production systems.

Next Steps

In the next chapter, we'll dive into vector stores—the systems that organize and retrieve these embeddings efficiently. You'll learn how to choose, implement, and optimize vector databases to complete your RAG pipeline.

## CHAPTER 7

Vector Stores: Organizing and Retrieving Your Knowledge

In our journey through RAG systems, we've explored how documents are loaded, split into manageable chunks, and transformed into vector embeddings. Now we arrive at a critical component that ties everything together: vector stores.

Vector stores are specialized databases designed to store, organize, and retrieve vector embeddings efficiently. Think of them as intelligent librarians who not only shelve your books (embeddings) in a systematic way but also know precisely how to find the most relevant ones when you come with a query. Unlike traditional databases that excel at exact matches, vector stores specialize in similarity searches—finding vectors that are "close" to your query vector in a high-dimensional space.

As RAG applications scale to handle thousands or millions of documents, the efficiency and capabilities of your vector store become increasingly important. A well-chosen vector store can mean the difference between a responsive, accurate system and one that's slow or returns irrelevant results.

© Ranajoy Bose 2025
R. Bose, *Mastering Retrieval-Augmented Generation*,
https://doi.org/10.1007/979-8-8688-1808-0_7

In this chapter, we'll explore

- The fundamental concepts behind vector stores

- Different types of vector stores and their unique strengths

- How to select the right vector store for your specific needs

- Implementation strategies for both local development and production environments

- Advanced techniques for optimizing vector retrieval

Let's begin by understanding what makes vector stores special and how they differ from traditional databases.

Companion Notebook: A comprehensive companion notebook for this section is available in the book's repository (**RAG_Chapter7_01_Vector_Stores.ipynb**). The notebook contains detailed, executable examples of all concepts covered here, including: Setup instructions for required packages, Sample data for experimentation, Implementation of various vector stores (FAISS, Chroma, Qdrant), Common vector store operations, Integration with RAG pipelines, and Advanced usage patterns. We recommend running the notebook alongside your reading to reinforce the concepts with practical experience. Some sections require API keys for services like OpenAI, but alternative free options are provided where possible.

Understanding Vector Databases: Beyond Traditional Storage

Traditional databases excel at storing and retrieving structured data through exact matches. If you query for a product with ID "12345" or customers named "Smith," these databases quickly find the exact records. However, they struggle with semantic similarity, finding conceptually related text that doesn't contain the exact search terms.

This is where vector databases shine. Storing data as vector embeddings in a multidimensional space enables similarity-based retrieval that captures semantic relationships between concepts.

The Fundamental Difference

Consider how you might find information about "renewable energy technologies":

In a traditional database:

- You'd need to search for exact keywords like "renewable energy" AND "technologies."

- Results containing synonyms like "sustainable power solutions" might be missed.

- You'd rely on carefully constructed Boolean queries or full-text search with various limitations.

In a vector database:

- Your query gets converted to a vector embedding.

- The database finds vectors that are "closest" in the embedding space.

- Results capture semantic similarity, returning relevant content even when keywords differ.

- The search inherently understands that solar panels, wind turbines, and geothermal systems are all related to your query.

Core Components of Vector Stores

Modern vector stores typically consist of:

1. **Storage Layer**: Where the vector embeddings and associated metadata are physically stored

2. **Index Layer**: Optimized data structures that enable efficient similarity search

3. **Query Interface**: Methods to perform various types of vector searches

4. **Metadata Filtering**: Capability to combine vector similarity with traditional filtering

See Figure 7-1 for an overview of the Vector Stores.

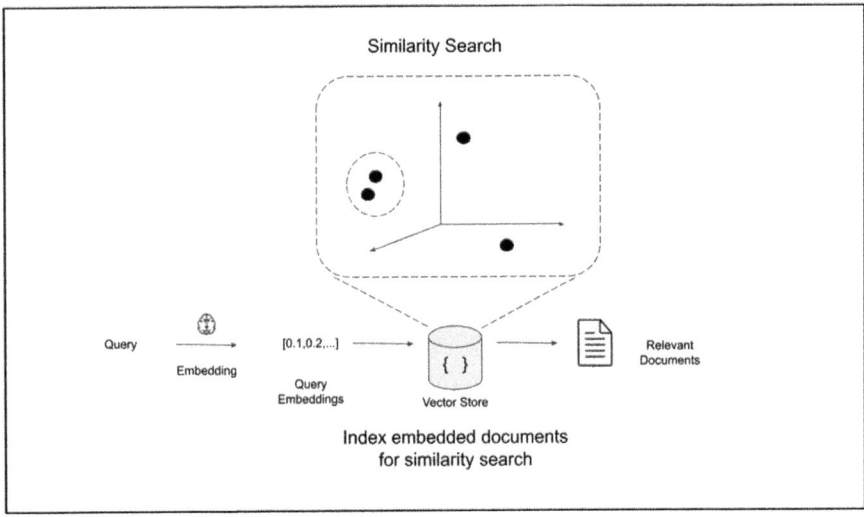

Figure 7-1. *Vector Stores: overview*

Vector Similarity Metrics

Vector databases rely on distance or similarity metrics to determine how "close" vectors are to each other:

- **Cosine Similarity**: Measures the cosine of the angle between vectors, ranging from -1 (exactly opposite) to 1 (exactly the same)

- **Euclidean Distance**: The straight-line distance between two points in the vector space

- **Dot Product**: A simple multiplication of corresponding elements, often used when vectors are normalized

- **Manhattan Distance**: The sum of absolute differences between vector components

Different vector stores may use different metrics, and some allow you to choose which one to use based on your specific embedding model and use case.

In the next section, we'll explore the various types of vector stores available and how to choose the right one for your RAG application.

Types of Vector Stores: Choosing Your Knowledge Repository

The vector store landscape offers a variety of options, each with distinct characteristics suited to different use cases. Let's explore the major categories and specific implementations to help you make an informed choice for your RAG application.

Local File-Based Vector Stores

These solutions store vectors directly on your local file system, making them ideal for development, testing, and smaller-scale applications.

FAISS (Facebook AI Similarity Search)

Developed by Facebook Research, FAISS is a high-performance library for efficient similarity search. It's particularly well-suited for dense vector embeddings.

```
from langchain.vectorstores import FAISS
from langchain.embeddings import OpenAIEmbeddings

# Initialize the embedding model
embeddings = OpenAIEmbeddings()

# Create a FAISS vector store from documents
vectorstore = FAISS.from_documents(documents, embeddings)

# Save to disk
vectorstore.save_local("faiss_index")

# Load from disk
loaded_vectorstore = FAISS.load_local("faiss_index", embeddings)
```

Chroma

Chroma is an open-source embedding database designed specifically for AI applications. It offers a simple API and is a great choice for getting started quickly.

```python
from langchain.vectorstores import Chroma

# Create a Chroma vector store
vectorstore = Chroma.from_documents(
    documents,
    embeddings,
    persist_directory="./chroma_db"
)

# Later, you can load the persisted database
persistent_vectorstore = Chroma(
    persist_directory="./chroma_db",
    embedding_function=embeddings
)
```

In-Memory Vector Stores

These solutions keep all vectors in RAM, offering extremely fast retrieval at the cost of scalability limitations.

Qdrant

Qdrant can be used both in-memory and as a persistent solution. It's especially powerful for filtering operations combined with vector search.

```python
from langchain.vectorstores import Qdrant

# Create an in-memory Qdrant instance
vectorstore = Qdrant.from_documents(
    documents,
    embeddings,
    location=":memory:",  # Use in-memory storage
    collection_name="my_documents"
)
```

Cloud-Based/Managed Vector Stores

These solutions offer scalability, managed infrastructure, and enterprise features.

Pinecone

Pinecone is a fully managed vector database for machine learning applications with real-time, high-dimensional vector search capabilities.

```
from langchain.vectorstores import Pinecone
import pinecone

# Initialize Pinecone
pinecone.init(
    api_key="your-api-key",
    environment="your-environment"
)

# Create a Pinecone index if it doesn't exist
index_name = "langchain-demo"
if index_name not in pinecone.list_indexes():
    pinecone.create_index(
        name=index_name,
        dimension=1536,  # Dimension for OpenAI embeddings
        metric="cosine"
    )

# Create the vector store
index = pinecone.Index(index_name)
vectorstore = Pinecone.from_documents(documents, embeddings,
index_name=index_name)
```

Specialized Vector Stores

These solutions address specific use cases or performance characteristics.

Milvus

Milvus is an open-source vector database designed for scalability and billions of vectors.

```
from langchain.vectorstores import Milvus

# Connect to Milvus
vectorstore = Milvus.from_documents(
    documents,
    embeddings,
    connection_args={"host": "localhost", "port": "19530"},
    collection_name="document_collection"
)
```

PGVector

PGVector extends PostgreSQL with vector similarity search capabilities, ideal if you're already using PostgreSQL.

```
from langchain.vectorstores import PGVector

# Connection string for PostgreSQL
connection_string = "postgresql://username:password@
localhost:5432/vector_db"

# Create a PGVector store
vectorstore = PGVector.from_documents(
    documents,
    embeddings,
```

```
    connection_string=connection_string,
    collection_name="document_collection"
)
```

In the next section, we'll dive deeper into how to select the right vector store for your specific use case, considering factors like scale, performance, and deployment constraints.

Selecting the Right Vector Store: Decision Factors

Choosing an appropriate vector store is critical to the success of your RAG application. Let's explore the key factors that should influence your decision.

Scale Requirements

The volume of data you need to store and search is perhaps the most significant factor in your decision:

Small-Scale (thousands of documents)

- Local solutions like FAISS or Chroma may be sufficient.

- In-memory stores provide excellent performance.

- Simpler deployment and maintenance.

Medium-Scale (hundreds of thousands of documents)

- Consider self-hosted solutions like Qdrant or Weaviate

- May require dedicated infrastructure but offer more control

- Balance between performance and cost

Large-Scale (millions+ documents)

- Cloud-based solutions like Pinecone or Milvus become necessary.

- Distributed architecture for horizontal scaling.

- Need for specialized indexing techniques.

Performance Considerations

Different vector stores optimize for different aspects of performance:

Query Latency

- FAISS is designed for high-speed retrieval, especially with optimized indices.

- In-memory solutions generally provide the lowest latency.

- Consider approximate nearest neighbor (ANN) algorithms for massive datasets.

Insertion Speed

- If your data changes frequently, evaluate the cost of index updates.

- Some stores require periodic reindexing for optimal performance.

- PGVector offers good balance for write-heavy workloads.

Memory Footprint

- FAISS and other specialized libraries offer memory-efficient indexing.

- Cloud solutions abstract this concern but impact pricing.

- Consider disk-based options like Chroma for memory-constrained environments.

Feature Requirements

Vector stores offer varying capabilities beyond basic similarity search:

Metadata Filtering

- Qdrant and Weaviate excel at combining vector search with metadata filtering

- Useful for constraining searches to specific categories, dates, or sources

Hybrid Search

- Some stores can combine vector similarity with keyword matching.

- Weaviate and Pinecone offer built-in hybrid search capabilities.

Multi-Tenancy

- If serving multiple users or applications, consider isolation capabilities.

- Cloud solutions typically offer better multitenant support.

Access Controls

- Enterprise deployments may require fine-grained permissions.

- Milvus and Pinecone offer comprehensive security features.

Deployment Constraints

Your operational environment will influence your options:

Local Development

- FAISS and Chroma are excellent for laptop-based development.

- Simple persistence to disk and version control.

Self-Hosted Production

- Consider Docker-ready solutions like Qdrant or Weaviate.

- Evaluate resource requirements and scaling options.

Cloud-Native

- Managed services reduce operational overhead.

- Consider integration with existing cloud services.

- Evaluate pricing models carefully (transaction vs. storage-based).

Cost Structure

Different vector stores have varying cost implications:

Open-Source vs. Commercial

- Many powerful options (FAISS, Milvus) are fully open-source.

- Commercial solutions typically offer better support and managed services.

Usage-Based Pricing

- Cloud solutions often charge based on data volume, query volume, or both.

- Understand the cost implications as your application scales.

Infrastructure Costs

- Self-hosted solutions require resources to operate.

- Consider both storage and compute requirements.

In the next section, we'll explore how to implement and interact with vector stores in a RAG application, focusing on the common operations and best practices.

Implementing Vector Stores in RAG Applications

Now that we understand the landscape of vector stores, let's dive into practical implementation. We'll explore common patterns and operations that apply across most vector store implementations.

Basic Vector Store Operations

Most vector stores in RAG applications support a common set of operations:

Creating a Vector Store from Documents

The most common pattern is to create a vector store from a collection of documents:

```
from langchain.vectorstores import Chroma
from langchain.embeddings import OpenAIEmbeddings
```

```
# Initialize embedding model
embeddings = OpenAIEmbeddings()

# Create a vector store from documents
vectorstore = Chroma.from_documents(
    documents,  # List of Document objects
    embeddings,
    persist_directory="./chroma_db"
)
```

Adding Documents to an Existing Store

You can incrementally add documents to most vector stores:

```
# Add new documents to the vector store
vectorstore.add_documents(new_documents)

# For stores that require explicit persistence
vectorstore.persist()
```

Similarity Search

The fundamental operation in any vector store is similarity search:

```
# Basic similarity search
results = vectorstore.similarity_search("renewable energy
innovations", k=5)

# Search with score (similarity metric)
results_with_scores = vectorstore.similarity_search_with_score(
    "renewable energy innovations",
    k=5
)
```

```
# Process the results
for doc, score in results_with_scores:
    print(f"Content: {doc.page_content[:100]}...")
    print(f"Metadata: {doc.metadata}")
    print(f"Similarity Score: {score}\n")
```

Metadata Filtering

Many vector stores support filtering based on document metadata:

```
# Search with metadata filter
results = vectorstore.similarity_search(
    "solar panel efficiency",
    k=5,
    filter={"source": "research_paper", "year": {"$gte": 2020}}
)
```

The exact syntax for filters varies by vector store implementation, but most support basic comparison operators.

Security Considerations for Persisted Indices

Many vector store implementations use Python's pickle serialization format to save and load data. While convenient, this introduces potential security risks since pickle files can be modified to execute arbitrary code when loaded.

When loading a previously saved vector store, modern implementations require an explicit acknowledgment of this risk:

```
# Loading a FAISS index from disk with security flag
vectorstore = FAISS.load_local(
    "faiss_index",
    embeddings,
    allow_dangerous_deserialization=True  # Security flag
)
```

Important   Only use the allow_dangerous_deserialization
flag with files you've created yourself or from trusted sources. Never
load vector stores from untrusted sources without proper security
validation.

For production environments, consider

- Using cloud-based vector stores that handle
  serialization securely

- Implementing validation checks before loading local
  vector stores

- Creating separate, isolated environments for loading
  potentially untrusted vector stores

Integration with Retrievers

Vector stores are typically wrapped in retrievers to integrate with the RAG
pipeline:

```
# Create a retriever from the vector store
retriever = vectorstore.as_retriever(
    search_type="similarity",
    search_kwargs={"k": 5}
)

# Use the retriever in a chain
from langchain.chains import RetrievalQA
from langchain.llms import OpenAI
```

```
qa_chain = RetrievalQA.from_chain_type(
    llm=OpenAI(),
    chain_type="stuff",
    retriever=retriever
)

# Query the chain
response = qa_chain.run("What are the latest innovations in
solar energy?")
print(response)
```

Advanced Vector Store Operations

As your RAG application matures, you may need more sophisticated vector store capabilities that go beyond basic similarity search. These advanced techniques can significantly improve the quality and relevance of retrieved information.

Maximum Marginal Relevance (MMR) Search

Standard similarity search often returns highly similar documents that contain redundant information. Consider a query about "renewable energy solutions" that returns five nearly identical documents about solar panels—this provides little value to the user.

Maximum Marginal Relevance (MMR) addresses this problem by optimizing for both relevance and diversity. It works through a two-step process:

1. First, retrieve a larger initial pool of relevant documents (the fetch_k parameter).

2. Then, iteratively select documents that balance relevance to the query with diversity from already-selected documents.

The `lambda_mult` parameter controls this balance: values closer to 1 prioritize relevance, while values closer to 0 prioritize diversity.

```
# MMR search to get diverse but relevant results
diverse_results = vectorstore.max_marginal_relevance_search(
    "renewable energy solutions",
    k=5,               # Number of results to return
    fetch_k=20,        # Initial pool of results to consider
    lambda_mult=0.5    # Diversity parameter (0 = max diversity,
                         1 = max relevance)
)
```

In practice, MMR might return a mix of documents covering solar, wind, hydro, and geothermal energy rather than five variations on the same solution. This provides the LLM with a richer context for generating comprehensive responses.

Let's visualize how MMR works. Imagine documents as points in space, where proximity represents similarity:

1. **Standard search**: Returns the 5 closest points to the query

2. **MMR search**: First selects the closest point, then balances closeness with distance from previously selected points

This diversity is particularly valuable when

- Exploring a broad topic with multiple facets

- Building comprehensive knowledge bases

- Avoiding "echo chamber" effects in information retrieval

Hybrid Search

Vector search excels at semantic understanding but sometimes misses exact keyword matches. Conversely, traditional keyword search catches exact matches but misses semantic connections. Hybrid search combines both approaches for more robust retrieval.

Hybrid search works by

1. Performing both vector similarity search and keyword search separately

2. Combining the results using a weighted scoring mechanism

3. Returning the highest-scoring documents based on both methods

The alpha parameter determines the balance between the two approaches:

```
# Using hybrid search (implementation varies by vector store)
results = vectorstore.hybrid_search(
    "photovoltaic efficiency improvements",
    k=5,
    alpha=0.5  # Weight between vector and keyword search (0 =
    all keyword, 1 = all vector)
)
```

Consider a query like "photovoltaic efficiency improvements":

- Vector search might find documents discussing general solar panel efficiency, even if they don't contain the exact word "photovoltaic."

- Keyword search would find documents containing "photovoltaic" and "efficiency," even if they use these terms in different contexts.

- Hybrid search combines these strengths, finding semantically relevant documents that also contain key terms.

This approach is particularly valuable for

- Technical or specialized domains with precise terminology

- Queries containing specific product names, people, or places

- Applications where both conceptual understanding and precise term matching are important

Not all vector stores support hybrid search natively. For those that don't, you can implement a custom solution by

1. Running separate vector and keyword searches

2. Assigning scores to each result

3. Combining and reranking based on a weighted formula

Deleting Documents

As your knowledge base evolves, you may need to remove outdated or incorrect documents from your vector store. Most modern vector stores provide methods for selective document deletion:

```
# Delete documents by IDs
vectorstore.delete(["doc1", "doc2", "doc3"])

# Some stores support deletion by metadata filter
vectorstore.delete(filter={"source": "outdated_website"})
```

Efficient document deletion is crucial for maintaining the accuracy and relevance of your RAG system over time. When implementing a document deletion strategy, consider

- Tracking document IDs systematically for future reference

- Using descriptive metadata to enable bulk deletions when information sources change

- Implementing periodic review processes to identify outdated content

In the next section, we'll explore advanced techniques for optimizing vector stores to improve retrieval quality and performance in production RAG systems.

Optimizing Vector Stores for Performance and Quality

As RAG applications scale from experimental prototypes to production systems, vector store optimization becomes increasingly important. In this section, we'll explore advanced strategies to enhance search performance and retrieval quality and explain the underlying principles in detail.

Companion Notebook: A comprehensive companion notebook for this section is available in the book's repository (**RAG_Chapter7_02_ Vector_Store_Optimization.ipynb**). The notebook contains detailed, executable examples of all concepts covered here, including HNSW and IVF indexing with parameter tuning, vector compression techniques, including quantization, semantic chunking comparison, re-ranking and advanced retrieval methods, and performance analysis with visualizations. You can run these directly in Google Colab to get hands-on experience.

Indexing Strategies for Large-Scale Vector Stores

When your vector collection grows beyond a few thousand entries, brute-force comparison becomes impractical. At this scale, Approximate Nearest Neighbor (ANN) algorithms become essential. These algorithms trade a small amount of accuracy for dramatic performance improvements, often making searches hundreds or thousands of times faster.

HNSW (Hierarchical Navigable Small World)

HNSW is one of the most powerful indexing algorithms in modern vector search systems, consistently delivering excellent performance across various benchmarks.

How HNSW Works:

Imagine a multilayered graph where each node (vector) connects to its neighbors. HNSW organizes this graph hierarchically:

1. The bottom layer contains all vectors, each connected to its nearest neighbors.

2. Upper layers contain progressively fewer nodes, forming "express lanes" for traversal.

3. Connections between nodes follow small-world network principles, allowing for efficient navigation.

During search, the algorithm

- Starts at a random entry point in the top layer

- Greedily navigates to the nearest neighbor of the query

- Once it can't get any closer at the current layer, it descends to the next lower layer

- This process repeats until reaching the bottom layer, where fine-grained search happens

This approach creates a balance of "long-range" connections (for jumping quickly to the right neighborhood) and "short-range" connections (for precise final navigation).

```
# Example: Creating a FAISS index with HNSW
import faiss

# Assume vectors is a numpy array of shape (n_vectors,
dimension)
dimension = vectors.shape[1]

# Create an HNSW index
index = faiss.IndexHNSWFlat(dimension, 32)  # 32 connections
per layer
index.hnsw.efConstruction = 200  # Controls build-time accuracy
index.hnsw.efSearch = 128        # Controls search-time
                                   accuracy

# Add vectors to the index
index.add(vectors)
```

The key parameters for tuning HNSW indices require careful consideration:

- M: The number of connections per node (default 32 in our example).

  - Higher values create more connections, improving accuracy but increasing memory usage and build time.

  - Lower values save memory but may require more hops to reach the target.

  - Most applications find optimal performance in the 16-64 range.

- `efConstruction`: Controls index build quality.

  - Higher values create a more optimal graph structure but slow down construction.

  - Values of 100–500 typically offer a good balance for most applications.

  - Critical for one-time index builds that will serve many queries.

- `efSearch`: Controls search precision.

  - Higher values examine more potential paths during search.

  - Directly impacts the accuracy-speed trade-off at query time.

  - Can be adjusted dynamically based on query requirements.

HNSW excels when

- Query speed is critical

- You need high recall rates (>95%)

- Memory availability is not severely constrained

- Your vectors have moderate to high dimensionality (32-1536)

IVF (Inverted File Index)

While HNSW creates a connected graph, IVF takes a fundamentally different approach by partitioning the vector space into clusters. This "divide and conquer" strategy dramatically reduces the search space.

How IVF Works:

1. **Clustering Phase:** During index building, IVF applies k-means clustering to partition all vectors into clusters (called "cells").

2. **Assignment Phase:** Each vector is assigned to its nearest cluster centroid.

3. **Search Phase:** At query time, IVF

   - Identifies the closest cluster centroids to the query vector

   - Searches only within the most promising clusters

   - Returns the best matches from these selected clusters

This approach uses a two-level search: first finding relevant clusters, then searching within those clusters.

```
# Example: Creating a FAISS IVF index
dimension = vectors.shape[1]

# Choose number of clusters - a critical parameter
# Rule of thumb: sqrt(N) to 4*sqrt(N) where N is the number
of vectors
n_clusters = int(4 * np.sqrt(vectors.shape[0]))

# Create a quantizer for clustering
quantizer = faiss.IndexFlatL2(dimension)

# Create an IVF index
index = faiss.IndexIVFFlat(quantizer, dimension, n_clusters)

# Training is required for IVF (unlike HNSW)
# This establishes the cluster centroids
index.train(vectors)
```

```
# Add vectors to the index
index.add(vectors)

# Set the number of clusters to probe during search
index.nprobe = 10  # Higher = more accurate, slower search
```

Critical parameters for IVF include

- n_clusters: The number of partitions

  - Too few clusters mean each contains too many vectors, slowing the search.

  - Too many clusters can lead to suboptimal partitioning and reduced accuracy.

  - The square root of your dataset size is often a good starting point.

- nprobe: The number of clusters to search

  - This parameter offers a direct accuracy–speed trade-off at query time.

  - Searching 1 cluster is fastest but least accurate.

  - Searching all clusters equals a brute-force search (accurate but slow).

  - Typical values range from 1 to 10% of total clusters.

IVF is particularly well-suited for

- Very large datasets (millions to billions of vectors)

- Cases where you can afford a training phase before search

- Applications where search parameters need runtime adjustment

- Scenarios where memory efficiency is a priority

Managing Vector Store Size and Memory Usage

As vector collections grow, memory and storage requirements become substantial. Advanced compression techniques can dramatically reduce these requirements with minimal impact on quality.

Scalar Quantization

Vector quantization tackles the memory problem by reducing the precision of each vector component. It's analogous to converting high-resolution images to lower resolutions—you save space while preserving most of the critical information.

How Scalar Quantization Works:

1. Instead of storing each vector component as a 32-bit floating-point value, quantization maps it to a smaller integer representation.

2. This mapping uses a learned or predefined codebook that preserves the overall distribution of values.

3. During search, quantized values are compared directly, or can be reconstructed to approximations of the original vectors.

```
# Example: Creating a quantized FAISS index
dimension = vectors.shape[1]

# Create a quantized index (8 bits per component)
# Reduces memory by 4x compared to 32-bit floats
index = faiss.IndexScalarQuantizer(dimension, faiss.
ScalarQuantizer.QT_8bit)
```

```
# Or for more aggressive compression (4 bits per component)
# Reduces memory by 8x compared to 32-bit floats
index = faiss.IndexScalarQuantizer(dimension, faiss.
ScalarQuantizer.QT_4bit)

# Add vectors to the index
index.add(vectors)
```

The key trade-off is precision vs. memory:

- 8-bit quantization (256 distinct values per component) typically reduces accuracy by only 1–3%.

- 4-bit quantization (16 distinct values per component) might reduce accuracy by 5–15%.

- For embeddings used in RAG, 8-bit precision is often sufficient.

Scalar quantization shines when

- You need to reduce memory usage with minimal impact on quality.

- Your vectors have a relatively uniform distribution of values.

- Fast search is more important than perfect reconstruction of original vectors.

Product Quantization (PQ)

While scalar quantization treats each dimension independently, product quantization recognizes that dimensions in embedding vectors are often correlated. By exploiting these correlations, PQ achieves much higher compression rates.

How Product Quantization Works:

1. Each vector is split into several subvectors (e.g., a 768-dimension vector might be split into 8 subvectors of 96 dimensions each).

2. For each subvector position, a small codebook of representative values is learned from the training data.

3. Each original subvector is replaced by the index of its nearest codebook entry.

4. During search, distances are computed using a precomputed distance table between codebook entries.

This approach achieves remarkable compression because

- Instead of storing all dimensions for each vector, we only store codebook indices.

- With 256 codebook entries per subvector (8 bits), an entire subvector is represented by a single byte.

- The distance calculation becomes a set of lookups rather than floating-point operations.

```
# Example: Creating a FAISS index with product quantization
dimension = vectors.shape[1]
m = 8  # Number of subvectors (must divide dimension evenly)

# Create a PQ index
index = faiss.IndexPQ(dimension, m, 8)  # 8 bits per subvector

# Train the index on representative data
# This learns the optimal codebooks
index.train(vectors)
```

```
# Add vectors to the index
index.add(vectors)
```

Key parameters for tuning PQ:

- m: Number of subvectors

  - More subvectors means higher precision but larger codebooks

  - Fewer subvectors means greater compression but less precision

  - Must divide the vector dimension evenly

- Bits per subvector (typically 8)

  - Determines the codebook size (2^bits entries per subvector)

  - 8 bits (256 entries) balances precision and memory efficiency

The compression ratio is remarkable: a 1536-dimensional vector that normally requires 6KB (with 32-bit floats) might only need 8–32 bytes with PQ, a 200-750x reduction!

PQ is especially valuable when

- Memory or storage constraints are severe

- You have a very large collection (millions to billions of vectors)

- You can accept a moderate reduction in retrieval quality

- You have a representative training dataset

Retrieval Quality Optimization

While performance optimizations focus on speed and efficiency, retrieval quality optimizations ensure that the right documents are found regardless of how they're stored.

Semantic Chunking

Text splitting is often treated as a simple mechanical process, but the boundaries where we split text can dramatically impact retrieval quality.

How Semantic Chunking Works:
Rather than splitting documents based solely on character count, semantic chunking aims to preserve the natural units of meaning in the text:

1. It respects document structure by preferentially splitting at major boundaries like headings, paragraphs, and sentences.

2. It maintains context by ensuring chunks contain complete semantic units.

3. It allows strategic overlap so important context isn't lost between chunks.

```
from langchain.text_splitter import
RecursiveCharacterTextSplitter

# Advanced splitter that respects semantic boundaries
splitter = RecursiveCharacterTextSplitter(
    chunk_size=1000,
    chunk_overlap=200,
    # The order of separators matters - it tries each in sequence
    separators=["\n## ", "\n### ", "\n#### ", "\n\n", "\n",
    " ", ""]
)
```

```
# Create semantically coherent chunks
chunks = splitter.split_documents(documents)
```

The separators parameter is particularly important:

- The splitting process tries each separator in order.

- If a separator would create chunks of appropriate size, it's used.

- If not, it proceeds to the next separator.

- This creates a hierarchy from "prefer splitting at major section headings" down to "split between words if necessary."

Semantic chunking improves retrieval quality by

- Keeping related information together in the same chunk

- Preserving the natural flow and context of the original document

- Reducing the likelihood of retrieving fragments that require additional context to understand

For maximum effectiveness, consider document-specific splitting strategies:

- Academic papers might split by section, subsection, and paragraph.

- Code documentation might split by function, class, or module.

- Legal documents might split by article, clause, or provision.

Re-ranking

Initial results may not be optimally ordered even with the best vector search. Re-ranking applies a more sophisticated, typically more computationally intensive scoring model to refine the initial retrieval.

How Re-ranking Works:

1. The initial vector search retrieves a candidate set of documents (typically more than needed).

2. A re-ranking model evaluates each document-query pair more thoroughly.

3. Documents are reordered based on these more precise relevance scores.

4. Only the top-ranked documents after re-ranking are returned.

This two-stage approach combines the speed of vector search with the precision of more complex models.

```
from langchain.retrievers import ContextualCompressionRetriever
from langchain.retrievers.document_compressors import LLMRerank

# Create a retriever with LLM-based re-ranking
llm = OpenAI(temperature=0)
reranker = LLMRerank.from_llm(
    llm=llm,
    k=5  # Number of documents to return after re-ranking
)

rerank_retriever = ContextualCompressionRetriever(
    base_compressor=reranker,
```

```
    base_retriever=vectorstore.as_retriever(
        search_kwargs={"k": 20}  # Retrieve more docs initially
        for re-ranking
    )
)

# Get re-ranked documents
reranked_docs = rerank_retriever.get_relevant_documents(
    "What advances have been made in perovskite solar cells?"
)
```

In this implementation

1. The vector store retrieves 20 candidate documents.

2. Each document is evaluated by the LLM for relevance to the query.

3. The top 5 documents after re-ranking are returned.

Re-ranking can dramatically improve precision because

- It can utilize more powerful models than the initial retrieval.

- It can perform deeper semantic analysis of the full text.

- It can consider document-query interactions rather than just similarity.

- It can incorporate additional signals like document structure, authority, or recency.

Types of re-rankers include

- **LLM-based**: Using language models to evaluate relevance (as shown above)

- **Cross-encoders**: Models that process query and document together rather than separately

- **Hybrid**: Combining multiple signals, including sparse (keyword) and dense (semantic) relevance

Re-ranking works best when

- Precision of the top results is critical
- You can afford the additional computation time
- Your initial retrieval has high recall but imperfect precision
- Complex queries require nuanced understanding of relevance

Query Transformations

Sometimes the query itself needs transformation to better match the way information is stored in your documents.

How Query Transformation Works:
Query transformation techniques modify or expand the original query to improve retrieval:

1. **Query Expansion**: Adding related terms or synonyms
2. **Query Reformulation**: Rephrasing the query to better match document language
3. **Query Decomposition**: Breaking complex queries into simpler sub-queries

```
from langchain.retrievers import MultiQueryRetriever

# Create a retriever that generates multiple query variations
retriever = MultiQueryRetriever.from_llm(
    retriever=vectorstore.as_retriever(),
    llm=OpenAI()
)
```

```
# The retriever will generate variations of the query,
# perform separate searches, and combine the results
results = retriever.get_relevant_documents(
    "How do newer types of solar cells compare to silicon?"
```

This approach might transform the original query into variations like

- "What are the differences between newer solar cell technologies and traditional silicon cells?"

- "Comparison of perovskite, quantum dot, and silicon solar cells"

- "Advantages and disadvantages of next-generation solar cells versus silicon"

By searching with multiple formulations, the retriever can find relevant documents that might not match the original phrasing, significantly improving recall.

Query transformation is particularly valuable for

- Handling ambiguous or underspecified queries

- Bridging vocabulary gaps between queries and documents

- Finding information that might be expressed in different ways

- Complex information needs that span multiple aspects of a topic

Monitoring Vector Store Performance

In production environments, systematic monitoring helps identify and address performance issues before they impact users.

```python
import time
from contextlib import contextmanager

@contextmanager
def timing(label):
    start_time = time.time()
    try:
        yield
    finally:
        end_time = time.time()
        print(f"{label}: {end_time - start_time:.3f} seconds")

# Example usage
with timing("Vector search"):
    results = vectorstore.similarity_search("renewable
    energy", k=5)
```

A comprehensive monitoring strategy should track

Performance Metrics:

- Query latency (average, percentiles, distribution)

- Index build/update time

- Memory usage patterns

- Throughput (queries per second)

Accuracy Metrics:

- Precision@k (relevance of top k results)

- Recall@k (percentage of relevant items found in top k)

- Mean Reciprocal Rank (MRR)

- User engagement with results (clicks, time spent)

Resource Utilization:

- CPU usage during queries and updates

- Memory consumption

- Disk I/O for persistence operations

- Network traffic for distributed systems

By establishing baselines and alerting on deviations, you can maintain reliability while continuously improving your vector search performance.

In the next section, we'll explore how to effectively deploy vector stores in production environments, addressing considerations like scalability, availability, and data consistency.

Deploying Vector Stores in Production Environments

Moving a RAG system from development to production introduces new challenges and considerations. While the core principles remain the same, production environments demand attention to scalability, reliability, consistency, and security. In this section, we'll explore key considerations for effectively deploying vector stores in production systems.

Companion Notebook: To help you implement the deployment patterns discussed in this section, we've created a comprehensive companion notebook—RAG_Chapter7_03_Vector_Store_Deployment. ipynb. This notebook provides detailed code examples for scalability configurations, data synchronization strategies, high availability setups, security implementations, and monitoring tools.

Scalability Considerations

As your knowledge base grows and user traffic increases, your vector store must scale accordingly. Let's explore the two fundamental approaches to scaling and their implications.

Vertical vs. Horizontal Scaling

Vertical Scaling involves increasing the resources of a single instance—essentially getting a bigger machine. This approach works well up to a point and offers several advantages:

- **Simplicity**: No need to coordinate across multiple nodes

- **Performance**: Avoids network overhead between nodes

- **Consistency**: Data is centralized and always in sync

Here's how you might configure a vertically scaled Qdrant instance:

```
# Connect to a powerful single instance
client = QdrantClient(
    url="http://localhost:6333",
    timeout=120  # Increased timeout for larger operations
)

# Create a collection optimized for a powerful machine
client.recreate_collection(
    collection_name="documents",
    vectors_config=VectorParams(size=1536, distance=Distance.
    COSINE),
```

```
optimizers_config={
    "default_segment_number": 2,  # Fewer segments for
                                    more RAM
    "indexing_threshold": 50000,  # Higher threshold for
                                    batch processing
},
shard_number=1  # Single shard for vertical scaling
)
```

This configuration takes advantage of a powerful server by using fewer, larger segments and processing more vectors in each batch operation.

However, vertical scaling eventually hits physical limits. Even the most powerful servers have finite memory and processing capacity. When your vector collection reaches billions of vectors or your query volume exceeds what a single machine can handle, you'll need another approach.

Horizontal Scaling distributes your vector collection across multiple machines. This brings different advantages:

- **Virtually unlimited capacity**: Add more nodes as needed.

- **Fault tolerance**: The system can continue operating if some nodes fail.

- **Cost efficiency**: Can use commodity hardware instead of specialized servers.

A horizontally scaled configuration might look like this:

```
# Create a collection with multiple shards
collection = Collection(
    name="documents",
    schema=schema,
    shards_num=3,  # Distribute across 3 shards
    properties={"collection.ttl.seconds": 0}
)
```

The challenge with horizontal scaling lies in coordination. When vectors are spread across multiple machines, the system must decide which nodes to query and how to combine results. This introduces complexity in three key areas:

1. **Sharding**: How to divide vectors across nodes (by ID ranges, hash partitioning, or semantic clusters)

2. **Replication**: How to maintain copies for redundancy and load distribution

3. **Consistency**: How to handle updates across multiple nodes

Modern distributed vector databases implement sophisticated approaches to these challenges. For instance, Milvus uses a coordinator service to manage sharding policies and maintains a global view of the system, while Qdrant employs consensus algorithms to ensure updates are consistent across replicas.

Cloud-Native Deployments

For most production RAG systems, cloud-based vector stores offer significant advantages over self-hosted solutions. These managed services abstract away much of the operational complexity.

Here's how you might configure a cloud-based vector store like Pinecone:

```
# Initialize Pinecone with API key
pinecone.init(
    api_key="your-api-key",
    environment="gcp-starter"
)
```

```
# Create a distributed index with optimal configuration
pinecone.create_index(
    name="production-index",
    dimension=1536,
    metric="cosine",
    pods=3,  # Number of compute units
    pod_type="p2",  # Performance tier
    replicas=2  # For high availability and throughput
)
```

This configuration provides immediate benefits:

- Distributed architecture with multiple compute pods

- Built-in redundancy with replicas

- No infrastructure to manage

- Automatic scaling capabilities

The trade-off comes in flexibility and potential vendor lock-in. Self-hosted solutions give you complete control over the infrastructure and configuration but require expertise to operate effectively.

Data Consistency and Updates

Production RAG systems rarely contain static knowledge bases. As your source documents evolve, your vector store must stay in sync.

Handling Document Updates

When documents change, their vector representations must be updated accordingly. This seemingly simple requirement introduces several challenges:

1. **Identity Management**: To update a document, you must be able to identify its existing representation in the vector store. This requires maintaining a consistent ID scheme.

2. **Atomicity**: Updates should be atomic to prevent inconsistent states.

3. **Indexing Impact**: Most vector indices aren't designed for frequent updates.

The simplest approach is the delete-then-insert pattern:

```
# Add documents with explicit IDs
document_ids = [get_document_id(doc) for doc in documents]
vectorstore = Chroma.from_documents(
    documents,
    embeddings,
    ids=document_ids
)

# When documents change:
updated_documents = [updated_doc1, updated_doc2]
updated_ids = [get_document_id(doc) for doc in updated_
documents]

# Delete old versions
vectorstore.delete(updated_ids)

# Add updated versions
vectorstore.add_documents(updated_documents, ids=updated_ids)
```

This works well for occasional updates but becomes inefficient for frequent changes. More sophisticated vector stores offer "upsert" operations that handle this pattern internally with optimized index updates.

Synchronization Strategies

Keeping your vector store synchronized with your source data requires a systematic approach. Three primary strategies exist, each with different trade-offs:

1. **Event-driven updates** process changes immediately when source documents change:

```python
def document_change_handler(document_id, new_content,
operation_type):
    """Handle updates to documents in real-time."""
    if operation_type == "CREATE" or operation_type == "UPDATE":
        # Create embedding and update vector store
        doc_object = Document(
            page_content=new_content,
            metadata={"doc_id": document_id}
        )
        vector_id = f"doc_{document_id}"
        vectorstore.add_documents([doc_object],
        ids=[vector_id])

    elif operation_type == "DELETE":
        # Remove from vector store
        vector_id = f"doc_{document_id}"
        vectorstore.delete([vector_id])
```

   This approach minimizes the time between a document update and its availability in search results. It works well when

   - Real-time consistency is critical

   - Updates are relatively infrequent

   - Your source system can reliably emit change events

The challenge lies in handling high-volume updates and ensuring reliability when event processing fails.

2. **Batch synchronization** updates the vector store at regular intervals:

```python
def scheduled_synchronization():
    """Synchronize vector store with source system
    periodically."""
    # Get all document IDs from source system
    source_doc_ids = source_system.get_all_document_ids()

    # Get all document IDs from vector store
    vector_ids = vectorstore.get_all_ids()

    # Find documents to add/remove/update through set
    operations
    docs_to_add = [id for id in source_doc_ids
                    if f"doc_{id}" not in vector_ids]

    ids_to_remove = [id for id in vector_ids
                      if id.startswith("doc_") and id[4:] not in
                      source_doc_ids]
    # Process additions and removals
    # ...
```

This approach is more efficient for high-volume changes and provides more predictable system load. The downside is that users may see outdated information between synchronization runs.

3. **Change data capture (CDC)** tracks changes at the database level, offering a middle ground between the previous approaches:

```python
def process_database_change_log(change_log):
    """Process database change log entries."""
    for change in change_log:
        document_id = change['document_id']
        vector_id = f"doc_{document_id}"

        if change['operation'] == 'INSERT' or
        change['operation'] == 'UPDATE':
            # Get latest document content
            document = source_system.get_document(document_id)

            # Update vector store
            vectorstore.add_documents([document],
            ids=[vector_id])

        elif change['operation'] == 'DELETE':
            # Remove from vector store
            vectorstore.delete([vector_id])
```

CDC works by monitoring database transaction logs to identify changes without polling or explicit events. This approach shines when your source data resides in a database that supports CDC.

In practice, many production systems combine these approaches—using event-driven updates for critical changes while relying on scheduled batch synchronization as a fallback to ensure eventual consistency.

High Availability and Disaster Recovery

For production systems, downtime isn't just an inconvenience—it can have significant business impact. A robust vector store deployment must continue operating despite component failures and provide mechanisms to recover from catastrophic events.

Replication

Replication maintains multiple copies of your vector data, serving two primary purposes:

Redundancy ensures that if one copy becomes unavailable, others can take its place. Here's how you might configure replication in a system like Weaviate:

```
# Create a class (collection) with replication configuration
class_obj = {
    "class": "Document",
    "vectorizer": "none",  # We'll provide our own vectors
    "vectorIndexConfig": {
        "distance": "cosine",
        # Index parameters...
    },
    "replicationConfig": {
        "factor": 2  # Each vector exists on 2 nodes
    }
}

# Create the class with replication
client.schema.create_class(class_obj)
```

With this configuration, every vector is stored on two different nodes, providing resilience against individual node failures.

Load distribution uses multiple replicas to handle more queries simultaneously. This pattern works well when read operations significantly outnumber writes (common for RAG systems).

The key consideration with replication is consistency—how quickly changes to one replica propagate to others. Different vector stores offer various consistency models, from eventual consistency (changes propagate asynchronously) to strong consistency (all replicas update simultaneously).

Backup and Restore

Even with replication, regular backups remain essential. They protect against scenarios where replicas might simultaneously fail or become corrupted.

For file-based vector stores like FAISS, backups can be straightforward:

```python
import faiss
import time

# Create a timestamped backup
backup_path = f"backups/faiss_index_{int(time.time())}.bin"

# Write the index to disk
faiss.write_index(faiss_index, backup_path)

# Later, to restore:
restored_index = faiss.read_index(backup_path)
```

For cloud-based stores, you might need to implement custom backup logic, or use their built-in backup features if available:

```python
def backup_pinecone_index(index_name, bucket_name, prefix):
    """Back up a Pinecone index to S3."""
    # Get the index
    index = pinecone.Index(index_name)

    # Define a batch size for retrieval
    batch_size = 1000

    # Process in batches
    # ...
```

A comprehensive backup strategy includes regular full backups, incremental backups for recent changes, and validated restore procedures with defined recovery time objectives.

Security Considerations

Vector stores often contain sensitive information extracted from internal documents, customer data, or proprietary knowledge bases. Securing this information requires a multilayered approach.

Data Encryption

Protecting vector data requires encryption in two states:

Encryption in transit ensures that data moving between components cannot be intercepted:

```
# Connect using TLS for encryption in transit
client = QdrantClient(
    url="https://your-server:6333",  # HTTPS endpoint
    timeout=60,
    api_key="your-api-key"  # Authentication
)
```

Encryption at rest protects data stored on disk or in databases from unauthorized access, typically implemented at the storage or database level.

While vector embeddings themselves might not seem sensitive (they're mathematical representations without obvious meaning), they can still leak information when analyzed. More importantly, the metadata associated with vectors often contains explicit sensitive information.

Access Controls

Implementing proper authentication and authorization ensures that only authorized users and systems can access your vector store:

```
# Connect with authentication
connections.connect(
    alias="default",
    host="localhost",
    port="19530",
    user="your-username",  # Authentication
    password="your-password"
)
```

Many RAG applications require sophisticated access controls that reflect the permissions on the original documents. For example, if a user doesn't have access to a particular document in the content management system, they shouldn't receive results from that document through the RAG interface either.

Monitoring and Maintenance

Ongoing monitoring helps identify and address issues proactively, before they impact users.

Performance Metrics

A comprehensive monitoring strategy should track key metrics across query performance, index health, and system resources. Here's a simple wrapper that can help capture operation metrics:

```
@contextmanager
def monitor_operation(operation_name, metadata=None):
    """Monitor the performance of vector store operations."""
    start_time = time.time()
    error = None
    try:
        yield
```

```
    except Exception as e:
        error = e
        raise
    finally:
        duration = time.time() - start_time
        log_entry = {
            "operation": operation_name,
            "duration_seconds": duration,
            "success": error is None,
            "metadata": metadata or {}
        }
        logging.info(f"Vector Store Metric: {log_entry}")

# Usage example
with monitor_operation("similarity_search", {"query":
"renewable energy", "k": 5}):
    results = vectorstore.similarity_search("renewable
    energy", k=5)
```

This monitoring approach allows you to track performance trends over time, establishing baselines and identifying potential issues before they become critical.

Health Checks

Regular health checks complement metrics by actively verifying that the system works as expected:

```
def vector_store_health_check(vectorstore):
    """Perform basic health check on vector store."""
    try:
        # Try a simple query
        start_time = time.time()
```

```python
vectorstore.similarity_search("test query", k=1)
query_time = time.time() - start_time

# Check if query time is within acceptable range
if query_time > 1.0:   # More than 1 second is slow
    return {
        "status": "warning",
        "message": f"Slow query response: {query_
        time:.2f} seconds",
        "query_time": query_time
    }

return {
    "status": "healthy",
    "message": "Vector store is responding normally",
    "query_time": query_time
}

except Exception as e:
    return {
        "status": "unhealthy",
        "message": f"Vector store error: {str(e)}",
        "error": str(e)
    }
```

Health checks like this can be integrated into monitoring systems to trigger alerts when problems arise. More sophisticated health checks might include

- **Canary queries** that verify search quality with known expected results

- **Indexing checks** that confirm new documents appear in search results within expected time frames

- **Connectivity checks** that validate all components of a distributed system are communicating properly

When health checks fail, the monitoring system should trigger appropriate alerts and potentially initiate automated remediation steps.

Operational Considerations

Beyond the technical aspects, successfully operating vector stores in production requires operational discipline.

Capacity planning ensures you have sufficient resources to handle growth:

```
# Example capacity planning calculation
docs_per_month = 10000  # Expected document growth rate
avg_embeddings_per_doc = 5  # Average chunks per document
embedding_dimension = 1536  # Dimension of OpenAI embeddings
bytes_per_dimension = 4  # 32-bit float

# Monthly storage growth (raw vectors, not including indices)
monthly_storage_gb = (docs_per_month * avg_embeddings_per_doc *
                      embedding_dimension * bytes_per_dimension)
                      / (1024**3)

print(f"Expected monthly storage growth: {monthly_storage_
gb:.2f} GB")
```

This type of capacity planning helps you provision appropriate resources ahead of time, rather than reacting to resource constraints after they impact users.

Update windows schedule potentially disruptive maintenance during low-traffic periods. For vector stores that require periodic optimization or reindexing, these windows should be carefully planned:

```
# Example scheduled optimization (pseudocode)
def scheduled_optimization(vectorstore, start_time=datetime.
time(2, 0)):  # 2 AM
    """Perform index optimization during off-hours."""
    # Wait until scheduled time
    now = datetime.datetime.now().time()
    if now > start_time:
        # If optimization is CPU-intensive, consider throttling
        vectorstore.optimize_index(max_cpu_percent=70)
        logging.info("Index optimization complete")
```

Documentation captures the knowledge needed to operate the system effectively, including architecture diagrams, configuration parameters, troubleshooting guides, and recovery procedures.

By addressing these operational aspects alongside the technical implementation, you can ensure that your vector store remains reliable, performant, and secure as your RAG application evolves.

Vector Store Evaluation and Benchmarking

Selecting the right vector store for your RAG application requires objective evaluation and benchmarking. In this section, we'll explore methodologies for measuring vector store performance and quality, helping you make informed decisions based on your specific requirements.

Companion Notebook: To help you implement the evaluation techniques discussed in this section, we've created a comprehensive benchmarking notebook—RAG_Chapter7_04_Vector_Store_ Evaluation. This notebook includes detailed code for performance testing, quality evaluation, and comparative analysis of vector store implementations.

Measuring Vector Store Performance

Performance evaluation must consider multiple dimensions beyond simple query speed:

Latency Benchmarking

Query latency—the time from submission to result retrieval—is perhaps the most visible performance metric. When evaluating latency, consider these key measurements:

- **Mean latency**: The average response time

- **Median latency**: The typical response time (less affected by outliers)

- **p95 latency**: The worst response time for 95% of queries

- **Variability**: The difference between minimum and maximum latencies

```
# Simple latency benchmark example
def test_latency(vector_store, query, iterations=100):
    latencies = []
    for _ in range(iterations):
        start_time = time.time()
        vector_store.similarity_search(query, k=5)
        latencies.append(time.time() - start_time)
    return statistics.mean(latencies), sorted(latencies)
    [int(iterations * 0.95)]
```

In production RAG systems, consistent performance is often more important than raw speed. A vector store with higher but consistent latency may provide a better user experience than one with lower average but highly variable latency. This can be particularly important for interactive applications where unpredictable response times frustrate users.

Different vector stores exhibit distinct latency profiles. FAISS and Qdrant typically offer excellent query speeds due to their optimized C++ cores, while some cloud-based solutions may introduce additional network latency that adds to overall response time.

Throughput Testing

Throughput measures how many queries a vector store can handle simultaneously—critical for multiuser applications. Unlike latency, which focuses on individual query performance, throughput testing subjects the vector store to concurrent load to identify bottlenecks and saturation points.

Effective throughput testing requires

- Simulating multiple concurrent users

- Using a representative query mix

- Running tests long enough to identify performance degradation

- Measuring both query rate (queries per second) and latency under load

A well-designed vector store should scale near-linearly with hardware resources up to a certain point. Beyond that point, you'll observe diminishing returns where adding more concurrent queries causes both throughput to plateau and latency to increase dramatically.

When interpreting throughput results, pay attention to two key indicators:

1. **Maximum sustainable throughput**: The highest query rate before latency becomes unacceptable

2. **Graceful degradation**: How the system behaves beyond its capacity limit

Some vector stores degrade gracefully under load, maintaining reasonable latency for all queries while processing fewer per second. Others might maintain throughput at the expense of dramatically increased latency for some queries, creating a poor experience for those users.

Scalability Analysis

Understanding how performance scales with data volume is crucial for long-term planning. As your vector collection grows, both index build time and query latency will change—sometimes dramatically.

Key scalability metrics to measure include

- **Index build time**: How long it takes to create the index from scratch

- **Incremental update time**: How long it takes to add new documents

- **Query latency vs. collection size**: How search speed changes as data grows

- **Memory usage vs. collection size**: How resource requirements scale

The ideal scalability pattern shows sublinear growth in both query time and memory usage as data volume increases. Vector stores with efficient indexing algorithms like HNSW typically show logarithmic rather than linear scaling in query time, meaning that a 10x increase in data might only result in a 2x increase in search time.

Different vector stores exhibit vastly different scalability characteristics:

- FAISS excels at handling millions of vectors with efficient memory usage.

- Qdrant and Weaviate offer good performance for both reads and writes at scale.

- Cloud solutions like Pinecone can scale to billions of vectors through distributed architecture.

Retrieval Quality Metrics

Performance means little if the retrieved results aren't relevant. Measuring retrieval quality requires different methodologies that assess how effectively a vector store returns the most relevant information.

Precision and Recall

The standard information retrieval metrics of precision and recall are applicable to vector stores:

- **Precision**: The fraction of retrieved documents that are relevant

- **Recall**: The fraction of relevant documents that are retrieved

- **F1 Score**: The harmonic mean of precision and recall, providing a single metric to balance both

```
# Basic precision and recall calculation
precision = len(relevant_retrieved) / len(retrieved)
recall = len(relevant_retrieved) / len(relevant)
f1 = 2 * (precision * recall) / (precision + recall)
```

The challenge with these metrics is creating reliable relevance judgments. For domain-specific applications, experts may need to manually evaluate which documents are relevant for each test query. This creates a "ground truth" set that serves as the basis for evaluation.

In RAG applications, precision often matters more than recall, as providing a few highly relevant results is typically better than returning many marginally relevant ones. This is especially true when those results will be processed by an LLM, which can be overwhelmed by too much irrelevant context.

Ranked Retrieval Metrics

For RAG applications, the order of results matters significantly. Several metrics capture this aspect of retrieval quality:

Mean Average Precision (MAP) provides a more nuanced view that considers the order of results:

- Calculates precision at each position where a relevant document appears

- Averages these precision values for each query

- Averages across all queries to get the final score

Normalized Discounted Cumulative Gain (nDCG) goes further by

- Assigning graded relevance scores (e.g., 0-3) rather than binary relevance

- Discounting the value of documents that appear lower in the results

- Normalizing against the ideal ordering of results

These ranked metrics better reflect how RAG applications use vector stores in practice. Since LLMs typically process the top-k results in order, having the most relevant information ranked higher directly impacts the quality of the generated response.

296

Comparative Analysis Methodology

When evaluating multiple vector stores, a structured methodology helps ensure fair and comprehensive comparison:

1. **Establish a common dataset**: Use the same documents and queries across all vector stores.

2. **Standardize embeddings**: Use identical embedding models to isolate vector store performance.

3. **Control compute resources**: Run tests on equivalent hardware configurations.

4. **Balance evaluation metrics**: Consider both performance and quality metrics.

5. **Document configuration details**: Record all parameters and settings for each vector store.

A typical comparative analysis might include a scoring rubric with weighted criteria:

- **Performance (40%)**: Latency, throughput, and scalability

- **Quality (30%)**: Precision, recall, MAP, and nDCG

- **Features (15%)**: Filtering capabilities, update operations, etc.

- **Operability (15%)**: Ease of deployment, monitoring, backup/restore

The weights should reflect your specific priorities. For interactive applications, low latency might be critical, while batch processing systems might prioritize throughput and scalability.

Testing RAG Retrieval End-to-End

While isolated vector store benchmarks provide valuable insights, ultimately what matters is how the vector store performs within your complete RAG pipeline. End-to-end testing evaluates the entire retrieval and generation process:

1. **Query selection**: Choose representative queries from your target domain.

2. **Retrieval evaluation**: Measure the relevance of retrieved context.

3. **LLM output assessment**: Evaluate the quality of generated responses.

4. **Human judgment**: Collect feedback on overall system quality.

Metrics for end-to-end evaluation might include

- **Response accuracy**: How factually correct the generated responses are

- **Context relevance**: How useful the retrieved context was for answering the query

- **Hallucination rate**: How often the LLM generates information not in the retrieved context

- **Response completeness**: How thoroughly the response addresses all aspects of the query

This holistic approach recognizes that a vector store that excels in isolation may not necessarily produce the best results when integrated with your specific LLM and prompt strategy.

Practical Benchmarking Strategy

For most RAG applications, a progressive benchmarking strategy works well:

1. **Start with quick comparisons**: Use a small representative dataset to narrow down options.

2. **Benchmark promising candidates**: Conduct detailed performance and quality tests on 2–3 top contenders.

3. **Prototype integration**: Build minimal RAG pipelines with the finalists.

4. **Test at scale**: Validate that performance holds with realistic data volumes.

5. **Simulate production load**: Ensure the system meets performance requirements under realistic query loads and usage patterns.

Remember that benchmarking is not a one-time activity. As your data grows and usage patterns evolve, periodically re-evaluate your vector store's performance and consider whether adjustments or alternatives might better serve your needs.

By applying these evaluation methodologies, you can select a vector store that not only performs well in benchmarks but actually delivers the retrieval quality and performance your RAG application requires in production.

Future Trends in Vector Stores

As RAG systems continue to evolve, vector stores are advancing rapidly to meet growing demands for scale, performance, and capabilities. In this section, we'll explore emerging trends that are likely to shape the future of vector databases and how they might influence your RAG architecture decisions.

Multimodal Vector Databases

While current RAG applications primarily focus on text, the future of retrieval is increasingly multimodal—combining text, images, audio, video, and other data types within a unified vector space.

Cross-Modal Retrieval

Multimodal vector stores allow queries in one modality to retrieve content in another:

```
# Example of cross-modal retrieval
# Query with text, retrieve relevant images
text_query = "a serene mountain landscape at sunset"
image_results = multimodal_vectorstore.cross_modal_search(
    text_query,
    source_modality="text",
    target_modality="image",
    k=5
)
```

This capability enables powerful applications like

- Finding images that match textual descriptions

- Retrieving documents relevant to an uploaded image

- Discovering videos related to audio queries

The key innovation is the alignment of different embedding spaces. Models like CLIP, ImageBind, and FLAVA establish joint embeddings where semantically similar content across modalities occupies nearby positions in the vector space.

Unified Storage Architecture

Multimodal vector stores require sophisticated architecture to handle the diverse characteristics of different data types:

- **Modality-specific preprocessing**: Images need resizing and normalization, text requires tokenization, audio needs frequency analysis.

- **Specialized embedding models**: Different encoders optimize for specific data types.

- **Heterogeneous vector dimensions**: Text embeddings might be 1536-dimensional while image embeddings could be 512-dimensional.

- **Variable indexing strategies**: Different modalities may benefit from different ANN algorithms.

Leading vector databases are beginning to incorporate these capabilities, with specialized collections or namespaces for different modalities while maintaining the ability to search across them.

Domain-Specific Vector Optimizations

General-purpose embedding models like those from OpenAI are being complemented by domain-specialized embeddings that capture nuanced relationships within specific fields.

Domain-Adapted Vector Stores

Vector stores optimized for specific domains can dramatically improve retrieval quality:

```
# Example of a domain-specialized vector store for legal
documents
legal_vectorstore = DomainVectorStore.from_documents(
    legal_documents,
    model="legal-embeddings-v2",  # Domain-specific model
    citation_graph=True,  # Legal-specific feature
    precedence_awareness=True  # Understanding case hierarchy
)
```

Domain adaptations appear across several dimensions:

- **Specialized embedding models**: Fine-tuned on domain-specific corpora (medical literature, legal documents, financial reports)

- **Custom distance metrics**: Tailored similarity functions that better capture domain-specific relationships

- **Metadata-aware retrieval**: Integrating domain knowledge (e.g., citation relationships in academic papers)

- **Domain-specific filtering**: Specialized facets relevant to particular fields

These optimizations are especially valuable in fields with specialized terminology and unique relationship types, such as healthcare, law, finance, and scientific research.

Emerging Distance Metrics and Algorithms

Traditional cosine or Euclidean distance metrics are being supplemented by more sophisticated approaches to measuring vector similarity.

Learned and Adaptive Metrics

Rather than using fixed distance functions, learned metrics adapt to specific data characteristics:

```
# Hypothetical example of a learned metric
vectorstore = VectorStore(
    metric_type="adaptive",
    metric_learning={
        "training_data": relevance_judgments,
        "optimization": "precision_at_k",
        "regularization": 0.01
    }
)
```

These advanced approaches include

- **Mahalanobis distance**: Accounts for correlations between different dimensions

- **Metric learning**: Automatically learning optimal distance functions from labeled examples

- **Contextual metrics**: Distance functions that adapt based on query context

- **Hybrid similarity**: Combining multiple distance functions with learned weights

Meanwhile, new indexing algorithms continue to push the performance envelope:

- **Graph-based indices** beyond HNSW, with adaptive connectivity

- **Learned index structures** that optimize organization based on query patterns

- **Quantization techniques** with minimal accuracy loss

- **Hardware-accelerated search** using GPUs and specialized chips

The most promising approaches combine algorithmic innovations with hardware acceleration to enable larger indices and faster search times.

Edge Deployment of Vector Stores

While cloud-based vector stores dominate current deployments, edge computing is driving a new wave of lightweight, distributed vector databases.

On-Device Vector Search

Compact vector stores running directly on end user devices enable new capabilities:

```
# Example of loading a compact vector store on a mobile device
edge_vectorstore = LiteVectorDB.load(
    "model_assets/knowledge_base.lvdb",
    max_memory_mb=50,
    quantization="int8"
)
```

```
# Perform search locally without network latency
results = edge_vectorstore.search("how to troubleshoot
connection issues")
```

This approach offers several advantages:

- **Privacy preservation**: Sensitive queries never leave the device

- **Offline operation**: Retrieval works without Internet connectivity

- **Reduced latency**: No network round-trips for common queries

- **Lower operational costs**: Reduced cloud compute and bandwidth usage

Enabling this trend are techniques like

- **Model distillation**: Creating smaller embedding models that approximate larger ones

- **Extreme quantization**: Reducing vector precision with minimal quality loss

- **Progressive loading**: Fetching only relevant portions of the vector database as needed

- **Incremental updates**: Efficiently updating on-device indices without full rebuilds

Edge vector stores won't replace cloud solutions for large-scale applications, but will create a complementary tier for privacy-sensitive, latency-critical, or bandwidth-constrained use cases.

Federated Vector Search

Beyond individual edge deployments, federated approaches enable search across distributed vector stores while preserving data sovereignty.

Cross-Silo Architecture

Federated vector search allows querying across organizational boundaries:

```
# Example of federated vector search
results = federated_vectorstore.search(
    "treatment protocols for rare diseases",
```

```
sources=["hospital_a", "hospital_b", "research_institute"],
aggregation_method="reciprocal_rank_fusion"
)
```

This approach enables

- **Data sovereignty**: Each organization maintains control of its data

- **Cross-organizational knowledge**: Querying across institutional boundaries

- **Privacy preservation**: Raw data never leaves its origin

- **Regulatory compliance**: Respecting data residency requirements

Implementing federated vector search requires sophisticated orchestration:

- **Query routing**: Determining which nodes should receive each query

- **Result aggregation**: Combining and ranking results from multiple sources

- **Distributed embedding**: Standardizing or aligning embedding spaces across nodes

- **Security protocols**: Ensuring appropriate access controls across boundaries

This model is particularly relevant for industries with strong data protection requirements like healthcare, finance, and government.

Self-improving Vector Stores

The next generation of vector stores will continuously learn and adapt based on usage patterns.

Relevance Feedback Loops

Self-improving systems incorporate user interactions to enhance retrieval quality:

```
# Recording user feedback for system improvement
vectorstore.record_interaction(
    query="carbon capture techniques",
    shown_documents=[doc1, doc2, doc3, doc4, doc5],
    clicked_documents=[doc2, doc4],
    dwell_times={doc2: 45, doc4: 120}  # Seconds spent viewing
)
```

These feedback loops enable

- **Personalized retrieval**: Adapting to individual user preferences

- **Query understanding**: Learning from rephrasing and refinements

- **Implicit relevance signals**: Incorporating clicks, dwell time, and conversational context

- **Continuous improvement**: Systems that get better with more usage

The most sophisticated implementations might even adjust their embedding spaces or distance metrics based on observed user behavior, creating truly adaptive retrieval systems.

Integration with Reasoning Engines

Vector retrieval is increasingly complemented by symbolic reasoning capabilities that enhance the quality and reliability of retrieved information.

Hybrid Retrieval Systems

Next-generation systems combine vector search with knowledge graphs and reasoning:

```
# Example of reasoning-enhanced retrieval
results = hybrid_retriever.retrieve(
    query="What medications might interact with ramipril?",
    reasoning_steps=["identify_entity_type", "expand_related_
    concepts", "apply_medical_constraints", "verify_facts"]
)
```

These hybrid approaches offer several benefits:

- **Factual verification**: Cross-checking retrieved information against structured knowledge

- **Logical consistency**: Ensuring retrieved context forms a coherent set

- **Causal reasoning**: Understanding relationships beyond mere similarity

- **Explainability**: Providing clear reasoning paths for retrieved information

By combining the flexibility of vector search with the precision of symbolic systems, these hybrid approaches aim to overcome the limitations of each method used in isolation.

Conclusion

The vector store landscape is evolving rapidly, with innovations addressing current limitations and enabling entirely new capabilities. As you design and implement RAG systems, consider not only current requirements but how these emerging trends might influence your architecture choices.

The most forward-looking implementations will build in the flexibility to incorporate these advances as they mature, ensuring your RAG applications can evolve alongside the underlying technology.

Key Takeaways and Best Practices

Let's consolidate our understanding of vector stores with practical guidance for implementing effective solutions in your RAG applications.

Vector Store Selection Framework

When choosing a vector store, consider these key factors:

Scale requirements:

- **Small scale (thousands of documents)**: FAISS, Chroma

- **Medium scale (hundreds of thousands)**: Qdrant, Weaviate, Pinecone

- **Large scale (millions+)**: Milvus, distributed Qdrant, OpenSearch Serverless

Update patterns:

- **Static data**: Any solution, optimize for query performance

- **Occasional updates**: Prefer solutions with efficient upsert operations

- **Frequent updates**: Weaviate, Qdrant, Milvus, PGVector
- **Streaming updates**: Solutions with real-time indexing capabilities

Query characteristics:

- **Low latency critical**: FAISS (smaller datasets), Qdrant
- **Complex filtering needs**: Qdrant, Weaviate, Pinecone
- **Hybrid search required**: Weaviate, Elasticsearch, OpenSearch

Operational constraints:

- **Self-hosted only**: FAISS, Chroma, Qdrant, Weaviate, Milvus
- **Minimal maintenance**: Managed services like Pinecone
- **Cost-sensitive**: Open-source solutions like FAISS, PGVector

The right choice ultimately depends on your specific requirements. Benchmark candidate solutions with realistic data volumes and query patterns before committing to a production deployment.

Implementation Best Practices

Data Preparation:

- Implement semantic chunking that respects document structure.
- Create consistent document IDs based on stable properties.

- Define a rich metadata schema to support filtering.
- Clean and normalize text before embedding.

Vector Store Configuration:

- Match index parameters to your dataset size and query patterns.
- Choose distance metrics appropriate for your embedding model.
- Configure memory/performance trade-offs based on hardware constraints.
- Set appropriate dimensions based on your embedding model.

Synchronization Strategy:

- Implement efficient document update detection.
- Use batch operations for initial loading and bulk updates.
- Consider eventual consistency for high-throughput systems.
- Keep track of embedding model versions to prevent drift.

Query Optimization:

- Find the optimal retrieval count (k) for your specific use case.
- Use metadata filters to narrow search space.
- Consider hybrid retrieval for balancing semantic and keyword matching.
- Implement caching for common queries.

Common Pitfalls to Avoid

Technical Pitfalls:

- Inconsistent vector dimensions when mixing embedding models

- Poor chunking strategies that break semantic coherence

- Suboptimal index parameters causing performance issues

- Inadequate error handling for API failures

Architectural Pitfalls:

- Underestimating resource requirements for production loads

- Neglecting backup and recovery procedures

- Missing monitoring for performance and quality

- Ignoring security and access control requirements

Integration Pitfalls:

- Misaligned query and document formulations

- Excessive retrieval counts overwhelming LLMs

- Missing feedback loops for quality improvement

- Insufficient testing with realistic query patterns

Recommended Configurations for Common Use Cases

Enterprise Knowledge Base: A distributed vector database like Weaviate or Milvus with access control metadata, regular incremental synchronization, and hybrid search capabilities.

Customer Support Bot: A managed service like Pinecone with query caching, real-time monitoring, and metadata filtering by product and user segment.

Personal Knowledge Assistant: Lightweight solutions like FAISS or Chroma with incremental updates, semantic chunking, and potential local deployment for privacy.

Research Tool: Vector stores with strong filtering and re-ranking capabilities, rich metadata for citations, and hybrid search for precise retrieval.

By applying these principles and avoiding common pitfalls, you can create vector store implementations that effectively meet the needs of your RAG applications while ensuring performance, reliability, and maintainability.

Summary and Next Steps

This chapter covered the fundamentals of vector stores and their role in RAG systems:

- **Core concepts**: How vector stores enable efficient similarity search

- **Implementation**: Local (FAISS, Chroma) vs. cloud (Pinecone) solutions

- **Advanced techniques**: Hybrid search, MMR, and production optimizations like HNSW/IVF indexing

Key Takeaways:

1. Match your vector store choice to scale, latency, and feature needs (e.g., metadata filtering).

2. Prioritize retrieval quality with re-ranking and query transformations.

3. Plan for production requirements (scalability, updates, and security).

Next Steps

In Chapter 8, we'll explore retrievers—the intelligent components that leverage vector stores to fetch the most relevant context for LLMs. You'll learn advanced retrieval strategies and their integration into end-to-end RAG systems.

CHAPTER 8

Retrievers: Finding the Most Relevant Information

In the fast-evolving landscape of Retrieval-Augmented Generation (RAG) systems, retrievers play a critical role. They are the intelligent research assistants that sift through vast repositories of knowledge to find the most relevant information for a given query. Unlike vector stores, which primarily focus on storing and indexing documents, retrievers are responsible for the actual retrieval process, implementing sophisticated strategies to ensure the most pertinent information is surfaced.

Retrievers serve as the bridge between a user's natural language query and the knowledge base, transforming ambiguous human questions into precise search operations. Their effectiveness directly impacts the quality of the generated responses in a RAG system—even the most advanced language model can only work with the given context.

In this chapter, we'll explore the fundamentals of retrievers, their various types, and advanced techniques to optimize their performance. We'll examine how retrievers can be customized, combined, and enhanced to overcome common challenges in information retrieval. By the end of this chapter, you'll have a comprehensive understanding of retrievers and the practical skills to implement them effectively in your RAG applications.

© Ranajoy Bose 2025
R. Bose, *Mastering Retrieval-Augmented Generation*,
https://doi.org/10.1007/979-8-8688-1808-0_8

Note   This chapter is accompanied by a companion Google Colab notebook containing complete code examples and implementations of the concepts discussed here. The code snippets in this chapter are simplified extracts from that notebook to illustrate key concepts.

Fundamentals of Retrieval

At its core, a retriever is a component that takes a query (typically a string) and returns a list of relevant documents. While this interface is straightforward, the underlying mechanisms can vary significantly in complexity and sophistication.

The Retriever Interface

In LangChain and most RAG frameworks, retrievers implement a straightforward interface:

- **Input**: A query (string)

- **Output**: A list of documents (standardized Document objects)

This simple interface belies the complexity beneath. A retriever might employ vector similarity search, keyword matching, or even leverage LLMs to interpret queries and find relevant information.

The Retrieval Pipeline

Let's break down the typical retrieval pipeline:

1. **Query Processing**: The user's raw query is processed, which may involve cleaning, tokenization, and transformation.

2. **Search Execution**: The processed query is used to search through indexed documents using various algorithms.

3. **Result Ranking**: Retrieved documents are ranked based on relevance criteria.

4. **Result Filtering**: Irrelevant results may be filtered out based on predetermined thresholds.

5. **Result Formatting**: The final results are formatted as Document objects containing both the content and metadata.

Evaluation Metrics for Retrieval Quality

To assess retriever performance, several metrics are commonly used:

- **Precision**: The fraction of retrieved documents that are relevant

- **Recall**: The fraction of relevant documents that are retrieved

- **F1 Score**: The harmonic mean of precision and recall

- **Mean Reciprocal Rank (MRR)**: Measures where the first relevant document appears in the result list

- **Mean Average Precision (MAP)**: Measures the average precision across multiple queries

- **Normalized Discounted Cumulative Gain (NDCG)**: Evaluates the usefulness of results based on their position

Common Challenges in Retrieval

Several challenges can impact retrieval effectiveness:

- **Vocabulary mismatch**: When the query uses different terminology than the documents

- **Ambiguity**: When a query has multiple possible interpretations

- **Context sensitivity**: When the meaning of a query depends on context not provided

- **Scale**: Efficiently searching through millions or billions of documents

- **Relevance decay**: When information becomes less relevant over time

- **Query complexity**: Handling complex, multipart queries effectively

Let's look at a simple example of implementing a custom retriever that addresses some of these challenges:

```
from langchain.schema import BaseRetriever, Document
from typing import List

class CustomKeywordRetriever(BaseRetriever):
    """A simple custom retriever with similarity scores."""
```

```python
def __init__(self, documents: List[Document]):
    self.documents = documents

def _get_relevant_documents(self, query: str) ->
List[Document]:
    results = []
    query_terms = set(query.lower().split())

    for doc in self.documents:
        # Count matching terms
        doc_terms = set(doc.page_content.lower().split())
        match_count = len(query_terms.
        intersection(doc_terms))

        # Add similarity score to metadata
        if match_count > 0:
            doc_with_score = Document(
                page_content=doc.page_content,
                metadata={**doc.metadata, "similarity_
                score": match_count / len(query_terms)}
            )
            results.append(doc_with_score)

    # Sort by similarity score
    results.sort(key=lambda x: x.metadata["similarity_
    score"], reverse=True)
    return results[:5]  # Return top 5 matches
```

This simple custom retriever demonstrates several key concepts:

- Implementing the standard retriever interface

- Adding similarity scores to retrieval results

- Basic keyword matching and result ranking

In practice, retrievers are much more sophisticated, often leveraging advanced algorithms and pre-built components, which we'll explore in subsequent sections.

Vector Store Retrievers

Vector stores have become a cornerstone of modern RAG systems, offering efficient similarity search capabilities. When used as retrievers, they transform the search process from traditional keyword matching to semantic matching based on vector embeddings.

Using Vector Stores As Retrievers

Converting a vector store into a retriever is typically straightforward. Most vector store implementations in frameworks like LangChain offer an as_retriever() method to create a retriever interface:

```
from langchain.vectorstores import Chroma
from langchain.embeddings import OpenAIEmbeddings

# Create a vector store
embeddings = OpenAIEmbeddings()
vectorstore = Chroma.from_documents(documents, embeddings)

# Convert to a retriever
retriever = vectorstore.as_retriever()

# Use the retriever
docs = retriever.invoke("What is semantic search?")
```

This pattern works across various vector database providers, creating a consistent interface regardless of the underlying technology.

Common Vector Similarity Methods

Vector retrievers rely on different similarity metrics to determine relevance:

- **Cosine Similarity**: Measures the cosine of the angle between vectors (1 for identical, 0 for orthogonal)

- **Dot Product**: Simple multiplication of vector components (higher for similar vectors with large magnitudes)

- **Euclidean Distance**: Measures the straight-line distance between vectors (smaller for more similar vectors)

Each method has its strengths and typical use cases. For example, cosine similarity is often preferred when vector magnitude should not impact similarity assessments.

Generating Multiple Embeddings per Document

A powerful technique to improve retrieval quality is to generate multiple embeddings for each document. This approach is beneficial when documents contain diverse topics or when different sections might be relevant to different queries.

```
from langchain.retrievers import MultiVectorRetriever
from langchain.storage import InMemoryStore

# Storage for original documents
docstore = InMemoryStore()

# Create embeddings for document summaries and sections
id_key = "doc_id"  # Metadata field to connect pieces
```

```
retriever = MultiVectorRetriever(
    vectorstore=vectorstore,
    docstore=docstore,
    id_key=id_key,
)
```

In the companion notebook, we'll implement a complete example that

1. Splits documents into chunks

2. Generates embeddings for each chunk

3. Maintains references to parent documents

4. Retrieves complete documents when chunks
   match a query

Retrieving Whole Documents for Chunks

When working with chunked documents, retrieving the original document rather than just the matching chunks is often valuable. This provides more context for the language model:

```
from langchain.retrievers import ParentDocumentRetriever
```

```
retriever = ParentDocumentRetriever(
    vectorstore=child_vectorstore,
    docstore=parent_docstore,
    child_splitter=text_splitter,
)
```

```
# Retrieves parent documents when child chunks match
docs = retriever.get_relevant_documents("Tell me about quantum
computing")
```

This approach addresses the context fragmentation problem that occurs when documents are split into small chunks for indexing. The companion notebook demonstrates a complete implementation with document loading, chunking, indexing, and retrieval operations.

Enhancing Retrieval Quality

Even the best vector stores and similarity algorithms can struggle with certain types of queries. In this section, we'll explore techniques to enhance retrieval quality.

Generating Multiple Queries

One effective approach is query expansion—transforming a single query into multiple related queries to improve recall:

```python
from langchain.retrievers.multi_query import
MultiQueryRetriever
from langchain.chat_models import ChatOpenAI

# Initialize LLM for query generation
llm = ChatOpenAI(temperature=0)

# Create a retriever that generates multiple queries
retriever = MultiQueryRetriever.from_llm(
    retriever=base_retriever,
    llm=llm
)

# Original query: "How do batteries work?"
# Might generate additional queries like:
# - "Explain the chemical processes in batteries"
# - "What is the mechanism of energy storage in batteries?"
```

The LLM generates variations of the original query, each potentially matching different documents in the knowledge base. The results are then combined to provide a more comprehensive response.

Contextual Compression

Contextual compression filters and condenses retrieved documents to focus on the most relevant portions:

```
from langchain.retrievers import ContextualCompressionRetriever
from langchain.retrievers.document_compressors import
LLMChainExtractor

# Create a compressor that extracts relevant information
compressor = LLMChainExtractor.from_llm(llm)

# Wrap the base retriever with contextual compression
compression_retriever = ContextualCompressionRetriever(
    base_compressor=compressor,
    base_retriever=base_retriever
)
```

This technique reduces noise in the retrieved context, focusing the language model's attention on the most pertinent information. The companion notebook demonstrates several compression strategies, from simple length-based filtering to sophisticated LLM-based extractors.

Mitigating the "Lost in the Middle" Effect

Research has shown that information in the middle of a document list tends to receive less attention from language models—a phenomenon known as the "lost in the middle" effect. To mitigate this, we can reorder retrieved results based on relevance:

```
from langchain.retrievers import ReorderRetriever

retriever = ReorderRetriever(
    base_retriever=vector_retriever,
    # Document reordering logic
    reordering_function=lambda docs, query: sorted(
        docs,
        key=lambda doc: doc.metadata.get("relevance_score", 0),
        reverse=True
    )
)
```

The companion notebook explores various reordering strategies, including

- Relevance-based ordering
- Time-weighted ordering
- Source credibility weighting
- Hybrid approaches that combine multiple factors

Generating Metadata Filters

Another powerful technique is to use metadata to filter retrieved documents. This can be done manually or with an LLM:

```
from langchain.retrievers.self_query.base import
SelfQueryRetriever

# Create a retriever that generates metadata filters
retriever = SelfQueryRetriever.from_llm(
    llm,
    vectorstore,
    document_content_description="Scientific articles about
    climate change",
```

```
metadata_field_info=[
    {"name": "year", "description": "Publication year"},
    {"name": "author", "description": "First author's
    last name"},
    {"name": "topic", "description": "Main research topic"}
]
)
```

This approach enables more precise results filtering based on metadata fields such as dates, authors, categories, or custom attributes. The LLM analyzes the query to determine appropriate metadata constraints, enhancing retrieval precision.

In the companion notebook, we'll implement and test all these enhancement techniques, allowing you to see firsthand how they improve retrieval quality in different scenarios.

Specialized Retrieval Approaches

As RAG systems mature, specialized retrievers have emerged to address specific use cases and challenges. These specialized approaches go beyond basic vector similarity search to incorporate temporal awareness, hybrid techniques, and domain-specific optimizations.

Time-Weighted Retrieval

Information relevance often changes over time. A time-weighted retriever assigns higher importance to more recent documents, which is particularly valuable for news, research, and rapidly evolving domains:

```
from langchain.retrievers import
TimeWeightedVectorStoreRetriever
import datetime
```

```
# Create a time-weighted retriever
retriever = TimeWeightedVectorStoreRetriever(
    vectorstore=vectorstore,
    decay_rate=0.01,  # Controls how quickly relevance decays
    with time
    search_kwargs={"k": 10}  # Number of documents to retrieve
)

# Add document with timestamp
retriever.add_documents([
    Document(
        page_content="Latest research on fusion energy",
        metadata={"created_at": datetime.datetime.now().
        timestamp()}
    )
])
```

The time-weighted retriever applies a decay function to relevance scores based on document age, ensuring newer content is prioritized when appropriate. The companion notebook demonstrates different decay functions and their effects on retrieval results.

Hybrid Vector and Keyword Retrieval

Keyword-based search and vector-based semantic search each have their strengths. Hybrid retrievers combine both approaches:

```
from langchain.retrievers import BM25Retriever,
EnsembleRetriever

# Create keyword-based retriever
bm25_retriever = BM25Retriever.from_documents(documents)
bm25_retriever.k = 10  # Number of documents to retrieve
```

```
# Create vector-based retriever
vector_retriever = vectorstore.as_retriever(search_
kwargs={"k": 10})

# Combine both retrievers
hybrid_retriever = EnsembleRetriever(
    retrievers=[bm25_retriever, vector_retriever],
    weights=[0.5, 0.5]  # Equal weighting
)
```

This hybrid approach excels at handling both keyword-specific queries and semantic queries. For example, a query containing a specific product code might work better with keyword search, while a conceptual query would benefit from semantic search.

Self-Querying Retrievers

Self-querying retrievers use an LLM to convert natural language queries into structured queries with filters:

```
from langchain.retrievers.self_query.base import
SelfQueryRetriever

retriever = SelfQueryRetriever.from_llm(
    llm,
    vectorstore,
    document_content_description="Technical documentation",
    metadata_field_info=[
        {"name": "version", "description": "Software version"},
        {"name": "topic", "description": "Documentation topic"}
    ]
)
```

```
# The query "Show me v2.0 deployment guides" might be
converted to:
# - semantic search for "deployment guides"
# - with filter: version == "2.0"
```

This approach is particularly effective for knowledge bases with rich metadata, allowing users to naturally express filtering criteria without learning a query syntax. The companion notebook shows how self-querying retrievers handle complex queries that combine semantic search with precise filtering conditions.

Contextual Retrievers

Contextual retrievers maintain awareness of conversation history, allowing for more relevant document retrieval in multiturn interactions:

```
from langchain.memory import ConversationBufferMemory
from langchain.chains import ConversationalRetrievalChain

memory = ConversationBufferMemory(
    memory_key="chat_history",
    return_messages=True
)

# Create a retrieval chain with conversation context
retrieval_chain = ConversationalRetrievalChain.from_llm(
    llm=llm,
    retriever=retriever,
    memory=memory
)

# First query: "What is RAG?"
# Follow-up query: "How does it compare to fine-tuning?"
# The retriever understands "it" refers to RAG
```

This approach is essential for conversational AI applications, where queries often reference previous exchanges. The companion notebook demonstrates techniques for managing conversation context and resolving references for more accurate retrieval.

Combining Retrieval Strategies

Individual retrieval strategies each have their strengths and weaknesses. By combining multiple approaches, we can create more robust and effective retrieval systems.

Ensemble Retrievers

Ensemble retrievers combine results from multiple retrievers, similar to ensemble methods in machine learning:

```
from langchain.retrievers import EnsembleRetriever

# Create an ensemble of retrievers
ensemble_retriever = EnsembleRetriever(
    retrievers=[
        keyword_retriever,
        vector_retriever,
        time_weighted_retriever
    ],
    weights=[0.3, 0.5, 0.2]  # Weighted by importance
)
```

This approach allows for weighted combinations of different retrieval strategies, leveraging each of their strengths. The companion notebook explores optimal weighting strategies and evaluation methods to tune ensemble retrievers.

Fusion Techniques

Beyond simple weighted combinations, more sophisticated fusion techniques can improve retrieval quality:

```
from langchain.retrievers.merger_retriever import MergerRetriever

# Create a retriever that merges results using RRF
retriever = MergerRetriever(
    retrievers=[retriever1, retriever2, retriever3],
    # Using Reciprocal Rank Fusion
    merger="rrf"  # Alternative: "interleave" or custom
    function
)
```

Reciprocal Rank Fusion (RRF) is particularly effective, giving higher weight to documents that appear at top positions across multiple retrievers. The companion notebook implements several fusion algorithms, including

- Reciprocal Rank Fusion

- CombSUM (sum of normalized scores)

- Interleaving (alternating results from different retrievers)

- MaxScore (taking the maximum score for each document)

Re-ranking Strategies

Re-ranking involves applying a secondary scoring pass to refine the initial retrieval results:

```
from langchain.retrievers import ContextualCompressionRetriever
from langchain.retrievers.document_compressors import LLMRerank
```

```
# Create a re-ranker
reranker = LLMRerank.from_llm(
    llm=llm,
    query_key="query",   # Key for the query in the prompt
    doc_key="doc",       # Key for the document in the prompt
    top_n=5              # Number of documents to return after
                         re-ranking
)
```

Re-ranking is particularly effective when the initial retrieval casts a wide net, and a more sophisticated (but potentially more computationally expensive) evaluation can be applied to a smaller set of candidates. The companion notebook shows how re-ranking with cross-encoders or LLMs can significantly improve retrieval precision.

Implementing and Optimizing Retrievers

Having explored various retriever types and strategies, let's now focus on practical implementation aspects, optimizations, and performance considerations.

Popular Retriever Frameworks and Libraries

Several frameworks provide robust retriever implementations:

- **LangChain**: Offers a comprehensive suite of retrievers with a unified interface

- **FAISS**: Meta's library for efficient similarity search and clustering

- **Elasticsearch**: Powerful search engine with retrieval capabilities

- **Pinecone, Weaviate, Chroma, Milvus**: Vector database systems with built-in retrieval mechanisms

For most RAG applications, using a higher-level framework like LangChain provides the simplest path to production, with abstractions that unify diverse retrieval approaches.

Step-by-Step Implementation Example

Let's implement a practical retriever pipeline combining several techniques we've discussed:

```python
from langchain.retrievers import
ContextualCompressionRetriever, MultiQueryRetriever
from langchain.retrievers.document_compressors import
DocumentCompressorPipeline
from langchain.text_splitter import
RecursiveCharacterTextSplitter

# 1. Create base vector store retriever
base_retriever = vectorstore.as_retriever(search_kwargs=
{"k": 20})

# 2. Add multi-query capability
multi_query_retriever = MultiQueryRetriever.from_llm(
    retriever=base_retriever,
    llm=llm
)

# 3. Add contextual compression
# First: filter by relevance score
relevance_filter = RelevanceFilter(threshold=0.7)
# Second: use LLM to extract relevant parts
extractor = LLMChainExtractor.from_llm(llm)
# Create a pipeline of compressors
compressor_pipeline = DocumentCompressorPipeline(
    transformers=[relevance_filter, extractor]
)
```

```
# Final retriever with compression
final_retriever = ContextualCompressionRetriever(
    base_compressor=compressor_pipeline,
    base_retriever=multi_query_retriever
)
```

This implementation combines multiple query generation for improved recall with contextual compression for improved precision. The companion notebook implements this pipeline and evaluates its performance on various query types.

Configuration and Optimization Parameters

Optimal retriever configuration depends on your specific use case and requirements. Key parameters to consider include

- **k**: Number of documents to retrieve (affects recall vs. processing time)

- **Score thresholds**: Minimum relevance score for inclusion

- **Chunk size**: For document splitting (affects context granularity)

- **Embedding model**: Quality and dimensionality of embeddings

- **Search algorithm**: Exact vs. approximate nearest neighbors

- **Retrieval weights**: For ensemble methods

The companion notebook provides a parameter tuning section demonstrating how to optimize these settings for different objectives.

Performance Considerations

As your system scales, several performance factors become critical:

- **Latency**: How quickly results are returned

- **Throughput**: Number of queries handled per second

- **Memory usage**: RAM requirements for index
  and runtime

- **Storage efficiency**: Disk usage for document storage

- **Update performance**: Speed of adding or modifying
  documents

Optimizing these factors often involves trade-offs. For example, approximate nearest neighbor algorithms trade perfect accuracy for faster retrieval, which is usually acceptable in RAG applications.

```
# Example of configuring FAISS for performance
from langchain.vectorstores import FAISS

# Create an index with performance-optimized parameters
# 1. nlist: number of clusters (affects index build time and
search speed)
# 2. nprobe: number of clusters to search (affects accuracy
vs. speed)
vectorstore = FAISS.from_documents(
    documents,
    embeddings,
    index_kwargs={"nlist": 256},
    search_kwargs={"nprobe": 20}
)
retriever = vectorstore.as_retriever()
```

The companion notebook explores these performance parameters and provides benchmarking examples to help you find the right balance for your application.

Real-world Applications and Case Studies

Retrievers play diverse roles across different applications, each with unique requirements and challenges.

Enterprise Search Applications

Enterprise search often involves heterogeneous document types, access controls, and domain-specific terminology:

```
# Example of enterprise search retriever with security
filtering
retriever = vectorstore.as_retriever(
    search_kwargs={
        "k": 10,
        "filter": {
            "access_level": {"$lte": user.access_level},
            "department": {"$in": user.departments}
        }
    }
)
```

Enterprise applications typically require

- Integration with identity management systems
- Support for document-level access controls
- Handling of domain-specific file formats
- Federated search across multiple repositories

Conversational Retrieval Systems

Conversational systems must maintain context across multiple turns and handle ambiguous references:

```python
from langchain.chains import ConversationalRetrievalChain

# Create a conversational retrieval chain
chain = ConversationalRetrievalChain.from_llm(
    llm=llm,
    retriever=retriever,
    get_chat_history=lambda h: h,  # Function to format
chat history
    verbose=True
)

# Maintain conversation state
chat_history = []
query = "What are the key features of RAG?"
result = chain.invoke({"question": query, "chat_history":
chat_history})
chat_history.append((query, result["answer"]))

# Follow-up question using context
follow_up = "How does it compare with fine-tuning?"
# The system understands "it" refers to RAG from chat history
```

The companion notebook implements a complete conversational RAG system and demonstrates techniques for handling context, references, and clarification requests.

Compliance and Security Considerations

When implementing retrievers for sensitive domains like healthcare, finance, or legal applications, several compliance factors must be considered:

- **Data residency**: Ensuring data remains in approved jurisdictions

- **Audit trails**: Logging retrieval requests and responses

- **PII handling**: Detecting and redacting personally identifiable information

- **Explainability**: Providing rationale for why documents were retrieved

- **Bias mitigation**: Ensuring balanced representation in retrieval results

The companion notebook includes a section on implementing compliance-aware retrievers with appropriate safeguards for sensitive applications.

Conclusion

Retrievers are the vital research assistants determining what information reaches your language model. Their effectiveness directly impacts the quality, relevance, and trustworthiness of your RAG system's outputs.

As we've explored in this chapter, retrievers have evolved from simple similarity search mechanisms to sophisticated systems that can

- Generate multiple queries to improve recall

- Compress and refine results for better precision

- Combine different retrieval strategies through ensembling and fusion

- Incorporate temporal awareness and domain knowledge

- Handle conversation context and complex filtering requirements

By understanding and implementing these advanced retrieval techniques, you can significantly enhance your RAG application's performance. The companion notebook provides hands-on experience with all the concepts covered in this chapter, allowing you to experiment and adapt these approaches to your specific use cases.

As retrieval technology continues to evolve, the integration of hybrid methods, contextual understanding, and domain-specific optimizations will likely become standard practice. By mastering the fundamentals and advanced techniques presented in this chapter, you'll be well-equipped to incorporate these innovations into your RAG systems.

Implementation Guidelines and Future Directions

Best Practices for Retriever Implementation

Based on our exploration of retrievers, here are key best practices for effective implementation:

Select the Right Retriever Architecture

Choose your retriever architecture based on your specific requirements:

- **For general-purpose semantic search**: Vector store retrievers are often sufficient

- **For queries with specific terms or codes**: Consider hybrid retrievers or keyword-first approaches

- **For evolving knowledge bases**: Implement time-weighted or recency-aware approaches

- **For conversational systems**: Use retrievers with conversation context awareness

- **For complex filtering needs**: Implement self-querying or metadata-aware retrievers

Optimize the Chunking Strategy

Document chunking significantly impacts retrieval quality:

```python
from langchain.text_splitter import
RecursiveCharacterTextSplitter

# Create a strategic text splitter
text_splitter = RecursiveCharacterTextSplitter(
    chunk_size=500,          # Shorter chunks for precise
                             retrieval
    chunk_overlap=50,        # Overlap to maintain context
                             across chunks
    separators=["\n\n", "\n", " ", ""],   # Respect document
                                          structure
    length_function=len      # Character count (alternatively
                             use tokens)
)

# Split while preserving metadata
split_docs = text_splitter.split_documents(documents)
```

Consider these chunking guidelines:

- Match chunk size to your use case (smaller for precise QA, larger for summarization).

- Ensure sufficient overlap to maintain context across chunk boundaries.

- Respect natural document boundaries (sections, paragraphs) when possible.

- Preserve metadata during splitting for later filtering and provenance tracking.

Implement Comprehensive Evaluation

Evaluate retrievers using diverse metrics and test sets:

```
from langchain.evaluation import EvaluatorType, evaluate_retriever

# Create evaluation dataset
eval_dataset = [
    {"query": "What is RAG?", "relevant_docs": [...],
    "irrelevant_docs": [...]},
    {"query": "How do transformers work?", "relevant_docs":
    [...], "irrelevant_docs": [...]}
]

# Evaluate the retriever
evaluation_results = evaluate_retriever(
    retriever,
    eval_dataset,
    metrics=[
        EvaluatorType.RETRIEVAL_PRECISION,
        EvaluatorType.RETRIEVAL_RECALL,
        EvaluatorType.RETRIEVAL_FAITHFULNESS
    ]
)
```

The companion notebook includes a comprehensive evaluation framework for measuring retriever performance across various dimensions.

Monitor and Adapt

Retrieval systems should adapt to changing requirements and user patterns:

- Implement logging to track query patterns and retrieval performance

- Use feedback loops to identify and address retrieval failures

- Periodically re-evaluate and tune retrieval parameters

- Consider A/B testing when making significant retrieval changes

Future Directions in Retrieval Technology

As retrieval technology continues to evolve, several promising directions are emerging:

Multimodal Retrievers

Future retrievers will work seamlessly across text, images, audio, and video:

```
# Conceptual example of a multimodal retriever
multimodal_retriever = MultiModalRetriever(
    text_retriever=vectorstore.as_retriever(),
    image_retriever=image_vectorstore.as_retriever(),
    fusion_strategy="cross_attention"
)
```

```
# Query could be text, image, or both
docs = multimodal_retriever.get_relevant_documents({
    "text": "Show me examples of architectural designs with
    curved facades",
    "image": uploaded_image   # Reference image
})
```

Multimodal retrievers will enable more natural and comprehensive information access, particularly valuable for domains like design, medicine, and education.

Reasoning-Enhanced Retrievers

Retrievers incorporating step-by-step reasoning will improve precision and relevance:

```
# Conceptual example of a reasoning-enhanced retriever
reasoned_retriever = ReasoningRetriever(
    base_retriever=vectorstore.as_retriever(),
    reasoning_llm=llm,
    max_reasoning_steps=3
)

# The retriever might:
# 1. Analyze the query to identify key information needs
# 2. Determine if the query should be decomposed
# 3. Formulate a retrieval plan with filtering criteria
# 4. Execute retrieval with dynamic adjustments
```

By incorporating reasoning capabilities, these retrievers can handle complex, multi-hop questions that require synthesizing information from multiple sources.

Adaptive and Learning Retrievers

Future retrievers will learn from user interactions and adapt to specific domains:

```
# Conceptual example of an adaptive retriever
adaptive_retriever = AdaptiveRetriever(
    base_retriever=vectorstore.as_retriever(),
    learning_rate=0.01,
    adaptation_strategy="relevance_feedback"
)

# Learning from feedback
adaptive_retriever.update_from_feedback(
    query="What are the latest treatments for Alzheimer's?",
    retrieved_docs=[doc1, doc2, doc3],
    positive_feedback_docs=[doc1, doc3],
    negative_feedback_docs=[doc2]
)
```

These systems will improve over time based on explicit feedback and implicit signals, becoming increasingly aligned with specific user needs and domain-specific requirements.

Privacy-Preserving Retrievers

As privacy concerns grow, retrievers that protect sensitive information will become essential:

```
# Conceptual example of a privacy-preserving retriever
private_retriever = PrivacyPreservingRetriever(
    base_retriever=vectorstore.as_retriever(),
    pii_detection_model=pii_detector,
```

```
    anonymization_strategy="differential_privacy",
    epsilon=0.1  # Privacy budget
)
```

These systems will use techniques like differential privacy, federated learning, and secure multiparty computation to enable effective retrieval while safeguarding sensitive information.

Summary and Next Steps

Retrievers are the vital research assistants that determine what information reaches your language model. Their effectiveness directly impacts the quality, relevance, and trustworthiness of your RAG system's outputs.

Key Takeaways

As we've explored in this chapter, retrievers have evolved from simple similarity search mechanisms to sophisticated systems that can

- Generate multiple queries to improve recall
- Compress and refine results for better precision
- Combine different retrieval strategies through ensembling and fusion
- Incorporate temporal awareness and domain knowledge
- Handle conversation context and complex filtering requirements

By understanding and implementing these advanced retrieval techniques, you can significantly enhance your RAG application's performance. The companion notebook provides hands-on experience with all the concepts covered in this chapter, allowing you to experiment and adapt these approaches to your specific use cases.

Next Steps

1. **Experiment with Hybrid Retrievers**: Combine keyword-based (BM25) and vector-based retrievers to balance precision and recall.

2. **Evaluate Performance**: Use metrics like MRR, NDCG, and recall@k to compare different retriever configurations.

3. **Explore Specialized Retrievers**: Implement time-weighted or self-querying retrievers for domain-specific applications.

4. **Optimize for Scale**: Benchmark latency and throughput for large-scale deployments.

As retrieval technology continues to evolve, the integration of hybrid methods, contextual understanding, and domain-specific optimizations will likely become standard practice. By mastering the fundamentals and advanced techniques presented here, you'll be well-equipped to build robust retrieval systems that power effective RAG applications.

PART III

Advanced Implementation

Prompt Templates: The Communication Experts That Structure Interactions with the LLM

In the intricate machinery of a Retrieval-Augmented Generation (RAG) system, prompt templates are the vital communication layer that bridges retrieved information with language models. Think of them as expert translators, converting raw knowledge into precisely formatted instructions that Large Language Models (LLMs) can understand and act upon effectively.

What Are Prompt Templates?

Prompt templates are structured formats that organize instructions, context, and queries when communicating with an LLM. In RAG systems, these templates are particularly crucial as they determine how retrieved

© Ranajoy Bose 2025
R. Bose, *Mastering Retrieval-Augmented Generation*,
https://doi.org/10.1007/979-8-8688-1808-0_9

349

information gets presented to the model, significantly influencing the
quality, relevance, and accuracy of generated outputs.

At their simplest, a prompt template might look like

```
Answer the question based on the context below.

Context: {retrieved_documents}

Question: {user_query}

Answer:
```

But this simplicity is deceptive. The art and science of prompt
engineering has evolved dramatically since the early days of LLMs,
transforming from basic text prompts to sophisticated, programmatically
constructed templates that can adapt dynamically to different queries,
contexts, and needs.

The Evolution of Prompt Engineering

The field of prompt engineering has undergone several key
evolutionary stages:

1. **Basic Prompting**: Early interactions with language
   models relied on simple instructions with minimal
   structure.

2. **Few-Shot Learning**: Practitioners discovered that
   providing examples within prompts dramatically
   improved performance.

3. **Structured Templates**: The development of
   consistent formats with clear delineation between
   instructions, context, and queries.

4. **Dynamic Templates**: Modern approaches that programmatically construct prompts based on retrieval results, query analysis, and available context window.

5. **Chain-of-Thought and Multistep Reasoning**: Advanced techniques that guide models through explicit reasoning steps.

Why Prompt Templates Matter in RAG

In traditional LLM applications, prompt design impacts performance. In RAG systems, it becomes critical for several reasons:

1. **Context Integration**: RAG systems must seamlessly blend retrieved information with user queries, requiring careful formatting to maintain coherence.

2. **Information Prioritization**: With limited context windows, templates must strategically arrange retrieved information to ensure the most relevant content gets proper attention.

3. **Hallucination Reduction**: Well-crafted prompts can significantly reduce the tendency of LLMs to generate incorrect information by clearly separating retrieved facts from areas requiring inference.

4. **Source Attribution**: Templates can instruct models to attribute information to sources, enhancing transparency and trustworthiness properly.

5. **Task Specification**: Different RAG applications (question answering, summarization, comparison) require specialized prompt structures to optimize performance.

The Impact of Template Quality

The difference between a mediocre and an excellent prompt template in RAG systems can be striking. Consider two approaches to the same RAG-based question answering system:

Basic Template:

```
Context: {documents}
Question: {query}
Answer:
```

Enhanced Template:

```
You are an expert assistant tasked with answering questions
based solely on the provided information.

CONTEXT:
{documents}

QUESTION:
{query}

INSTRUCTIONS:
1. Answer the question using only information from the context
2. If the context doesn't contain the answer, respond with "I
   don't have enough information to answer this question"
3. Cite relevant sources from the context using [Document X]
   notation
4. Use bullet points for multi-part answers

YOUR ANSWER:
```

The enhanced template typically produces more accurate, better-structured, and more reliable responses by providing clear guardrails and expectations for the model.

In the following sections, we'll explore the anatomy of effective prompt templates, examine different template types for various RAG applications, and learn practical techniques for creating, testing, and optimizing these crucial components of any successful RAG system.

Anatomy of an Effective Prompt Template

Every effective prompt template for RAG systems contains several key components that work harmoniously to guide the LLM toward generating high-quality responses. Understanding these components and how they interact is essential for designing templates that maximize the benefits of retrieved information.

Core Components of a Prompt Template

System Instructions

System instructions establish the foundational behavior and role for the LLM. They set the tone, define capabilities, and establish constraints for the interaction.

You are a specialized medical research assistant with expertise in cardiology. Your task is to provide accurate information based exclusively on the provided research papers. Do not draw on general knowledge beyond what is explicitly stated in the context.

System instructions are particularly important in RAG systems to

- Constrain the model to use only retrieved information

- Establish domain-specific personas for specialized
  applications

- Set appropriate levels of certainty and hedging based
  on information quality

Context Window

The context window contains retrieved documents or information
snippets that the model should use to generate its response. This is the
RAG-specific component that distinguishes these templates from general
LLM prompting.

CONTEXT:
[Document 1] A 2023 study by Chen et al. found that statin
therapy reduced major cardiovascular events by 24% in patients
with LDL cholesterol levels above 190 mg/dL, regardless of
prior cardiovascular disease history.

[Document 2] The JUPITER trial (2008) demonstrated that
rosuvastatin reduced the incidence of major cardiovascular
events by nearly 50% among individuals with elevated high-
sensitivity C-reactive protein levels but relatively low
cholesterol levels..

Effective context formatting typically includes

- Clear delineation between different documents

- Source identification for each information chunk

- Consistent formatting that makes it easy for the model
  to parse

Query Specification

The query specification presents the user's question or request in a clear, well-defined format.

```
QUESTION:
What is the evidence for statin efficacy in primary prevention
of cardiovascular disease?
```

This component should

- Clearly separate the query from context

- Preserve the user's original intent while potentially reformatting for clarity

- Include any query-specific parameters or constraints

Response Format Instructions

Response format instructions guide how the model should structure its answer, including level of detail, citation requirements, and formatting preferences.

```
RESPONSE GUIDELINES:
- Begin with a direct answer to the question
- Support your points with evidence from the provided studies
- Use [Document X] notation to cite your sources
- Include numerical data where available
- If the evidence is conflicting, acknowledge different
  perspectives
- Format lists using bullet points
- Limit your response to 3-4 paragraphs
```

These instructions are crucial for ensuring consistency and utility in RAG system outputs.

Template Structure Patterns

Several structural patterns have emerged as particularly effective for RAG
applications:

The Expert Advisor Pattern

```
You are an expert {domain} advisor with deep knowledge in
{specialty}.

CONTEXT:
{retrieved_documents}

QUESTION:
{user_query}

Provide a comprehensive answer based solely on the context
provided. Use the following format:
- Direct answer to the question
- Supporting evidence
- Any limitations or caveats
- Recommendations if applicable
```

This pattern works well for professional or academic domains where
authoritative, well-structured responses are valued.

The Documentary Evidence Pattern

```
Below are excerpts from reliable sources on {topic}.

SOURCE MATERIAL:
{retrieved_documents}

QUERY:
{user_query}
```

Based exclusively on the source material, please:
1. Answer the query directly
2. Cite each piece of information using [Source X] notation
3. If the information provided is insufficient, state
   this clearly

This pattern emphasizes source attribution and factual accuracy, making it ideal for research, legal, or compliance applications.

The Analytical Framework Pattern

ANALYTICAL TASK: {task_description}

RELEVANT DATA:
{retrieved_documents}

QUESTION FOR ANALYSIS:
{user_query}

Analyze the data provided to answer the question. Your
analysis should:
- Begin with your conclusion
- Present key evidence from the data
- Explain your reasoning step by step
- Note any data limitations or gaps

This pattern is optimized for scenarios requiring critical thinking and complex analysis of retrieved information.

Balancing Brevity and Completeness

One of the most challenging aspects of template design is finding the right balance between comprehensive instructions and token efficiency. Every token used for the template is one less available for retrieved content or model output.

Here are several strategies for maintaining this balance:

1. **Tiered Instruction Sets:** Use brief core instructions with more detailed sub-instructions only where necessary.

2. **Instruction Compression:** Refine verbose instructions into concise directives without losing meaning.

3. **Consistent Formatting:** Establish a consistent structure that requires less explicit instruction once the model recognizes the pattern.

4. **Symbolic Shorthand:** Develop compact notation for commonly used instructions (e.g., "[C]" for "cite your source").

Let's examine a template transformation that maintains functionality while reducing token count:

Original (Verbose):

```
You are a helpful assistant tasked with answering questions
based on the provided context. You should focus exclusively on
the information contained in the context and should not rely on
external knowledge. If the context doesn't contain sufficient
information to answer the question, please indicate this
clearly rather than attempting to generate an answer based on
what you think might be correct. When citing information from
the context, please use the document number in square brackets,
like [Document 1], to indicate your source.

Context:
{documents}
```

Question:
{query}

Please provide your answer below:

Optimized:

Answer using ONLY the context below. Say "Insufficient
information" if needed. Cite sources as [Doc X].

CONTEXT:
{documents}

QUESTION:
{query}

ANSWER:

The optimized template achieves the same functional goals with
significantly fewer tokens, allowing more room for retrieved documents in
the context window.

In the next section, we'll explore specific types of prompt templates
optimized for different RAG applications, from question answering to
summarization and beyond.

Types of Prompt Templates for RAG

Different RAG applications require specialized prompt templates to
optimize performance for specific tasks. In this section, we'll explore
various template types, their unique characteristics, and when to apply
each one.

Question-Answering Templates

Question-answering is perhaps the most common RAG application,
providing direct responses to user queries based on retrieved information.

```
qa_template = """
Answer the question based ONLY on the following context:

CONTEXT:
{context}

QUESTION:
{question}

ANSWER:
"""
```

Key Features:

- Clear instruction to use only provided context

- Distinct separation between context and question

- Simple structure optimized for direct answers

A robust question-answering template often includes additional
guardrails:

```
enhanced_qa_template = """
You are a precise question-answering system. Use ONLY the
provided context to answer.

CONTEXT:
{context}

QUESTION:
{question}
```

```
INSTRUCTIONS:
- Answer directly based on the context
- If the context doesn't contain the answer, say "The provided
  information doesn't answer this question"
- Do not introduce external knowledge
- Cite relevant parts using [Document X] notation

ANSWER:
"""
```

This enhanced template reduces hallucinations and improves attribution by adding specific constraints on the model's behavior.

Summarization Templates

Summarization templates condense multiple retrieved documents into coherent, concise summaries.

```
summarization_template = """
Summarize the following documents about {topic}:

DOCUMENTS:
{documents}

INSTRUCTIONS:
- Create a coherent summary of 3-5 paragraphs
- Capture the main points from all documents
- Maintain factual accuracy
- Include key statistics and findings
- Avoid introducing information not in the documents

SUMMARY:
"""
```

Key Features:

- Clear scope definition (topic)

- Specific length guidance

- Instructions to maintain factual accuracy

- Warning against introducing external information

For multidocument summarization with potential contradictions:

```
multi_doc_template = """
Create a comprehensive summary of these potentially conflicting
documents:

DOCUMENTS:
{documents}

INSTRUCTIONS:
- Synthesize a cohesive summary of 4-6 paragraphs
- Identify areas of consensus across documents
- Explicitly note contradictions or disagreements
- Weight information by recency and source credibility where
  indicated
- Maintain neutral tone throughout

SUMMARY:
"""
```

This template acknowledges potential contradictions in retrieved information and provides strategies for handling them.

Comparison Templates

Comparison templates help analyze similarities and differences between entities or concepts described in retrieved documents.

```
comparison_template = """
Compare and contrast the following {entities} based on the
provided information:

INFORMATION:
{documents}

ENTITIES TO COMPARE:
{entity_1}
{entity_2}

INSTRUCTIONS:
- Create a structured comparison covering key aspects
- Highlight similarities and differences
- Use a balanced approach giving equal attention to each entity
- Only compare aspects mentioned in the documents
- Use a two-column format for clarity

COMPARISON:
"""
```

Key Features:

- Clear identification of entities to compare

- Structured format suggestion

- Balance requirement between entities

- Constraint to only compare documented aspects

Reasoning Templates

Reasoning templates guide the model through explicit logical steps to reach conclusions based on retrieved information.

```
reasoning_template = """
Analyze the following information and reason through the
question step by step:

INFORMATION:
{documents}

QUESTION:
{question}

REASONING PROCESS:
1. Identify relevant facts from the information
2. Consider what these facts tell us about the question
3. Evaluate any assumptions or inferences needed
4. Reach a logical conclusion

YOUR ANALYSIS:
"""
```

Key Features:

- Explicit reasoning steps

- Structured analytical approach

- Clear separation between facts and inferences

For more complex reasoning, Chain-of-Thought (CoT) templates have proven particularly effective:

```
cot_template = """
Reason through this complex question using the provided
information:

CONTEXT:
{documents}

QUESTION:
{question}
```

```
REASONING STEPS:
1. Let's identify the key elements of this question
2. Now, let's extract relevant information from our context
3. Let's consider what additional reasoning is needed
4. Let's work through each logical step
5. Finally, let's formulate our answer based on this reasoning

DETAILED ANALYSIS:
"""
```

This approach guides the model to break down complex problems into manageable steps, often improving accuracy on difficult questions.

Task-Specific Templates

Beyond these general categories, specialized templates can be designed for domain-specific tasks:

Code Generation Template

```
code_template = """
Write code based on the following technical documentation:

DOCUMENTATION:
{documents}

TASK:
{task}

REQUIREMENTS:
- Language: {language}
- Include comments explaining key functionality
- Follow best practices for {language}
- Only use functions/methods documented in the provided context
```

CODE SOLUTION:
"""

Content Creation Template

```
content_template = """
Create {content_type} based on these information sources:

SOURCES:
{documents}

TOPIC:
{topic}

SPECIFICATIONS:
- Tone: {tone}
- Target audience: {audience}
- Length: {length}
- Key points to include: {key_points}
- Only use facts from the provided sources

CONTENT:
"""
```

Multi-Stage Templates

For complex RAG applications, multistage templates guide the model through sequential processing steps:

```
multistage_template = """
Process this query through multiple stages using the provided
context:

CONTEXT:
{documents}
```

```
QUERY:
{query}

STAGE 1: INFORMATION EXTRACTION
Identify and list all relevant facts from the context related
to the query.

STAGE 2: ANALYSIS
Analyze these facts, noting connections, patterns, and
implications.

STAGE 3: RESPONSE FORMULATION
Synthesize your analysis into a comprehensive response to
the query.

BEGIN PROCESSING:
"""
```

This approach is particularly useful for complex questions requiring multiple cognitive operations.

Template Selection Guidelines

When selecting a template type for your RAG application, consider

1. **Query Complexity**: Simple factual questions work well with basic QA templates, while complex analytical questions benefit from reasoning templates.

2. **Information Characteristics**: Contradictory or diverse sources may require comparison or multistage templates.

3. **Output Requirements**: Consider the desired
    format, length, and style of the response.

4. **Domain Specificity**: Technical or specialized
    domains often benefit from task-specific templates
    with domain-appropriate constraints.

In the next section, we'll explore how to make these templates dynamic
and adaptable to different retrieval results and query types.

Dynamic Prompt Construction

Static templates serve as excellent starting points, but the most
sophisticated RAG applications employ dynamic prompt construction—
adapting templates in real-time based on query characteristics, retrieval
results, and available context windows. This section explores techniques
for creating flexible, responsive prompting systems.

Variable Substitution: The Foundation
of Dynamic Prompts

At its core, dynamic prompt construction begins with simple variable
substitution. Most template libraries provide mechanisms to inject
variables into predefined templates:

```
from langchain.prompts import PromptTemplate

template = """
Answer the question based on the context below.

Context: {context}

Question: {question}
```

Answer:
"""

```python
prompt = PromptTemplate(
    input_variables=["context", "question"],
    template=template,
)
```

This basic approach can be extended to incorporate conditional logic, transformations, and complex variable processing.

Programmatic Template Selection

Different query types often benefit from different template structures. A robust RAG system might analyze the query to select the most appropriate template:

```python
def select_template(query: str) -> PromptTemplate:
    comparison_terms = ["compare", "difference", "versus",
    "vs", "similarities"]
    summarization_terms = ["summarize", "summary", "overview",
    "brief"]

    if any(term in query.lower() for term in comparison_terms):
        return templates["comparison"]
    elif any(term in query.lower() for term in
    summarization_terms):
        return templates["summarization"]
    else:
        return templates["qa"]
```

This approach allows your RAG system to seamlessly switch between different interaction modes based on query analysis.

Context Window Management

Perhaps the most critical aspect of dynamic prompt construction is managing the context window effectively. LLMs have token limits, and efficiently allocating those tokens between instructions, context, and response space requires careful planning.

```python
def fit_documents_to_token_limit(
    docs: List[Document],
    token_limit: int,
    model: str = "gpt-3.5-turbo"
) -> str:
    processed_docs = []
    current_tokens = 0

    for i, doc in enumerate(docs):
        doc_text = f"\n[Document {i+1}]: {doc.page_content}\n"
        doc_tokens = count_tokens(doc_text, model)

        if current_tokens + doc_tokens <= token_limit:
            processed_docs.append(doc_text)
            current_tokens += doc_tokens
        else:
            break

    return "".join(processed_docs)
```

This ensures you maximize the use of available context window while keeping within token limits.

Document Prioritization Strategies

Not all retrieved documents are equally relevant or important. Dynamic prompt construction should incorporate strategies for prioritizing content.

```python
def prioritize_documents(docs: List[Document]) ->
List[Document]:
    # Sort by relevance score (descending)
    sorted_docs = sorted(docs, key=lambda x: x.metadata.
    get('relevance_score', 0), reverse=True)

    # Group documents by relevance score
    high_relevance = [doc for doc in sorted_docs if doc.
    metadata.get('relevance_score', 0) > 0.7]
    medium_relevance = [doc for doc in sorted_docs if 0.4
    <= doc.metadata.get('relevance_score', 0) <= 0.7]

    # Combine with priority weighting
    return high_relevance + medium_relevance
```

Advanced implementations might incorporate semantic similarity, document recency, entity matching, and user preference history.

Content Transformation Techniques

Dynamic prompt construction often involves transforming retrieved content to better fit within templates:

```python
def extract_relevant_passages(doc: Document, query: str, max_
length: int = 200) -> str:
    # Simple approach: find query terms in text
    query_terms = set(query.lower().split())

    # Create overlapping windows of text
    window_size = min(max_length, len(doc.page_content))

    # Find passage with most query term matches
    # ... (implementation details in companion notebook)

    return best_passage
```

Common transformation techniques include passage extraction, information chunking, entity highlighting, and table restructuring.

Adaptive Instruction Calibration

The system instructions portion of templates can also adapt dynamically based on retrieval quality and query characteristics:

```
def calibrate_instructions(query: str, retrieved_docs:
List[Document]) -> str:
    # Base instructions
    instructions = "Answer based on the context provided."

    # Assess retrieval quality
    confidence = retrieval_confidence(retrieved_docs)

    if confidence < 0.5:
        instructions += " If the context doesn't contain
        sufficient information, clearly state what's missing."

    # Analyze query complexity
    if requires_reasoning(query):
        instructions += " Break down your reasoning process
        step by step."

    return instructions
```

This approach allows the system to provide more tailored guidance to the LLM based on the specific scenario.

Putting It All Together

To create a fully dynamic RAG prompt, we integrate all these components into a comprehensive workflow:

```python
def build_dynamic_rag_prompt(query: str, docs: List[Document],
max_tokens: int = 3500) -> str:
    # 1. Score and prioritize documents
    scored_docs = add_relevance_scores(docs, query)
    prioritized_docs = prioritize_documents(scored_docs)

    # 2. Transform documents
    transformed_docs = [transform_document(doc, query) for doc
    in prioritized_docs]

    # 3. Calibrate instructions
    system_instruction = calibrate_instructions(query,
    scored_docs)

    # 4. Fit within token limit
    # ... (implementation details in companion notebook)

    # 5. Construct the final prompt
    final_prompt = template_skeleton.replace("[PLACEHOLDER]",
    context_text)

    return final_prompt
```

Companion Notebook: Dynamic Prompt Construction

For complete implementations of all concepts covered in this section, refer to the companion Google Colab notebook: "Chapter9_DynamicPrompts. ipynb". This notebook includes

1. Complete code for all the examples shown in this section

2. Token counting utilities and document processing functions

3. Interactive examples demonstrating template selection logic

4. Document prioritization and transformation implementations

5. A comprehensive implementation that integrates all components

The notebook allows you to experiment with different query types, document sets, and token limitations to see how dynamic prompt construction works in practice. It's designed to run in Google Colab with minimal setup—just add your API key if you want to generate actual responses.

In the next section, we'll explore how to enhance templates further with few-shot learning examples to guide model behavior more precisely.

Few-Shot Learning and Examples in Templates

Few-shot learning is a powerful technique for enhancing prompt templates by including examples that demonstrate desired behavior. In the context of RAG systems, well-crafted examples can guide the LLM to follow specific patterns when processing retrieved information.

The Power of Examples in Prompts

Few-shot learning works by leveraging the model's ability to recognize patterns from demonstrations. We can guide the model to follow similar patterns when generating responses by including examples of ideal inputs and outputs.

```
few_shot_template = """
Answer the question based on the context. If the answer isn't
in the context, say "I don't have enough information."

Example 1:
Context: The Golden Gate Bridge was completed in 1937. It has a
total length of 8,981 feet.
Question: When was the Golden Gate Bridge completed?
Answer: The Golden Gate Bridge was completed in 1937.

Example 2:
Context: The Eiffel Tower is 330 meters tall and was completed
in 1889.
Question: How deep is the foundation of the Eiffel Tower?
Answer: I don't have enough information.

Now, answer the following:
Context: {context}
Question: {question}
Answer:
"""
```

This approach shows the model exactly how to handle both cases
where information is available and cases where it's missing.

Designing Effective Few-Shot Examples

The quality and design of examples significantly impact their effectiveness.
Here are key principles for creating optimal examples:

1. **Representativeness**: Examples should represent
   the range of expected inputs and desired outputs.

2. **Specificity**: Examples should demonstrate specific
   behaviors you want the model to emulate.

3. **Diversity**: Include a variety of scenarios to help the
    model generalize effectively.

4. **Complexity Gradient**: Order examples from simple
    to complex to help the model build understanding.

Let's look at a template with diverse examples for a RAG application:

```
diverse_examples_template = """
Answer based on the provided information.

Example 1: [Direct answer with citation]
Context: [Doc1] Einstein published the theory of relativity
in 1905.
Question: When did Einstein publish the theory of relativity?
Answer: Einstein published the theory of relativity in
1905 [Doc1].

Example 2: [Synthesis across documents]
Context: [Doc1] The Great Depression began with the stock
market crash in 1929.
[Doc2] The Great Depression lasted until about 1939.
Question: What was the Great Depression?
Answer: The Great Depression was an economic downturn that
began with the stock market crash in 1929 [Doc1] and lasted
until about 1939 [Doc2].

Context: {context}
Question: {question}
Answer:
"""
```

Example Categories for RAG Applications

For RAG systems, several categories of examples have proven particularly
valuable:

Source Attribution Examples

These demonstrate how to properly cite sources from retrieved
documents:

```
attribution_example = """
Context: [Doc1] The Python programming language was created by
Guido van Rossum in 1991.
Question: Who created Python?
Answer: Python was created by Guido van Rossum in 1991 [Doc1].
"""
```

Handling Incomplete Information

These show how to respond when retrieved documents don't contain
the answer:

```
incomplete_info_example = """
Context: [Doc1] JavaScript was created by Brendan Eich in 1995.
Question: When was TypeScript released?
Answer: I don't have enough information in the provided context
to answer when TypeScript was released.
"""
```

Resolving Contradictions

These demonstrate how to handle conflicting information in retrieved
documents:

```
contradiction_example = """
Context: [Doc1] Some studies suggest coffee may increase heart
disease risk.
[Doc2] Recent research indicates moderate coffee consumption
may reduce heart disease risk.
```

Question: Does coffee increase heart disease risk?
Answer: The information in the context presents conflicting views. According to [Doc1], some studies suggest coffee may increase heart disease risk, while [Doc2] indicates recent research shows moderate coffee consumption may actually reduce heart disease risk. The scientific consensus appears to be evolving on this topic.
"""

Dynamic Example Selection

For advanced RAG systems, we can programmatically select the most relevant examples based on the query type:

```python
def select_examples(query_type):
    example_library = {
        "factual": [
            {"context": "Rome is the capital of Italy.",
             "question": "What is the capital of Italy?",
             "answer": "Rome is the capital of Italy."}
        ],
        "comparative": [
            {"context": "Python is dynamically typed. Java is
            statically typed.",
             "question": "Compare Python and Java typing
            systems.",
             "answer": "Python is dynamically typed while Java
            is statically typed."}
        ],
        "not_found": [
            {"context": "The Earth orbits the Sun.",
             "question": "How many moons does Mars have?",
```

```
            "answer": "I don't have enough information to
            answer this question."}
    ]
}

# Return relevant examples based on query type
return example_library.get(query_type, example_
library["factual"])
```

Balancing Example Count with Context Limitations

A key challenge with few-shot examples is that they consume valuable token space that could otherwise be used for retrieved documents. Finding the right balance requires careful consideration:

```
def build_prompt_with_examples(query, docs, max_tokens=3500):
    # Detect query type
    query_type = detect_query_type(query)

    # Select appropriate examples (1-2 examples)
    examples = select_examples(query_type)[:2]

    # Format examples section
    examples_text = format_examples(examples)
    examples_tokens = count_tokens(examples_text)

    # Calculate remaining tokens for context
    base_template_tokens = count_tokens(template_without_examples)
    available_for_context = max_tokens - base_template_tokens -
    examples_tokens

    # Process documents to fit remaining space
    context = fit_documents_to_token_limit(docs, available_for_
    context)
```

```
# Build final prompt
prompt = template_without_examples.format(
    examples=examples_text,
    context=context,
    question=query
)

return prompt
```

Example Rotation Strategies

For systems that handle diverse query types, rotating examples can be an
effective strategy:

1. **Query-Based Rotation**: Select examples most like
   the current query.

2. **Performance-Based Rotation**: Track which
   examples yield better results and prioritize them.

3. **Diversity Rotation**: Ensure coverage of different
   response patterns and edge cases.

```
def rotate_examples(example_bank, query, rotation_
strategy="query_similarity"):
    if rotation_strategy == "query_similarity":
        # Select examples with highest semantic similarity
        to query
        # Implementation details in companion notebook
        return select_most_similar_examples(example_
        bank, query)
    elif rotation_strategy == "diversity":
        # Select examples that cover different patterns
        return select_diverse_examples(example_bank)
```

```
else:
    # Default strategy
    return example_bank[:2]  # Just take first two examples
```

Companion Notebook: Few-Shot Learning

To experiment with few-shot learning techniques in RAG templates, refer to our companion Google Colab notebook "Chapter9_FewShotExamples. ipynb" which builds on the previous section's dynamic prompt construction concepts. The notebook includes

1. A library of example templates for different RAG scenarios

2. Functions for dynamic example selection and rotation

3. Demonstration of token usage optimization with few-shot examples

4. Comparative performance analysis of zero-shot vs. few-shot prompting

5. Implementation of adaptive example selection based on query characteristics

With these few-shot learning techniques, you can significantly improve the consistency, accuracy, and alignment of your RAG system's responses. In the next section, we'll explore prompt template libraries and frameworks that can help streamline the implementation of these techniques.

Prompt Template Libraries and Frameworks

Building prompt templates from scratch can be time consuming and complex. Fortunately, several libraries and frameworks have emerged to simplify this process, providing standardized templates, composition tools, and best practices for RAG applications. This section explores popular template frameworks and how to leverage them effectively.

Overview of Popular Template Frameworks

Several frameworks have emerged to address the challenges of prompt engineering in RAG systems:

LangChain

LangChain provides a comprehensive ecosystem for building LLM-powered applications, with robust support for prompt templates and RAG workflows:

```
from langchain.prompts import PromptTemplate,
ChatPromptTemplate
from langchain.prompts.few_shot import FewShotPromptTemplate
from langchain_core.output_parsers import StrOutputParser

# Basic template
qa_template = PromptTemplate.from_template(
    """Answer the question based on the context.

    Context: {context}
    Question: {question}

    Answer:"""
)
```

```python
# Few-shot template
examples = [
    {"context": "Paris is the capital of France.", "question":
    "What is the capital of France?", "answer": "Paris"},
    {"context": "Berlin is the capital of Germany.",
    "question": "What is the capital of Germany?", "answer":
    "Berlin"}
]

example_prompt = PromptTemplate.from_template(
    """Context: {context}
    Question: {question}
    Answer: {answer}"""
)
```

LangChain excels at composing prompts into chains and integrating them with retrievers, making it ideal for complex RAG workflows.

LlamaIndex

LlamaIndex (formerly GPT Index) specializes in data ingestion, indexing, and retrieval with strong support for RAG prompting:

```python
from llama_index.prompts import PromptTemplate

# QA prompt template
qa_template = PromptTemplate(
    "Context information is below.\n"
    "---------------------\n"
    "{context_str}\n"
    "---------------------\n"
    "Given the context information and not prior knowledge, "
```

```
    "answer the query.\n"
    "Query: {query_str}\n"
    "Answer: "
)

# Refine prompt template (for iterative refinement)
refine_template = PromptTemplate(
    "The original query is as follows: {query_str}\n"
    "We have provided an existing answer: {existing_answer}\n"
    "We have the opportunity to refine the existing answer "
    "with some more context below.\n"
    "------------\n"
    "{context_msg}\n"
    "------------\n"
    "Given the new context, refine the original answer to
    better "
    "answer the query. If the context isn't useful, return the
    original answer."
)
```

LlamaIndex provides specialized templates for different RAG patterns, including query transformations and response synthesis.

Haystack by Deepset

Haystack focuses on production-ready, modular components for question answering and RAG systems:

```
from haystack.nodes import PromptNode, PromptTemplate

template = PromptTemplate(
    prompt="""
    Answer the question based on the given documents.
    Documents: {documents}
```

```
    Question: {query}
    Answer:
    """,
    output_parser=None
)

prompt_node = PromptNode(
    model_name_or_path="gpt-3.5-turbo",
    default_prompt_template=template
)
```

Haystack excels in pipeline construction and document processing for RAG workflows.

Standard Templates for Common RAG Tasks

These frameworks provide ready-to-use templates for common RAG scenarios:

Question Answering Templates

```
# LangChain QA template
qa_template = """
You are an assistant for question-answering tasks. Use the
following pieces of context to answer the question at the end.
If you don't know the answer, just say that you don't know,
don't try to make up an answer.

CONTEXT:
{context}

QUESTION:
{question}
```

ANSWER:
"""

Summarization Templates

```
# LlamaIndex summarization template
summarize_template = """
CONTEXT:
{context_str}

Using the above context, generate a concise summary that
captures the key points.
Focus on the main ideas and significant details.

SUMMARY:
"""
```

Comparison Templates

```
# Comparison template
compare_template = """
Compare and contrast the following entities based on the
provided information:

CONTEXT:
{context}

ENTITIES TO COMPARE:
{entity_1}
{entity_2}

Provide a structured comparison highlighting key similarities
and differences.
Use only information stated in the context.

COMPARISON:
"""
```

Building Custom Template Libraries

For domain-specific applications, creating a custom template library can
enhance consistency and performance:

```
class MedicalRAGTemplates:
    """Custom template library for medical RAG applications."""

    @staticmethod
    def diagnosis_template(context, symptoms):
        return f"""
        You are a medical assistant providing information based
        on medical literature.
        Analyze the following symptoms using only the provided
        medical context.

        MEDICAL CONTEXT:
        {context}

        REPORTED SYMPTOMS:
        {symptoms}

        Based strictly on the medical context, provide:
        1. Possible conditions consistent with these symptoms
        2. Important missing information that would help narrow
           the possibilities
        3. Appropriate next steps based on medical guidelines

        Include citations to specific documents using [Doc X]
        notation.
        Emphasize that this is informational only and not a
        diagnosis.
        """
```

```python
@staticmethod
def medication_info_template(context, medication):
    return f"""
    Provide information about the following medication
    using only the provided context.

    CONTEXT:
    {context}

    MEDICATION:
    {medication}

    Include information on:
    - Approved uses
    - Common side effects
    - Typical dosing
    - Major contraindications

    Cite sources as [Doc X] and include appropriate medical
    disclaimers.
    """
```

Template Versioning and Management

For production RAG systems, template versioning becomes essential:

```python
class TemplateRegistry:
    """Registry for managing and versioning prompt
    templates."""

    def __init__(self):
        self.templates = {}
        self.version_history = {}
```

```python
def register_template(self, name, template,
version="1.0.0"):
    """Register a new template version."""
    if name not in self.templates:
        self.templates[name] = template
        self.version_history[name] = {version: template}
    else:
        self.templates[name] = template
        self.version_history[name][version] = template

def get_template(self, name, version=None):
    """Get a template by name and optional version."""
    if name not in self.templates:
        raise KeyError(f"Template '{name}' not found")

    if version is None:
        return self.templates[name]

    if version not in self.version_history[name]:
        raise KeyError(f"Version '{version}' not found for
        template '{name}'")

    return self.version_history[name][version]

def list_versions(self, name):
    """List all versions of a template."""
    if name not in self.version_history:
        raise KeyError(f"Template '{name}' not found")

    return list(self.version_history[name].keys())
```

Integration with RAG Pipelines

Templates need to integrate seamlessly with the broader RAG pipeline.
Here's how they typically fit into the workflow:

```python
from langchain.chains import RetrievalQA
from langchain_openai import OpenAI
from langchain.retrievers import your_retriever

# Create custom prompt
custom_prompt = PromptTemplate(
    template=qa_template,
    input_variables=["context", "question"]
)

# Create retrieval chain
qa_chain = RetrievalQA.from_chain_type(
    llm=OpenAI(),
    chain_type="stuff",
    retriever=your_retriever,
    chain_type_kwargs={"prompt": custom_prompt}
)

# Use the chain
response = qa_chain.invoke({"query": "What is the capital of
France?"})
```

Companion Notebook: Prompt Template Libraries

For complete implementations of template libraries and their integration
with RAG systems, refer to our companion Google Colab notebook
"RAG_Chapter9_03_Template_Libraries.ipynb" which builds on the
concepts from previous sections. The notebook includes

1. Examples of using LangChain, LlamaIndex, and Haystack prompting capabilities

2. Implementation of a custom template registry with versioning

3. Demonstrations of template integration into full RAG pipelines

4. Comparison of different frameworks for various RAG tasks

These template libraries and frameworks can dramatically reduce development time and improve consistency in your RAG applications. In the next section, we'll explore advanced prompt engineering techniques specifically designed to enhance RAG system performance.

Prompt Engineering Techniques for RAG

Effective prompt templates form the foundation of a RAG system, but mastering specific prompt engineering techniques can significantly enhance performance. This section explores strategies to improve accuracy, reduce hallucinations, and handle complex information retrieval scenarios.

Strategies for Reducing Hallucinations

Hallucinations—when the LLM generates incorrect information despite retrieved context—are one of the most persistent challenges in RAG systems. Several prompt engineering techniques can help minimize this issue:

Explicit Constraints and Boundaries

Clearly delineating the boundaries of what the model should and
shouldn't do is crucial:

```
anti_hallucination_template = """
Answer the question based EXCLUSIVELY on the provided context.
If the context doesn't contain enough information to answer
completely, say "I don't have enough information" rather than
guessing.

CONTEXT:
{context}

QUESTION:
{question}

ANSWER:
"""
```

Knowledge Boundary Acknowledgment

Instructing the model to explicitly acknowledge the limits of the provided
knowledge:

```
boundary_template = """
Use ONLY the facts present in the context below to answer the
question.
Your knowledge is limited to this context - do not introduce
external information.

CONTEXT:
{context}

QUESTION:
{question}
```

Begin your response with a statement of what you know based on
the context, then provide your answer based strictly on that
information.

ANSWER:
"""

Source Attribution Requirements

Enforcing citation and attribution helps the model stick to retrieved facts:

```
attribution_template = """
Answer based solely on the provided documents.

DOCUMENTS:
{context}

QUESTION:
{question}

INSTRUCTIONS:
1. Every factual statement must cite its source document using
   [Doc X] notation
2. If information from multiple documents is used, cite
   each source
3. If the question cannot be answered from the documents, state
   this clearly

ANSWER:
"""
```

Techniques to Improve Relevance and Coherence

RAG systems must not only be factual but also provide relevant, coherent
responses to user queries:

Contextual Priming

Prime the model with contextual information about the query domain:

```
priming_template = """
You are answering a question about {domain}.
When working with {domain} information, it's important to
consider {key_considerations}.

CONTEXT:
{context}

QUESTION:
{question}

Using the context provided, give a well-structured answer
addressing the question.

ANSWER:
"""
```

Response Structuring

Guiding the structure of responses improves coherence and information
organization:

```
priming_template = """
You are answering a question about {domain}.
When working with {domain} information, it's important to
consider {key_considerations}.

CONTEXT:
{context}

QUESTION:
{question}
```

Using the context provided, give a well-structured answer
addressing the question.

ANSWER:
"""

Query-Focused Reasoning

Encouraging the model to reason step-by-step about the query before
answering:

```
reasoning_template = """
Answer the question using the provided information.

CONTEXT:
{context}

QUESTION:
{question}

REASONING PROCESS:
1. First, identify what the question is specifically asking for
2. Then, locate the relevant information in the context
3. Consider whether the context fully answers the question
4. Formulate your answer based on this analysis

ANSWER:
"""
```

Methods for Handling Contradictory Information

Retrieved documents often contain conflicting information. These
techniques help the model navigate contradictions:

Contradiction Acknowledgment

Explicitly instructing the model to identify and address contradictions:

```
contradiction_template = """
Answer based on the following information, which may contain
contradictions.

CONTEXT:
{context}

QUESTION:
{question}

INSTRUCTIONS:
- If you notice contradictory information, explicitly identify
  the contradiction
- Present both perspectives with their respective sources
- If possible, explain potential reasons for the contradiction
- Indicate which answer has stronger support, if applicable

ANSWER:
"""
```

Information Quality Assessment

Guiding the model to evaluate the quality and reliability of different
sources:

```
quality_assessment_template = """
Analyze the following information and answer the question.

CONTEXT:
{context}
```

QUESTION:
{question}

When answering:
1. Assess the reliability of each piece of information
   (recency, source credibility)
2. Prioritize information from more authoritative sources
3. Consider the consistency across multiple sources
4. Indicate confidence level in your final answer

ANSWER:
"""

Multi-perspective Synthesis

Encouraging balanced presentation of conflicting viewpoints:

multi_perspective_template = """
The following sources may present different perspectives on the
question.

SOURCES:
{context}

QUESTION:
{question}

Present a balanced answer that:
- Synthesizes the different perspectives
- Acknowledges areas of consensus and disagreement
- Avoids favoring one viewpoint without justification
- Helps the reader understand the full picture

ANSWER:
"""

Approaches for Source Attribution and Citation

Proper attribution is critical for transparency and verifiability in RAG systems:

Inline Citation Format

Standardizing how the model should cite sources within text:

```
inline_citation_template = """
Answer the question using the provided sources.

SOURCES:
{context}

QUESTION:
{question}

Cite your sources using the format [Source X] immediately after
each piece of information that comes from that source. Every
factual statement should have a citation.

ANSWER:
"""
```

Evidence Grading

Instructing the model to grade the strength of evidence for different claims:

```
evidence_grading_template = """
Answer the question based on the provided information.

INFORMATION:
{context}

QUESTION:
{question}
```

For each key point in your answer:
1. Cite the relevant source(s) [Source X]
2. Indicate evidence strength (Strong, Moderate, Limited)
3. Note if critical information is missing

ANSWER:
"""

Source Qualification

Encouraging the model to qualify sources while citing them:

source_qualification_template = """
Answer using the following information sources.

SOURCES:
{context}

QUESTION:
{question}

When citing sources, include relevant qualifiers:
- Recency: Note publication date when relevant [Source X, 2022]
- Type: Identify source type [Source X, Research Paper]
- Agreement: Mention if multiple sources confirm the
  information [Sources X, Y]

ANSWER:
"""

Companion Notebook: Prompt Engineering Techniques

For hands-on exploration of these prompt engineering techniques, refer to our companion Google Colab notebook "RAG_Chapter9_04_Prompt_Engineering.ipynb". The notebook includes

1. Implementation and testing of anti-hallucination strategies

2. Comparative evaluation of different attribution approaches

3. Experiments with contradiction handling techniques

4. Automated evaluation of various prompt engineering methods

5. Real-world examples demonstrating how prompt engineering improves RAG performance

These techniques represent the cutting edge of prompt engineering for RAG systems. By thoughtfully applying these approaches in your templates, you can significantly enhance the reliability, relevance, and overall quality of your RAG application's outputs.

Testing and Optimizing Prompt Templates

Creating effective prompt templates is an iterative process that requires systematic testing and optimization. This section explores methodologies for evaluating, comparing, and refining templates to enhance the performance of your RAG system.

Evaluation Metrics for Prompt Template Performance

Assessing prompt templates requires clear, measurable criteria that align with your application's goals:

Accuracy Metrics

The most fundamental metric is factual accuracy—whether the generated content correctly reflects the retrieved information:

```python
def evaluate_factual_accuracy(responses, ground_truth):
    """Evaluate factual accuracy of responses against ground
    truth."""
    scores = []
    for response, truth in zip(responses, ground_truth):
        # Simple exact match for facts
        facts_in_response = extract_facts(response)
        facts_in_truth = extract_facts(truth)

        correct_facts = [f for f in facts_in_response if f in
        facts_in_truth]
        if len(facts_in_response) > 0:
            accuracy = len(correct_facts) / len(facts_in_
            response)
        else:
            accuracy = 0.0

        scores.append(accuracy)

    return sum(scores) / len(scores) if scores else 0.0
```

For production systems, manual evaluation by subject matter experts often provides the most reliable accuracy assessment, especially for nuanced topics.

Citation Quality

In RAG applications, proper attribution is crucial. This metric assesses whether the model correctly cites the sources of information:

```python
def evaluate_citation_quality(responses, contexts):
    """Evaluate the quality of source citations in
    responses."""
    scores = []
    for response, context in zip(responses, contexts):
        # Extract citations from response
        citations = re.findall(r'\[(?:Document|Source|Doc)
        (\d+)\]', response)

        # Count total citations and valid citations
        total_citations = len(citations)
        valid_citations = sum(1 for c in citations if
        f"Document {c}" in context)

        # Calculate score
        if total_citations > 0:
            score = valid_citations / total_citations
        else:
            score = 0.0

        scores.append(score)

    return sum(scores) / len(scores) if scores else 0.0
```

Hallucination Rate

Measuring the model's tendency to generate information not present in
the retrieved documents:

```python
def estimate_hallucination_rate(responses, contexts):
    """Estimate the rate of hallucinated content."""
    scores = []
    for response, context in zip(responses, contexts):
        # Extract statements from response (simplified
        approach)
        statements = [s.strip() for s in response.split('.') if
        len(s.strip()) > 20]

        # Count statements not supported by context
        (simplified)
        unsupported = 0
        for statement in statements:
            # Check if key elements of the statement appear in
            the context
            keywords = extract_keywords(statement)
            if not any(all(k.lower() in context.lower() for k
            in keywords) for context in contexts):
                unsupported += 1

        hallucination_rate = unsupported / len(statements) if
        statements else 0.0
        scores.append(hallucination_rate)

    return sum(scores) / len(scores) if scores else 0.0
```

Relevance Score

Measures how well the response addresses the specific query:

```
def calculate_relevance_score(responses, queries):
    """Calculate relevance of responses to queries."""
    # This typically requires semantic similarity calculation
    # or human evaluation for reliable results
    # Here's a placeholder for the concept
    return "Relevance scoring typically requires semantic
analysis or human evaluation"
```

A/B Testing Different Template Structures

Systematic comparison of template variations helps identify the most effective approaches:

Setting Up A/B Tests

```
def run_ab_test(template_a, template_b, queries,
contexts, llm):
    """Run an A/B test comparing two templates."""
    results_a = []
    results_b = []

    for query, context in zip(queries, contexts):
        # Format prompts
        prompt_a = template_a.format(context=context,
        question=query)
        prompt_b = template_b.format(context=context,
        question=query)
```

```
# Generate responses
response_a = llm.invoke(prompt_a).content
response_b = llm.invoke(prompt_b).content

results_a.append(response_a)
results_b.append(response_b)

return results_a, results_b
```

Controlled Variable Testing

To isolate the impact of specific template elements, test variations that
differ in only one aspect:

```
# Base template
base_template = """
Answer the question based on the context.

CONTEXT:
{context}

QUESTION:
{question}

ANSWER:
"""

# Variant with explicit anti-hallucination instruction
variant_template = """
Answer the question based on the context.

CONTEXT:
{context}

QUESTION:
{question}
```

Use ONLY information from the context. If the context doesn't
contain the answer, say "I don't have enough information."

ANSWER:
"""

The difference between these templates is precisely one
instruction

Significance Testing

For reliable conclusions, statistical significance testing helps determine
whether observed differences are meaningful:

```python
def check_statistical_significance(metrics_a, metrics_b,
confidence_level=0.95):
    """Check if the difference between metrics is statistically
    significant."""
    # This is a placeholder - actual implementation would use
    # appropriate statistical tests like t-test or Mann-Whitney U
    import scipy.stats as stats

    t_stat, p_value = stats.ttest_ind(metrics_a, metrics_b)

    return {
        "t_statistic": t_stat,
        "p_value": p_value,
        "significant": p_value < (1 - confidence_level)
    }
```

Iterative Refinement Based on Output Quality

Systematic template improvement involves a cyclical process:

Error Analysis

Cataloging the types of errors in model outputs guides targeted
improvements:

```python
def analyze_errors(responses, contexts, queries):
    """Analyze common error patterns in responses."""
    error_types = {
        "hallucination": 0,
        "missing_information": 0,
        "incorrect_citation": 0,
        "irrelevant_content": 0,
        "contradiction": 0
    }

    for response, context, query in zip(responses, contexts,
    queries):
        # Simplified error detection logic
        if contains_unsupported_claims(response, context):
            error_types["hallucination"] += 1

        if not addresses_query(response, query):
            error_types["irrelevant_content"] += 1

        # Additional error checks would go here

    return error_types
```

Template Refinement Cycle

The process of iterative improvement follows a structured path:

1. Establish baseline performance with initial template

2. Identify error patterns through systematic analysis

3. Implement targeted template modifications

4. Test modified templates against the same
evaluation set

5. Measure improvement and identify
remaining issues

6. Repeat until performance meets targets or plateaus

Template Versioning

Maintaining a record of template iterations helps track progress and
insights:

```python
class TemplateVersion:
    def __init__(self, template, description, changes=None):
        self.template = template
        self.description = description
        self.changes = changes or []
        self.metrics = {}
        self.timestamp = datetime.now()

    def add_metric(self, name, value):
        self.metrics[name] = value

    def summary(self):
        return {
            "description": self.description,
            "changes": self.changes,
            "metrics": self.metrics,
            "timestamp": self.timestamp.isoformat()
        }

# Example usage
template_history = []
```

```
base = TemplateVersion(
    template=base_template,
    description="Basic QA template"
)
base.add_metric("accuracy", 0.72)
base.add_metric("hallucination_rate", 0.15)
template_history.append(base)

v2 = TemplateVersion(
    template=variant_template,
    description="Added anti-hallucination instruction",
    changes=["Added explicit instruction not to use external
    knowledge"]
)
v2.add_metric("accuracy", 0.75)
v2.add_metric("hallucination_rate", 0.08)
template_history.append(v2)
```

Automated Prompt Optimization Techniques

For advanced RAG systems, automated optimization can increase
efficiency:

Programmatic Template Generation

```
def generate_template_variants(base_template, components):
    """Generate multiple template variants by combining
    components."""
    variants = []

    # Create all possible combinations of optional components
    import itertools
```

```python
for r in range(1, len(components) + 1):
    for combo in itertools.combinations(components.
    items(), r):
        variant = base_template
        description = "Base template with: "
        changes = []

        for name, text in combo:
            variant += text
            description += name + ", "
            changes.append(f"Added {name}")

        variants.append(TemplateVersion(
            template=variant,
            description=description.rstrip(", "),
            changes=changes
        ))

    return variants
```

Gradient-Based Optimization

For continuous parameters in templates (like token allocations), gradient-based approaches can be effective:

```python
def optimize_token_allocation(base_allocation, eval_function,
steps=10, learning_rate=0.05):
    """Optimize token allocation between different prompt
    components."""
    current_allocation = base_allocation.copy()
    best_allocation = base_allocation.copy()
    best_score = eval_function(best_allocation)
```

```python
for step in range(steps):
    # Try small adjustments to each component
    for component in current_allocation:
        # Try increasing this component's allocation
        test_allocation = current_allocation.copy()
        test_allocation[component] += 10  # Add 10 tokens

        # Normalize to maintain total token count
        total = sum(test_allocation.values())
        target_total = sum(current_allocation.values())
        scaling_factor = target_total / total
        test_allocation = {k: v * scaling_factor for k, v
        in test_allocation.items()}

        # Evaluate
        score = eval_function(test_allocation)
        if score > best_score:
            best_allocation = test_allocation
            best_score = score

    # Update current allocation
    current_allocation = best_allocation.copy()

return best_allocation, best_score
```

Evolution-Based Approaches

Genetic algorithms can efficiently explore the space of possible templates:

```python
def evolve_templates(population, fitness_function,
generations=5):
    """Evolve a population of templates using genetic algorithm
    principles."""
    for generation in range(generations):
        # Evaluate fitness
```

```python
    fitness_scores = [fitness_function(template) for
    template in population]

    # Select parents (simplified)
    parents = [population[i] for i in
    sorted(range(len(fitness_scores)),
                        key=lambda i: fitness_scores[i],
                        reverse=True)[:len(population)//2]]

    # Create next generation (simplified)
    next_gen = parents.copy()
    while len(next_gen) < len(population):
        # Select two parents
        parent1, parent2 = random.sample(parents, 2)

        # Crossover (simplistic text splicing)
        split_point = len(parent1) // 2
        child = parent1[:split_point] +
        parent2[split_point:]

        # Mutation (random instruction addition/removal)
        child = mutate_template(child)

        next_gen.append(child)

    population = next_gen

return population
```

Companion Notebook: Testing and Optimization

For practical implementation of these evaluation and optimization techniques, refer to our companion Google Colab notebook "RAG_Chapter9_05_Template_Optimization.ipynb". The notebook includes

1. Complete code for automated evaluation of prompt templates

2. Implementation of A/B testing for template comparison

3. Template versioning and error analysis tools

4. Automated template optimization examples

5. Case studies showing the iterative refinement process

Through systematic testing and optimization, you can significantly enhance the performance of your RAG system's prompt templates, reducing hallucinations while improving relevance and overall output quality.

Advanced Topics in Prompt Templates

As RAG systems mature in sophistication, prompt templates must evolve to handle increasingly complex scenarios. This section explores advanced prompting techniques that push the boundaries of what's possible with retrieval-augmented generation.

Multimodal Prompt Templates

Modern LLMs are increasingly capable of processing multiple modalities, enabling RAG systems that can incorporate images, diagrams, and other visual elements:

Image-Aware RAG Templates

```
multimodal_template = """
Use both the text context and images provided to answer the
question.

TEXT CONTEXT:
{text_context}

IMAGE CONTEXT:
{image_descriptions}

QUESTION:
{question}

When answering:
1. Consider information from both text and images
2. If images contain relevant information not in the text,
   include it
3. Specify whether your information comes from text or images
4. If more visual details would help, mention this

ANSWER:
"""
```

These templates provide explicit instructions for the model to integrate information across modalities. The {image_descriptions} placeholder can contain either automatic image captions from a vision model or explicit descriptions of the images provided by the retrieval system.

Structured Output with Visual References

```
visual_reference_template = """
Answer the question using the provided text and image
information.
```

```
TEXT CONTEXT:
{text_context}

IMAGE DESCRIPTIONS:
{image_descriptions}

QUESTION:
{question}

INSTRUCTIONS:
- Provide your answer with references to both text [T1, T2,
  etc.] and images [I1, I2, etc.]
- If the answer requires visual information not described, say
  so clearly
- Format tables or structured data appropriately

ANSWER:
"""
```

This approach enables precise attribution across modalities, helping users understand which parts of the response are derived from textual vs. visual sources.

Templates for Complex Reasoning Chains

Advanced RAG applications often require multistep reasoning to synthesize information and draw conclusions:

Chain-of-Thought RAG

```
cot_rag_template = """
Analyze the provided information to answer the question.

CONTEXT:
{context}
```

QUESTION:
{question}

REASONING PROCESS:
1. Identify key facts and concepts from the context relevant to
   the question
2. Analyze relationships between these elements
3. Consider any constraints or conditions mentioned
4. Develop logical inferences step by step
5. Formulate a well-reasoned answer

Provide your complete reasoning process before giving the
final answer.

ANSWER:
"""

Chain-of-Thought prompting encourages the model to work through complex problems methodically, often leading to more accurate conclusions, especially for questions requiring multiple logical steps.

Tree of Thoughts for Decision-Making

tot_template = """
Explore multiple reasoning paths to answer this question.

CONTEXT:
{context}

QUESTION:
{question}

SOLUTION APPROACH:
1. Path A: Consider first that...
   a. Analyze the implications...

b. Evaluate whether this leads to...
c. Determine if this approach is sufficient...

2. Path B: Alternatively, consider that...
   a. Analyze these implications...
   b. Evaluate whether this approach...
   c. Determine if this path is better...

3. Compare the outcomes of different reasoning paths
4. Select the most complete and accurate answer

Explicitly show your exploration of different reasoning approaches.

ANSWER:
"""

This template encourages the model to explore multiple potential approaches to a problem before settling on the most appropriate solution, similar to how humans might weigh different methods when faced with complex questions.

Retrieval-Augmented Verification

```
verification_template = """
Answer the question and then verify your answer using the
provided context.
```

CONTEXT:
{context}

QUESTION:
{question}

ANSWER PROCESS:

1. First, generate your best answer to the question
2. Then, analyze the context for specific information that confirms or contradicts your answer
3. Look for potential gaps or assumptions in your initial answer
4. Correct any errors or omissions based on the context
5. Provide your final, verified answer with appropriate confidence

FINAL ANSWER:
"""

This approach first commits to an answer, then deliberately checks that answer against the retrieved information, improving accuracy by catching potential errors or oversights.

Self-Reflective Prompts and Self-Correction

One of the most powerful advanced techniques is to prompt the model to critique and improve its own outputs:

Self-Reflection Template

```
self_reflection_template = """
Answer the question using the provided information, then
reflect on your answer.

CONTEXT:
{context}

QUESTION:
{question}
```

INITIAL ANSWER:
[Provide your best answer based on the context]

REFLECTION:
- What assumptions did I make in my answer?
- Did I use all relevant information from the context?
- Are there alternative interpretations I should consider?
- How confident am I in each part of my answer?
- What additional information would strengthen my answer?

IMPROVED ANSWER:
[Provide a refined answer based on your reflection]
"""

Self-reflection helps the model identify potential weaknesses in its own reasoning and improve its answer accordingly.

Adversarial Self-Critique

adversarial_template = """
You will answer the question in two phases: first as an answer
generator, then as a critical reviewer looking for flaws.

CONTEXT:
{context}

QUESTION:
{question}

PHASE 1 - ANSWER:
[Generate your best answer based on the context]

PHASE 2 - CRITIQUE:
[Now adopt a skeptical perspective]
- Identify any unsupported claims in the answer
- Find any missing context or nuance

- Check if all relevant information was included
- Look for potential misinterpretations

FINAL ANSWER:
[Provide an improved answer addressing the critique]
"""

This approach creates an internal dialogue where the model deliberately tries to find flaws in its own reasoning, then addresses them.

Templates for Hybrid Human–AI Workflows

Advanced RAG systems often function as part of human-in-the-loop workflows, requiring templates designed for collaboration:

Uncertainty Highlighting Template

```
uncertainty_template = """
Answer the question based on the provided context.

CONTEXT:
{context}

QUESTION:
{question}

FORMAT YOUR ANSWER AS FOLLOWS:
1. Main answer with high-confidence statements
2. [UNCERTAIN: Areas where information is ambiguous or
   incomplete]
3. [MISSING: Key information needed to fully answer the question]
4. Suggested follow-up questions for clarification

ANSWER:
"""
```

This template explicitly marks areas of uncertainty or missing information, making it easier for human reviewers to focus their attention on parts that require verification.

Feedback-Ready Template

```
feedback_template = """
Answer the question based on the context. Structure your
response to facilitate expert feedback.

CONTEXT:
{context}

QUESTION:
{question}

ANSWER STRUCTURE:
1. Summary (key points in 1-2 sentences)
2. Detailed explanation with reasoning
   - Point A: [Evidence] → [Conclusion]
   - Point B: [Evidence] → [Conclusion]
3. Alternative interpretations if applicable
4. Confidence assessment (High/Medium/Low) with rationale
5. [FEEDBACK REQUESTED: Specific aspects where expert input
   would be valuable]

ANSWER:
"""
```

This approach structures the response to make it easy for human experts to provide targeted feedback on specific parts of the answer.

Companion Notebook: Advanced Prompt Templates

For implementation examples of these advanced techniques, refer to our companion Google Colab notebook "RAG_Chapter9_06_Advanced_Templates.ipynb". The notebook includes

1. Implementation of multimodal RAG templates with image description integration

2. Chain-of-Thought and Tree of Thoughts templates with evaluation metrics

3. Self-reflective and adversarial self-critique examples

4. Templates optimized for human–AI collaboration workflows

5. Performance comparisons between standard and advanced templates

These advanced techniques can significantly enhance the capabilities of your RAG system, enabling it to handle complex tasks that require sophisticated reasoning, multimodal integration, and effective human collaboration.

Best Practices

Prompt templates are the communication layer that brings together retrieved information and language models in RAG systems. Throughout this chapter, we've explored various techniques for designing, optimizing, and deploying effective templates. In this section, we'll summarize key principles, highlight common pitfalls, and provide a practical checklist for creating high-performance templates.

Summary of Key Principles for Effective Prompt Templates

Several fundamental principles have emerged that apply across different RAG applications:

Clarity and Structure

The most effective templates provide clear boundaries between different components—context, query, instructions, and answer—allowing the LLM to easily navigate the prompt structure:

```
CONTEXT:
{context}

QUESTION:
{question}

INSTRUCTIONS:
[specific guidance]

ANSWER:
```

This structured approach consistently outperforms more casual or conversational formats.

Explicit Constraints

Being explicit about what the model should and shouldn't do significantly improves reliability:

```
Use ONLY information from the provided context.
If the context doesn't contain the answer, say "I don't have
enough information."
Do not introduce external knowledge not present in the context.
```

Such constraints help minimize hallucinations and keep responses grounded in retrieved information.

Task-Specific Guidance

Templates should include specific guidance tailored to the task at hand:

```
For a question-answering task:
- Answer directly and concisely
- Cite specific documents using [Doc X] notation
- Note any uncertainty in your answer

For a summarization task:
- Capture key points from all documents
- Maintain neutrality between potentially conflicting sources
- Organize information logically rather than document-
  by-document
```

The more aligned your instructions are with the specific task, the better the results will be.

Dynamic Adaptability

The most sophisticated RAG systems dynamically adjust templates based on

- Query characteristics

- Retrieved document quality and quantity

- Available context window

- User preferences and domain

This adaptive approach ensures optimal template usage across diverse scenarios.

Balance Between Instruction Detail and Token Efficiency

While detailed instructions improve performance, they also consume valuable tokens that could otherwise be used for context. Striking the right balance is crucial:

```
# Verbose but potentially wasteful
Please provide an accurate, comprehensive response to the
question based on the provided context. Ensure that your answer
is factually correct and aligns with the information given.
If the context contains insufficient information to fully
answer the question, please indicate this clearly rather than
attempting to provide information beyond what is available.
It's important that you cite your sources properly using the
[Document X] notation.
```

```
# Concise but still effective
Answer using ONLY the context. Say "Insufficient information"
if needed. Cite sources as [Doc X].
```

For most applications, concise instructions that preserve key constraints while minimizing token usage provide the best overall performance.

Common Pitfalls to Avoid

Even well-designed RAG systems can fall short due to template-related issues:

Overly Generic Templates

Generic templates that don't adapt to query type or content characteristics often produce suboptimal results. Consider

```
# Too generic
Answer the question based on the context.
```

This provides too little guidance about how to handle edge cases, conflicting information, or missing data.

Confusing Model Identity

Templates that create confusion about whether the model should respond as itself or adopt a specific persona can lead to inconsistent results:

```
# Confusing identity
You are an expert in astronomy. Based on the articles provided,
answer the question.
```

Since the model isn't actually an astronomy expert and is limited to the retrieved context, this framing can be counterproductive.

Insufficient Citation Guidance

Without explicit citation instructions, the model may blend information from multiple sources without proper attribution:

```
# Missing citation guidance
Use the information from the documents to answer the question.
```

This makes it impossible for users to verify which parts of the response come from which sources.

Ignoring Token Limits

Templates that don't account for token constraints may lead to truncated context or responses:

```
# Too verbose, wasting tokens
I want you to carefully analyze the following documents,
considering all relevant aspects and nuances. Take your time
to think step by step about the implications of each piece
of information. Consider multiple perspectives and potential
interpretations. Ensure that you...
```

Such verbose instructions consume valuable tokens that could be better used for retrieved content.

Overlooking Multimodal Components

As RAG systems increasingly incorporate multiple modalities, templates must provide explicit guidance for handling different information types:

```
# Missing multimodal guidance
Answer the question based on the context.
```

Without specific instructions on how to integrate information from text, images, or other modalities, the model may ignore critical visual information.

Checklist for Creating and Evaluating Templates

When developing templates for your RAG application, use this practical checklist:

Design Phase

- Clearly define the task and expected output format.
- Include explicit constraints on using only retrieved information.

- Provide guidance for handling insufficient or contradictory information.

- Specify citation requirements appropriate to the application.

- Allocate token budget for different prompt components.

- Consider the appropriate reasoning approach (direct, CoT, self-reflective).

- Adapt the template to domain-specific requirements.

Testing Phase

- Evaluate performance across a diverse set of queries.

- Test with different types and qualities of retrieved documents.

- Compare against simpler/baseline templates.

- Measure factual accuracy, citation quality, and relevance.

- Identify common error patterns.

- Check token usage efficiency.

- Gather user feedback on response quality.

Optimization Phase

- Streamline instructions while maintaining key constraints.

- Remove redundant or overly verbose language.

- Test template variations through A/B testing.

- Implement dynamic selection or adjustment mechanisms.

- Develop specializations for different query types.

- Create fallback templates for edge cases.

Deployment Phase

- Implement template versioning and change tracking.

- Monitor performance metrics in production.

- Establish feedback collection mechanisms.

- Create a schedule for regular template review and updates.

- Document template design decisions and optimization history.

Companion Notebook: Template Best Practices

For a practical implementation of the best practices discussed in this chapter, refer to our final companion notebook "RAG_Chapter9_07_ Template_Best_Practices.ipynb". This notebook includes

1. A complete template evaluation framework that checks for common issues

2. Implementation of the checklist as executable code

3. Examples of before/after template optimization

4. Performance comparison across different
    template designs

5. A template management system that incorporates
    all best practices

With these resources, you're ready to harness the full power of prompt
templates as the crucial communication layer in your RAG applications.

Summary and Next Steps

Prompt templates are the linchpin of effective RAG systems, transforming
retrieved knowledge into structured, actionable outputs.

Key Takeaways

This chapter explored

- **Core Principles**: Clarity, explicit constraints, and task-
  specific guidance

- **Advanced Techniques**: Dynamic construction, few-
  shot learning, and anti-hallucination strategies

- **Evaluation**: Metrics like accuracy, citation quality, and
  token efficiency

Next Steps

1. **Implement Templates**: Start with the basic
    QA template (Section 9.3) and iteratively add
    complexity.

2. **Benchmark Variations**: Use A/B testing to compare
    template performance (Section 9.8).

3. **Optimize for Production**:

   • Reduce token waste with concise instructions.

   • Integrate dynamic query analysis for template selection.

4. **Explore Advanced Patterns**: Experiment with Chain-of-Thought (Section 9.7) or multimodal templates (Section 9.9).

Resources for Further Learning

1. **Academic Research**: Follow publications from Anthropic, AI2, and Stanford CRFM on prompt engineering.

2. **Open-Source Libraries**: Study implementations in LangChain and LlamaIndex.

3. **Community Repositories**: Explore prompt collections on GitHub (e.g., Awesome-Prompt-Engineering).

4. **Interactive Tools**: Experiment with playgrounds like OpenAI's Prompt Designer.

5. **Model Documentation**: Review guidelines from OpenAI, Anthropic, and Mistral for model-specific optimizations.

By combining these best practices with systematic testing, you'll create robust templates that elevate your RAG system's reliability and relevance.

## CHAPTER 10

RAG in Action: Advanced Patterns for Unstructured Data

In our journey through Retrieval-Augmented Generation, we've assembled a powerful toolkit of components. We've explored document loaders that ingest content from various sources, text splitters that segment information into manageable chunks, embedding models that transform text into mathematical representations, vector stores that organize these embeddings, retrievers that find the most relevant information, and prompt templates that structure interactions with language models.

Now it's time to put these components to work in real-world applications. Like a master chef who moves beyond learning individual techniques to creating complete dishes, we're ready to combine our RAG components into cohesive systems that solve specific problems.

This chapter focuses on implementing advanced RAG patterns for unstructured data—text, documents, websites, and other content without predefined structure. We'll explore techniques that extend the basic RAG pattern to handle more complex scenarios: interactive question answering, information extraction, conversational interfaces, and sophisticated query analysis.

© Ranajoy Bose 2025
R. Bose, *Mastering Retrieval-Augmented Generation,*
https://doi.org/10.1007/979-8-8688-1808-0_10

Each section includes practical implementation patterns with code examples from our companion notebook, which you can run in Google Colab to see these techniques in action. As we explore each pattern, we'll examine

1. The problem it solves

2. The architecture required

3. Implementation details with code examples

4. Best practices and optimization strategies

5. Common pitfalls and how to avoid them

Let's begin with one of the most common applications of RAG: building interactive question-answering systems that go beyond simple one-off queries to deliver rich, conversational experiences.

Interactive Q&A Systems with RAG

Question-answering is perhaps the most natural application of RAG—providing accurate, sourced answers to user queries by retrieving and synthesizing information from a knowledge base. However, truly interactive Q&A systems require more sophistication than our initial implementation in Chapter 3.

This section'll extend the basic RAG Q&A architecture to handle conversation history, stream responses for a better user experience, provide transparent source attribution, and personalize retrieval based on user context.

Incorporating Chat History for Contextual Understanding

Imagine asking a friend "When was it released?" without context. They'd likely respond with confusion: "When was what released?" But if you've been discussing a specific movie, they'd understand your question in that context. The same principle applies to RAG-powered Q&A systems—they need conversation history to make sense of contextual questions.

Let's examine how to implement chat history in a RAG system:

```python
from langchain.chains import ConversationalRetrievalChain
from langchain.memory import ConversationBufferMemory
from langchain_community.chat_models import ChatOpenAI
from langchain_community.vectorstores import Chroma
from langchain_openai import OpenAIEmbeddings

# Initialize memory to store conversation history
memory = ConversationBufferMemory(
    memory_key="chat_history",
    return_messages=True
)

# Create a conversational retrieval chain with memory
qa_chain = ConversationalRetrievalChain.from_llm(
    llm=ChatOpenAI(model="gpt-3.5-turbo"),
    retriever=vectorstore.as_retriever(),
    memory=memory
)

# First question with no prior context
first_response = qa_chain.invoke({"question": "What is RAG?"})
print(first_response['answer'])
```

```
# Follow-up question that relies on conversation history
follow_up_response = qa_chain.invoke({"question": "What are its
advantages?"})
print(follow_up_response['answer'])
```

This implementation uses `ConversationBufferMemory` to maintain the history of interactions, allowing the model to understand follow-up questions that reference previous context.

Balancing History Size with Relevance

Chat history is valuable for context, but it also consumes precious token space in your LLM's context window. As conversations grow longer, you need strategies to manage this trade-off. Here are three approaches, from simplest to most sophisticated:

1. **Fixed-size window**: Keep only the most recent N exchanges.

2. **Summarization**: Periodically condense older exchanges into summaries.

3. **Selective retention**: Keep exchanges most relevant to the current topic.

Here's an example of implementing a fixed-size window:

```
# Initialize memory with a fixed-size window of 5 exchanges
windowed_memory = ConversationBufferWindowMemory(
    memory_key="chat_history",
    k=5,  # Number of exchanges to keep
    return_messages=True
)
```

For more dynamic conversations, we can implement a summarization approach:

```python
from langchain.memory import ConversationSummaryMemory

# Memory that uses an LLM to summarize conversation history
summary_memory = ConversationSummaryMemory(
    llm=ChatOpenAI(model="gpt-3.5-turbo"),
    memory_key="chat_history",
    return_messages=True
)
```

The summarization approach keeps the context concise as the conversation grows longer, preserving key information while reducing token consumption.

Best Practices for Chat History

1. **Include timestamps** when conversations span multiple sessions.

2. **Preserve entity references** to maintain coherence about people, places, or things.

3. **Consider privacy implications** of storing conversation history.

4. **Monitor token usage** to stay within LLM context window limits.

Adding chat history transforms a basic Q&A system into a truly conversational experience, allowing users to ask follow-up questions naturally without repeating context in every query.

In the companion notebook, we provide a complete implementation that demonstrates these concepts with an interactive example. You'll see how the system correctly interprets follow-up questions like "Can you elaborate on that?" or "What are the disadvantages?" by contextualizing the conversation.

Streaming Responses for Better User Experience

When asking a question to a RAG system, waiting several seconds for a complete response can make the interaction feel sluggish and unnatural. Streaming provides a more engaging experience by delivering content incrementally as it's generated, watching someone type a response in a chat application.

Let's look at how to implement streaming in a RAG-powered Q&A system:

```
from langchain.chains import RetrievalQA
from langchain.callbacks.streaming_stdout import
StreamingStdOutCallbackHandler
from langchain_community.chat_models import ChatOpenAI
from langchain_core.runnables import RunnablePassthrough

# Initialize a streaming-enabled LLM
streaming_llm = ChatOpenAI(
    model="gpt-3.5-turbo",
    streaming=True,
    callbacks=[StreamingStdOutCallbackHandler()]
)

# Create a template for RAG
from langchain.prompts import ChatPromptTemplate
```

```
template = """Answer the question based on the following
context:

Context: {context}

Question: {question}
"""
prompt = ChatPromptTemplate.from_template(template)

# Create a streaming RAG chain
def format_docs(docs):
    return "\n\n".join(doc.page_content for doc in docs)

streaming_chain = (
    {"context": retriever | format_docs, "question":
    RunnablePassthrough()}
    | prompt
    | streaming_llm
)

# Invoke with streaming output
streaming_chain.invoke("What is retrieval augmented
generation?")
```

This implementation uses LangChain's streaming capabilities to display content as it's being generated, rather than waiting for the complete response.

Handling Partial Document Retrieval in Streaming

One subtlety with streaming in RAG systems is that document retrieval typically happens before any content generation begins. This means users still experience a brief delay before seeing output. For a better experience, we can split the process into phases:

1. First, display a "searching knowledge base..." message

2. Then, show the retrieved documents as they're found

3. Finally, stream the generated response

Here's how to implement this approach:

```python
import time
from typing import List
from langchain_core.documents import Document

def stream_retrieval_and_generation(query: str, retriever, llm,
prompt):
    # Phase 1: Notify user about search
    print("Searching knowledge base...\n")

    # Phase 2: Retrieve documents with visual feedback
    start_time = time.time()
    retrieved_docs = retriever.get_relevant_documents(query)
    retrieval_time = time.time() - start_time

    print(f"Found {len(retrieved_docs)} relevant documents in
    {retrieval_time:.2f} seconds:")
    for i, doc in enumerate(retrieved_docs, 1):
        preview = doc.page_content[:100] + "..." if len(doc.
        page_content) > 100 else doc.page_content
        print(f"Document {i}: {preview}\n")

    # Phase 3: Generate and stream the response
    print("\nGenerating response:")
```

```
# Prepare context from retrieved documents
context = "\n\n".join(doc.page_content for doc in
retrieved_docs)

# Create and invoke the streaming chain
chain = prompt | llm
response = chain.invoke({"context": context,
"question": query})

return response
```

This phased approach keeps users engaged throughout the process, providing visibility into the system's operation and reducing perceived waiting time.

Best Practices for Streaming

1. **Show intermediate states** like retrieval progress to indicate the system is working.

2. **Consider chunked streaming** for long responses, where content is batched into meaningful segments.

3. **Monitor client connection** status to detect if a user has disconnected during streaming.

4. **Implement a timeout mechanism** to handle cases where generation takes too long gracefully.

Streaming transforms the user experience from a "request and wait" interaction to a more conversational flow. This small change can significantly improve user engagement and satisfaction, especially for longer responses that might otherwise result in a frustrating wait.

Source Attribution and Transparency

One of RAG's key advantages is its ability to ground responses in specific source documents. Making these sources transparent to users builds trust and provides a path to verify information. Let's explore how to implement source attribution in RAG systems.

The most basic approach is to return the source documents alongside the generated response:

```
from langchain.chains import RetrievalQA

# Create a QA chain that returns source documents
qa_with_sources = RetrievalQA.from_chain_type(
    llm=llm,
    chain_type="stuff",
    retriever=retriever,
    return_source_documents=True
)

# Query the system
result = qa_with_sources({"query": "What are the components of
a RAG system?"})

# Access both the answer and the sources
answer = result["result"]
source_documents = result["source_documents"]

# Display answer and sources
print(f"Answer: {answer}\n")
print("Sources:")
for i, doc in enumerate(source_documents, 1):
    print(f"Source {i}: {doc.metadata.get('source',
    'Unknown')}")
    print(f"Content snippet: {doc.page_content[:150]}...\n")
```

This approach shows source documents separately from the answer. While functional, it doesn't directly connect specific claims in the answer to their sources.

In-Text Citations

For more precise attribution, we can instruct the model to include in-text citations. This requires a well-designed prompt template:

```
fcitation_prompt = """
Answer the question based solely on the following context.
Use [doc1], [doc2], etc. to indicate which document supports
each part of your answer.

Context:
{context}

Question: {question}

Answer with citations:
"""

# Modify the documents to include their index
def add_index_to_docs(docs):
    doc_string = ""
    for i, doc in enumerate(docs, 1):
        doc_string += f"[doc{i}]: {doc.page_content}\n\n"
    return doc_string

# Create the chain with citation instructions
citation_chain = (
    {"context": retriever | add_index_to_docs, "question":
    RunnablePassthrough()}
    | ChatPromptTemplate.from_template(citation_prompt)
    | llm
)
```

```
# Query with citations
response = citation_chain.invoke("What are the advantages
of RAG?")
print(response)
```

With this approach, the model will consist of citations like "[doc1]" directly in its response, connecting each claim to its source.

Building a Citation Lookup System

To make citations more useful, we can build a system that allows users to click on citations to view the full source:

```
def create_response_with_clickable_citations(query, response,
source_docs):
    # Extract citations like [doc1], [doc2] from the response
    import re
    citations = re.findall(r'\[doc(\d+)\]', response)

    # Create a mapping of citation numbers to documents
    citation_map = {}
    for citation in citations:
        doc_num = int(citation)
        if doc_num <= len(source_docs):
            doc = source_docs[doc_num-1]
            citation_map[f"doc{doc_num}"] = {
                "content": doc.page_content,
                "source": doc.metadata.get("source",
                "Unknown"),
                "url": doc.metadata.get("url", "")
            }
```

```
return {
    "response": response,
    "citations": citation_map
}
```

In a web application, you could use this mapping to create pop-up windows or expandable sections that show the full context when a user clicks on a citation.

Best Practices for Source Attribution

1. **Always include document metadata** such as titles, URLs, or page numbers to make sources traceable.

2. **Balance citation density** with readability—too many citations can interrupt the text flow.

3. **Handle conflicting information** transparently by acknowledging differences between sources.

4. **Consider confidence levels** for different sources based on their reliability.

Effective source attribution not only builds trust but also provides an educational dimension to RAG systems, allowing users to explore the knowledge base through citations and develop a deeper understanding of the topic.

User-Specific Retrieval Patterns

In many applications, the most relevant information depends on who's asking. A financial advisor RAG system should retrieve different investment information for a retired person vs. a young professional. Let's explore how to implement personalized retrieval in RAG systems.

There are three main approaches to user-specific retrieval:

1. **Query augmentation**: Enhance queries with user context

2. **Metadata filtering**: Retrieve only documents relevant to the user

3. **Personalized vector stores**: Maintain separate knowledge bases per user

Let's examine each approach:

Query Augmentation with User Context

The simplest approach is to enhance queries with relevant user information:

```python
def create_personalized_query(query: str, user_profile:
dict) -> str:
    """Augment the query with relevant user context."""

    # Extract relevant user attributes based on the query topic
    relevant_attributes = []

    if "investment" in query.lower() or "financial" in query.
lower():
        if "age" in user_profile:
            relevant_attributes.append(f"age {user_
            profile['age']}")
        if "risk_tolerance" in user_profile:
            relevant_attributes.append(f"risk tolerance: {user_
                profile['risk_tolerance']}")

    elif "health" in query.lower() or "medical" in query.lower():
        if "medical_conditions" in user_profile:
            conditions = user_profile["medical_conditions"]
```

```python
        relevant_attributes.append(f"medical conditions:
        {', '.join(conditions)}")

    # If we have relevant attributes, include them in the query
    if relevant_attributes:
        augmented_query = f"{query} [For a person with {';
        '.join(relevant_attributes)}]"
        return augmented_query

    # If no relevant attributes, return the original query
    return query

# Example usage
user_profile = {
    "age": 65,
    "risk_tolerance": "conservative",
    "medical_conditions": ["hypertension", "type 2 diabetes"]
}

original_query = "What investment strategies should I
consider?"
personalized_query = create_personalized_query(original_query,
user_profile)
print(f"Original: {original_query}")
print(f"Personalized: {personalized_query}")

# Use the personalized query with the retriever
docs = retriever.get_relevant_documents(personalized_query)
```

This technique helps retrieve documents that better match the user's specific circumstances without requiring any architectural changes to your RAG system.

Metadata Filtering by User Attributes

For more targeted retrieval, we can use metadata filters to select documents based on user characteristics:

```python
from langchain.retrievers import FilteredRetriever

def create_user_specific_filters(user_profile: dict) -> dict:
    """Create metadata filters based on user profile."""
    filters = {}

    # Add age range filter if user age is available
    if "age" in user_profile:
        age = user_profile["age"]
        if age < 30:
            filters["age_group"] = "young_adult"
        elif age < 60:
            filters["age_group"] = "middle_aged"
        else:
            filters["age_group"] = "senior"

    # Add risk profile filter if available
    if "risk_tolerance" in user_profile:
        filters["risk_profile"] = user_profile["risk_
        tolerance"]

    return filters

# Create a filtered retriever
user_filters = create_user_specific_filters(user_profile)
filtered_retriever = vectorstore.as_retriever(
    search_type="similarity",
    search_kwargs={"k": 4, "filter": user_filters}
)
```

```
# Retrieve documents matching both the query and user filters
results = filtered_retriever.get_relevant_
documents(original_query)
```

This approach requires your documents to have appropriate metadata tags that match user attributes, but it provides more precise control over retrieval.

Personalized Vector Stores

For the highest level of personalization, you can maintain separate vector stores for different users or user groups:

```python
def get_or_create_user_vectorstore(user_id: str, base_docs=None):
    """"Get or create a user-specific vector store."""
    store_path = f"./user_stores/{user_id}"

    # Check if the user's store already exists
    if os.path.exists(store_path):
        # Load existing store
        user_store = Chroma(
            persist_directory=store_path,
            embedding_function=embeddings
        )
        return user_store

    # Create new store with base documents plus user-specific docs
    if base_docs:
        user_store = Chroma.from_documents(
            documents=base_docs,
            embedding=embeddings,
            persist_directory=store_path
        )
        return user_store
```

```
    # Create empty store if no base documents
    user_store = Chroma(
        persist_directory=store_path,
        embedding_function=embeddings
    )
    return user_store

# Example usage
user_id = "user_123"
user_vectorstore = get_or_create_user_vectorstore(user_id,
base_docs=common_docs)

# Add user-specific documents
user_docs = load_user_documents(user_id)
if user_docs:
    user_vectorstore.add_documents(user_docs)

# Create a retriever from the user's store
user_retriever = user_vectorstore.as_retriever()
```

This approach is particularly useful for applications where users have personal documents (like emails or notes) that should be included in retrieval alongside common knowledge.

Security and Privacy Considerations

When implementing personalized retrieval, be mindful of these security and privacy considerations:

1. **Data segregation**: Ensure that one user cannot access another user's private data.

2. **Explicit consent**: Obtain clear permission to use personal data for retrieval enhancement.

3. **Data minimization**: Only store and use the necessary personal attributes.

4. **Secure storage**: Protect user profiles and personalized vector stores with appropriate encryption and access controls.

5. **Transparency**: Make it clear to users how their personal information influences retrieval results.

Best Practices for User-Specific Retrieval

1. **Start simple** with query augmentation before implementing more complex personalization.

2. **Use a hybrid approach** combining common knowledge with personalized documents.

3. **Implement feedback mechanisms** to learn which personalization strategies are most effective.

4. **Consider caching** personalized retrieval results to improve performance.

Personalized retrieval transforms RAG from a one-size-fits-all system to an experience tailored to each user's unique context and needs. When implemented correctly, it can significantly improve the relevance and usefulness of retrieved information.

Extraction Patterns in RAG Systems

While question-answering focuses on providing direct responses to queries, extraction is about systematically pulling structured information from unstructured content. RAG-powered extraction combines the contextual understanding of LLMs with retrieval to accurately identify and extract specific data points, entities, or patterns from large document collections.

Think of extraction as teaching your system to read documents the way a skilled analyst would—identifying key information, organizing it into a structured format, and presenting it in a way ready for further processing or analysis. This section explores three powerful patterns transforming your RAG system from merely answering questions to intelligently extracting and organizing knowledge.

Reference-Based Extraction

Have you ever tried to explain a task to someone, only to find that showing an example works much better than abstract instructions? Reference-based extraction applies this same principle to LLMs. Instead of relying on complex instructions, we show the model a clear example of what we want.

Imagine you're extracting information from scientific research papers. Each paper might use slightly different formatting or terminology, but they all contain similar core elements—titles, authors, publication dates, and findings.

```
from langchain.output_parsers import PydanticOutputParser
from pydantic import BaseModel, Field

# Define the structure we want to extract
class ResearchPaper(BaseModel):
    title: str = Field(description="The title of the
    research paper")
    authors: List[str] = Field(description="List of
    authors' names")
    publication_year: int = Field(description="Year the paper
    was published")
    abstract: str = Field(description="The paper's abstract")
```

The magic happens in how we craft our prompt. Rather than abstract instructions, we provide a concrete example that demonstrates both the input format and the expected output:

```
extraction_template = """
Extract structured information from the research paper below.

EXAMPLE INPUT:
Title: Advances in Neural Information Processing Systems
Authors: John Smith, Jane Doe
Publication: Conference on Neural Information Processing
Systems, 2022

EXAMPLE OUTPUT:
{
  "title": "Advances in Neural Information Processing Systems",
  "authors": ["John Smith", "Jane Doe"],
  "publication_year": 2022
}

INPUT PAPER: {paper_text}
OUTPUT: {format_instructions}
"""
```

This approach leverages the LLM's pattern recognition abilities. By showing it what a proper extraction looks like, you're guiding it to apply the same pattern to new documents. It's like giving the model a completed form and asking it to fill out similar forms for new inputs.

The reference-based approach is particularly powerful for domain-specific documents where the structure is consistent but might use specialized terminology. Legal contracts, financial reports, medical records follow patterns that become clearer with examples than with abstract rules.

When implementing reference-based extraction in production systems, including error handling and validation is wise. Even with clear examples, LLMs occasionally produce outputs that don't match your expected structure. A simple repair mechanism can significantly increase reliability:

```
def validate_and_repair(extracted_text, parser, llm):
    try:
        # First try direct parsing
        return parser.parse(extracted_text), True
    except Exception:
        # If that fails, ask the LLM to fix it
        repair_prompt = "The following JSON doesn't match the
        expected schema. Please fix it:"
        repair_chain = ChatPromptTemplate.from_template(repair_
        prompt) | llm
        repaired_text = repair_chain.invoke({"invalid_json":
        extracted_text}).content
        # Try parsing again
        return parser.parse(repaired_text), True
```

Handling Long-Form Content Extraction

Even the most advanced LLMs have context window limitations, making it challenging to process long documents like contracts, research papers, or annual reports. Let's explore three elegant strategies to overcome this limitation.

The Chunk-and-Summarize Approach divides and conquers. We split the document into manageable chunks, extract information from each one, and then consolidate the results:

```
# Create a map-reduce chain for processing financial metrics
in chunks
map_reduce_chain = MapReduceDocumentsChain(
    # Map: Extract from each chunk
    llm_chain=LLMChain(llm=llm, prompt=PromptTemplate.from_
    template(
        "Extract all financial metrics from this text: {text}"
    )),
    # Reduce: Combine all extractions
    reduce_documents_chain=StuffDocumentsChain(
        llm_chain=LLMChain(llm=llm, prompt=PromptTemplate.from_
        template(
            "Combine these financial metrics, removing
            duplicates: {text}"
        ))
    ),
    document_variable_name="text"
)
```

This approach mirrors how a human might tackle a lengthy document—read it section by section, take notes, and then synthesize the information at the end. It works particularly well for extracting distinct, factual information like financial metrics, key statistics, or entity mentions.

For narrative documents where context builds throughout the text, **Sequential Extraction with State Tracking** offers a more sophisticated approach. Instead of treating each chunk independently, we maintain a "state" of extracted information that evolves as we process each section:

```
def sequential_extraction(chunks, initial_state=None):
    """Process chunks while maintaining and updating state."""
    state = initial_state or {"entities": {}, "events": []}
```

```
for i, chunk in enumerate(chunks):
    prompt = f"""
Update our extraction based on this new information.

What we know so far:
Entities: {json.dumps(state["entities"])}
Events: {json.dumps(state["events"])}

New text (chunk {i+1} of {len(chunks)}):
{chunk.page_content}

Provide updates to our extraction:
"""

    # Extract from this chunk and update the state
    response = llm.invoke(prompt)
    updated_state = parse_updates(response.content)
    state = merge_states(state, updated_state)

return state
```

This method shines when handling documents like legal contracts where terms defined in early sections affect the interpretation of later sections. By maintaining state, we preserve context across chunk boundaries, creating a more coherent extraction.

For documents with predictable structures, **Targeted Extraction with Section Routing** provides an efficient alternative. Rather than processing the entire document blindly, we first identify relevant sections, then extract from each with specialized prompts:

```
# Define targeted sections for a financial report
financial_sections = {
    "FINANCIAL HIGHLIGHTS": "Extract revenue, profit, EPS, and
    growth metrics.",
    "REVENUE BY SEGMENT": "Extract revenue for each business
    segment with growth rates.",
```

```
    "OUTLOOK": "Extract projections and strategic initiatives
    for next year."
}

# First identify sections, then extract from each one
sections = identify_document_sections(financial_report)
extracted_data = {}

for section_name, instructions in financial_sections.items():
    if section_name in sections:
        section_text = extract_section_text(financial_report,
        sections[section_name])
        extracted_data[section_name] = extract_from_
        section(section_text, instructions)
```

This approach is like having specialized analysts for different document parts—one focuses on financial metrics, another on strategic plans, and each extracts exactly what they're trained to find. It's particularly effective for standardized documents like annual reports, academic papers, or technical documentation.

Function-Free Extraction Methods

While some LLM providers offer specialized function calling APIs for structured extraction, you might not always have access to these features. Function-free extraction methods allow you to implement structured extraction across any LLM provider.

Template-Based Extraction uses clear instructions and output formatting guidelines to guide the LLM:

```
response_schemas = [
    ResponseSchema(name="person_name", description="The full
    name of the person"),
    ResponseSchema(name="age", description="The age as an
    integer"),
```

```
    ResponseSchema(name="occupation", description="The person's
    job or primary occupation")
]
parser = StructuredOutputParser.from_response_schemas(response_
schemas)

template = """
Extract the following information about the person in the text.
{format_instructions}
TEXT: {text}
"""
```

The key is explicit instructions about how each field should be formatted. Rather than asking "What can you tell me about this person?"— we're providing a clear template for the response.

When you need to extract multiple fields with clear boundaries, **Delimiter-Based Extraction** offers an elegant solution:

```
delimiter = "####"
extraction_prompt = f"""
Extract information with these delimiter markers:
{delimiter} PERSON NAME
[full name]
{delimiter} AGE
[age as number]
{delimiter} OCCUPATION
[job title and field]

TEXT: {{text}}
"""
```

The delimiters create a structured format that's both clear to the LLM and easy to parse programmatically. It's like creating a fill-in-the-blank template that guides the model's response.

For complex data with nested structures, **JSON Schema-Guided Extraction** provides precise control:

```
json_schema = {
    "properties": {
        "person_name": {"type": "string"},
        "location": {
            "type": "object",
            "properties": {
                "city": {"type": "string"},
                "state": {"type": "string"}
            }
        },
        "skills": {
            "type": "array",
            "items": {"type": "string"}
        }
    }
}

json_prompt = f"""
Extract information according to this schema:
{json.dumps(json_schema, indent=2)}

Ensure your response is valid JSON.
TEXT: {{text}}
"""
```

This approach is especially powerful for hierarchical data where some fields contain nested objects or arrays. The schema acts as a blueprint that guides the LLM to produce output with exactly the structure you need.

When extracting information about multiple entities (like team members, products, or events), **Tabular Extraction** offers an intuitive format:

```python
def extract_as_table(text, columns):
    """Extract information into a markdown table format."""
    prompt = f"""
Extract information from the text into a markdown table.
Include columns for: {', '.join(columns)}
Format with proper markdown table syntax including header
and separator rows.

TEXT: {text}
"""

    table_markdown = llm.invoke(prompt).content
    return parse_markdown_table(table_markdown)
```

The tabular format naturally organizes information about multiple entities into rows and columns, making comparisons and analysis straightforward. It's particularly useful for summarizing information across similar items, like comparing features across products or qualifications across candidates.

These function-free extraction methods can be mixed and matched depending on your specific needs. For simple flat structures, delimiter-based extraction provides clarity. For complex nested data, JSON schema offers precision. For multiple entities, tabular extraction creates natural organization. By selecting the right approach for your use case, you can build robust extraction capabilities that work across any LLM provider.

Building Intelligent RAG-Powered Chatbots

RAG systems truly shine when implemented as interactive chatbots that can engage in meaningful, extended conversations. While our previous implementations focused on individual Q&A exchanges, a true

RAG-powered chatbot combines multiple capabilities: maintaining conversation history, adapting its retrieval strategy to the dialogue flow, integrating external tools when needed, and managing resources efficiently during long conversations.

Think of the difference between asking a librarian a single reference question vs. having an in-depth consultation about a research topic. The latter requires the librarian to remember what you've already discussed, connect new questions to previous context, know when to pull in special resources, and maintain coherence throughout an extended interaction. Our RAG chatbot aims to provide this richer experience.

In this section, we'll explore how to evolve a basic RAG implementation into a sophisticated conversational agent that can engage users in natural, informative dialogues grounded in your knowledge base. We'll build this intelligence layer by layer, starting with the foundation: memory management.

Memory Management Strategies

Effective memory management is the cornerstone of any conversational system. Without it, your chatbot will act like someone with short-term memory loss, forgetting the context of the conversation from one exchange to the next. Let's explore different approaches to memory management, each with its own trade-offs and use cases.

Short-Term vs. Long-Term Memory Architectures

Memory architectures in RAG chatbots fall into two broad categories: short-term approaches that maintain recent exchanges verbatim, and long-term approaches that compress or summarize older interactions.

Short-term memory is conceptually simple—keep a record of recent messages:

```
buffer_memory = ConversationBufferMemory(
    memory_key="chat_history",
    return_messages=True
)
```

This approach works well for brief interactions but becomes problematic as conversations grow. A simple solution is to implement a sliding window that keeps only the most recent exchanges:

```
window_memory = ConversationBufferWindowMemory(
    memory_key="chat_history",
    k=3,   # Number of exchanges to keep
    return_messages=True
)
```

For longer conversations, we need long-term memory strategies that compress older information while preserving important context. Summary-based approaches use the LLM itself to create concise summaries of past interactions:

```
summary_memory = ConversationSummaryMemory(
    llm=llm,
    memory_key="chat_history",
    return_messages=True
)
```

The ideal approach often combines these strategies—keeping recent exchanges verbatim for immediate context while summarizing older exchanges:

```
summary_buffer_memory = ConversationSummaryBufferMemory(
    llm=llm,
    memory_key="chat_history",
    max_token_limit=1000,
    return_messages=True
)
```

This hybrid approach gives your chatbot both the detailed recall of recent exchanges and the compressed wisdom of the broader conversation history.

Efficient Storage and Retrieval of Conversation History

As conversations grow longer, token usage becomes a critical concern. Most modern LLMs have token limits (typically 4,000 to 32,000 tokens), and every token used for conversation history is one less token available for retrieved context and generation.

One effective approach is to implement token-aware memory that automatically manages its size based on token limits:

```
token_memory = ConversationTokenBufferMemory(
    llm=llm,
    memory_key="chat_history",
    max_token_limit=1000,
    return_messages=True
)
```

This memory tracks token usage and trims older messages when needed to stay within the specified limit.

For production systems, you'll also want to implement persistent storage that maintains conversations across sessions. This involves serializing conversation history to a database or file system and loading it when a user returns:

```python
def save_conversation_history(user_id, history):
    """Save conversation history to storage"""
    # Convert message objects to serializable format
    serializable_history = []
    for msg in history:
        if isinstance(msg, HumanMessage):
            serializable_history.append({'type': 'human',
                'content': msg.content})
        elif isinstance(msg, AIMessage):
            serializable_history.append({'type': 'ai',
                'content': msg.content})

    # Save to file or database
    with open(f"{user_id}_conversation.json", 'w') as f:
        json.dump({'history': serializable_history}, f)
```

Prioritizing Relevance in Memory Utilization

Not all parts of a conversation are equally relevant to the current exchange. Rather than treating conversation history as a chronological queue, we can prioritize messages based on their relevance to the current query.

Consider a custom memory implementation that uses semantic similarity to retrieve the most relevant past exchanges:

```python
def get_relevant_history(query, history, k=3):
    """Retrieve messages most relevant to the current query"""
    query_embedding = embeddings.embed_query(query)

    # Get embeddings for all messages
    message_embeddings = []
    for msg in history:
        embed = embeddings.embed_query(msg.content)
        message_embeddings.append((msg, embed))
```

```
# Calculate similarities and sort
similarities = []
for msg, embed in message_embeddings:
    similarity = calculate_similarity(query_
    embedding, embed)
    similarities.append((msg, similarity))

# Return top k most relevant messages
sorted_history = sorted(similarities, key=lambda x: x[1],
reverse=True)
return [msg for msg, _ in sorted_history[:k]]
```

This relevance-based approach ensures that the most important context is preserved in the LLM's limited context window, even if it's not the most recent exchange.

By implementing these memory management strategies, your RAG chatbot will maintain a coherent understanding of the conversation over time. It will remember important details from earlier exchanges while efficiently managing token usage and prioritizing the most relevant context for each new query.

Contextual Retrieval for Natural Conversations

A truly intelligent RAG chatbot doesn't just remember past exchanges—it also understands how to use that history to improve its retrieval process. Contextual retrieval moves beyond treating each question in isolation, leveraging the full conversational context to find the most relevant information.

Query Formulation from Chat Context

In natural conversations, people often use pronouns, references to previously mentioned concepts, or shorthand queries that only make sense in context of the ongoing conversation. For instance, after discussing RAG systems, a user might simply ask "What are its limitations?" The word "its" is only meaningful in the context of the previous exchange.

To handle these contextual queries, we need to reformulate them into self-contained questions before passing them to our retriever:

```python
def rewrite_query_with_context(query, chat_history):
    """Rewrite a potentially contextual query into a standalone
    question"""
    if not chat_history:
        return query  # No context to consider

    # Format chat history for the LLM
    formatted_history = "\n".join([
        f"Human: {msg.content}" if isinstance(msg,
        HumanMessage) else
        f"AI: {msg.content}" for msg in chat_history
    ])

    rewrite_prompt = f"""
Given the conversation history below and the user's
latest query,
rewrite the query to be self-contained with all relevant
context.

Conversation history:
{formatted_history}

Latest query: {query}
```

```
Standalone query:
"""

response = llm.invoke(rewrite_prompt)
return response.content.strip()
```

This transformation gives your retriever a much better chance of finding relevant information. For example, "What are its limitations?" might be rewritten as "What are the limitations of Retrieval-Augmented Generation systems?"

The key insight is that query reformulation acts as a bridge between the natural flow of conversation and the more structured needs of the retrieval process.

Balancing Retrieval with Conversational Flow

Not every conversational turn requires retrieval. When a user says "Thank you" or "Can you explain that in simpler terms?"—triggering a retrieval process would be inefficient and potentially disruptive to the conversation flow.

A sophisticated RAG chatbot should dynamically determine when to retrieve information:

```
def should_retrieve(query, chat_history):
    """Determine if this query needs knowledge retrieval"""
    # Check for common conversational phrases that don't need
    retrieval
    conversational_patterns = [
        r"^(thank you|thanks|ok|great|awesome)",
        r"can you explain (that|this|it)",
        r"i (don't|do not) understand",
        r"could you (clarify|simplify)"
    ]
```

```
for pattern in conversational_patterns:
    if re.match(pattern, query.lower()):
        return False

# By default, retrieve for most queries
return True
```

This selective approach prevents unnecessary retrievals, making your chatbot more responsive and natural in conversation.

The actual implementation in a production system might be more sophisticated, using a classification model trained on conversation data to determine whether a given query requires retrieval.

Implementation Examples for Different Conversational Needs

Different conversational scenarios call for different retrieval strategies. Let's look at how we might adapt our retrieval process for different use cases:

For educational scenarios where detailed, comprehensive answers are valued:

```
def educational_retrieval(query, chat_history):
    """Retrieve with emphasis on comprehensive information"""
    standalone_query = rewrite_query_with_context(query,
    chat_history)

    # Retrieve more documents for comprehensive coverage
    docs = retriever.get_relevant_documents(
        standalone_query,
        search_kwargs={"k": 5, "fetch_k": 20}
    )

    # Sort by relevance but also try to include diverse sources
    return diversify_sources(docs)
```

For customer support scenarios where concise, direct answers are preferred:

```
def support_retrieval(query, chat_history):
    """Retrieve with emphasis on direct, actionable
    information"""
    standalone_query = rewrite_query_with_context(query, chat_
    history)

    # Add priority to recent or high-confidence documents
    filter_dict = {"confidence": {"$gte": 0.8}}

    # Retrieve fewer, more precisely targeted documents
    docs = retriever.get_relevant_documents(
        standalone_query,
        search_kwargs={"k": 2, "filter": filter_dict}
    )
```

The key principle is adapting your retrieval strategy to match the conversational context and user expectations. A research assistant might provide in-depth explorations of a topic, while a customer support bot focuses on quick, direct answers to specific questions.

By implementing contextual retrieval, your RAG chatbot will deliver more relevant information that fits naturally into the conversation flow. It will understand the implicit context in user queries, know when to retrieve vs. when to simply respond conversationally, and adapt its retrieval strategy to the specific needs of the conversation.

Tool Integration for Enhanced Capabilities

Even the most sophisticated RAG system with a comprehensive knowledge base will encounter questions it can't answer based solely on its indexed documents. By integrating external tools, your chatbot can dynamically extend its capabilities beyond the boundaries of its static knowledge base.

Combining RAG with External Tools

External tools expand your chatbot's capabilities in several
important ways:

1. **Accessing real-time information** that isn't in your
   knowledge base

2. **Performing calculations or transformations** on
   retrieved data

3. **Executing actions** based on user requests

4. **Searching specialized knowledge sources** for
   domain-specific queries

Let's implement a basic tool integration framework using
LangChain's tools:

```
# Define a knowledge base retriever tool
retriever_tool = create_retriever_tool(
    retriever,
    "KnowledgeBase",
    "Search the internal knowledge base for information about
    our products and services."
)

# Define a web search tool for current information
search_tool = Tool(
    name="WebSearch",
    func=DuckDuckGoSearchRun().run,
    description="Search the web for current information not in
    the knowledge base."
)
```

```
# Define a calculator tool for mathematical operations
calculator_tool = Tool(
    name="Calculator",
    func=lambda x: eval(x),
    description="Perform mathematical calculations. Input
    should be a valid mathematical expression."
)

# Create a list of available tools
tools = [retriever_tool, search_tool, calculator_tool]
```

With these tools defined, we can create an agent that knows when and how to use them:

```
agent = initialize_agent(
    tools,
    llm,
    agent=AgentType.CHAT_CONVERSATIONAL_REACT_DESCRIPTION,
    memory=memory,
    verbose=True
)
```

This agent can now dynamically decide whether to retrieve information from your knowledge base, search the web for current information, or perform calculations as needed.

Decision Frameworks for Tool Selection

The key to effective tool integration is knowing which tool to use for which query. Rather than overwhelming your chatbot with too many tools and letting it figure out the selection process through trial and error, you can implement a decision framework that guides tool selection:

```python
def select_tools(query, available_tools):
    """Select the appropriate tools for a given query"""

    tool_selection_prompt = f"""
Based on the user's query, select the most appropriate
tool(s) from the following options:

{[f"{t.name}: {t.description}" for t in available_tools]}

User query: {query}

Return only the names of the tools to use, separated by
commas, or "none" if no tools are needed.
    """
```

This approach uses the LLM's understanding of the query and tool capabilities to make a more directed selection, reducing unnecessary tool calls and improving efficiency.

Code Patterns for Seamless Tool Incorporation

When incorporating tools into your RAG chatbot, you want the integration to feel seamless to the user. Here's a pattern that combines RAG with tool usage in a natural way:

```python
def process_query_with_tools(query, chat_history):
    """Process a query using tools as needed, with RAG as the
    foundation"""

    # 1. Determine if tools are needed
    selected_tools = select_tools(query, available_tools)

    # 2. If tools are selected, use them
    if selected_tools:
        tool_response = agent.run(input=query,
            tools=selected_tools)
```

```
    return tool_response

# 3. Otherwise, fall back to standard RAG
rewritten_query = rewrite_query_with_context(query, chat_
history)
docs = retriever.get_relevant_documents(rewritten_query)
context = "\n\n".join(doc.page_content for doc in docs)

prompt = f"""
Answer the question based on the following context and chat
history.

Context: {context}

Question: {query}
"""

    return llm.invoke(prompt).content
```

This pattern starts with tool selection but falls back to standard RAG when appropriate. The user doesn't need to request tool usage explicitly—it happens automatically when needed.

Tools dramatically expand your chatbot's capabilities, enabling it to answer questions beyond the scope of its static knowledge base. By implementing thoughtful tool selection and seamless integration, you create a more capable assistant to draw on multiple information sources and capabilities while maintaining a consistent conversational experience.

Optimizing for Extended Conversations

Real-world chatbot interactions often span dozens or even hundreds of exchanges. As conversations grow longer, maintaining coherence, managing resources efficiently, and preserving important context become increasingly challenging. Let's explore strategies for optimizing your RAG chatbot for extended conversations.

The code snippets below provide a simplified view of the core concepts. For complete implementations of all the techniques discussed in this section, refer to the companion notebook in the book's repository, which contains detailed, executable code examples.

Strategies for Managing Growing Chat Histories

As conversations grow longer, the token count for storing the entire history quickly exceeds most LLMs' context windows. Here are several strategies to manage this growth:

The simplest approach is implementing a sliding window that retains only the most recent exchanges:

```
def get_recent_history(history, k=5):
    """Retrieve the most recent k exchanges"""
    return history[-2*k:] if len(history) > 2*k else history
```

For a more nuanced approach, you can implement token-based truncation that respects conversation boundaries while staying within token limits:

```
def truncate_history_by_tokens(history, max_tokens=2000):
    """Truncate history to stay within token limits"""
    # Estimate tokens for each message and keep as many as
      possible
    # while staying under the token limit, prioritizing recent
      messages
```

This approach keeps as much recent context as possible while respecting the LLM's context window limitations.

Summarization and Compression Techniques

Rather than simply truncating history, summarization offers a way to compress older exchanges while preserving their essential meaning:

```python
def compress_history(messages, max_tokens=1000):
    """Compress conversation history using summarization"""
    # Keep recent exchanges intact
    recent_messages = messages[-6:] if len(messages) > 6 else
    messages
    old_messages = messages[:-6] if len(messages) > 6 else []

    # If older messages exist, summarize them
    if old_messages:
        summary = create_summary_of_messages(old_messages)
        return [SystemMessage(content=summary)] + recent_
        messages
    return recent_messages
```

This hybrid approach gives you the best of both worlds: a concise summary of older exchanges that preserves key information, combined with the full detail of recent exchanges for immediate context.

Maintaining Coherence Across Conversation Sessions

Many real-world applications involve conversations that span multiple sessions, sometimes with days or weeks between interactions:

```python
class PersistentConversationManager:
    """Manages conversations that persist across multiple
    sessions"""

    def __init__(self, user_id, storage_path="./
    conversations"):
        self.user_id = user_id
        self.storage_path = storage_path
        self.current_session = []
```

```python
def start_new_session(self):
    """Start a new conversation session"""
    # Load previous history and compress if needed
    previous_history = self.load_conversation_history()
    if previous_history:
        self.current_session = compress_history(previous_
        history)
        return self.current_session
    else:
        return []
```

By implementing these optimization techniques, your RAG chatbot can maintain coherent, meaningful conversations that span many exchanges and multiple sessions. The chatbot will remember important context from earlier in the conversation, efficiently manage its memory usage, and provide a consistent experience that feels like talking to a knowledgeable assistant with good recall rather than a series of disconnected Q&A exchanges.

Bringing It All Together: A Complete RAG Chatbot

Let's now combine all these components into a comprehensive RAG-powered chatbot that showcases the full potential of the architecture:

```python
class ComprehensiveRAGChatbot:
    """A complete RAG chatbot with memory, tools, and
    contextual understanding"""

    def __init__(self, retriever, memory_type="summary_buffer",
    max_tokens=2000):
        # Initialize core components
        self.retriever = retriever
```

```python
        self.llm = ChatOpenAI(model="gpt-3.5-turbo",
        temperature=0.3)
        self.memory = self._initialize_memory(memory_type,
        max_tokens)
        self.tools = self._initialize_tools()
        self.agent = self._initialize_agent()

    def chat(self, query):
        """Process a user query and return a response"""
        # Determine if we should use tools or direct RAG
        use_tools = self._should_use_tools(query)

        if use_tools:
            # Use the agent with tools
            result = self.agent.invoke({"input": query})
            return result["output"]
        else:
            # Use direct RAG with context
            context = self._get_retrieval_context(query)
            response = self._generate_response(query, context)
            self._update_memory(query, response)
            return response
```

This simplified implementation illustrates the core architecture of our comprehensive RAG chatbot. It includes memory management, contextual retrieval, tool integration, and conversation optimization—all working together to create a seamless user experience.

By thoughtfully implementing these advanced patterns, you'll create RAG chatbots that maintain coherent, meaningful conversations across many exchanges, remember important context, retrieve relevant information, and seamlessly integrate external tools when appropriate.

Advanced Query Analysis and Optimization

In the world of Retrieval-Augmented Generation, the quality of your retrieval largely determines the quality of your responses. You can have the most sophisticated embedding model, the best-structured vector store, and the most powerful LLM, but if your queries aren't effectively formulated, the entire system falters. In this section, we'll explore advanced techniques to enhance query analysis and optimization in RAG systems, moving beyond basic retrieval to robust, intelligent querying.

Think of query optimization as the art of translation—converting a user's natural language question into the most effective search possible. This translation process must address ambiguities, incorporate context, handle edge cases, and adapt to the specific knowledge domain of your system.

We'll examine five key areas of query optimization: prompt engineering with examples, handling edge cases, multi-query architectures, advanced filtering, and managing high-cardinality variables. Each technique builds upon the last to create increasingly sophisticated retrieval capabilities.

The code examples in this section are simplified for clarity. For complete, executable implementations, refer to the companion notebook in the book's repository which includes detailed code for all the techniques discussed here.

Prompt Engineering with Examples

When users interact with a RAG system, their queries often lack specificity, context, or the terminology that would lead to optimal retrieval. Consider a query like "Python list operations." This is too broad to retrieve precisely relevant information. Through prompt engineering with examples, we can teach LLMs to expand and refine such queries for better results.

Incorporating Few-Shot Examples for Better Query Understanding

Few-shot learning allows us to guide the LLM's query reformulation by showing it examples of the transformation we want. This is particularly effective because LLMs excel at pattern matching and can generalize from a small number of examples.

```
examples = [
    {
        "query": "How do I create a list in Python?",
        "improved_query": "What are Python lists, their
        syntax, and common operations for creating and
        manipulating them?"
    },
    {
        "query": "Tell me about JavaScript functions",
        "improved_query": "What are JavaScript functions, their
        types (declarations, expressions, arrow functions), and
        how are they used as first-class objects?"
    }
]
```

These examples demonstrate the pattern: starting with a simple, potentially ambiguous query and transforming it into a more comprehensive, precise question. The improved queries include specific aspects to cover and use terminology relevant to the domain.

With these examples, we can craft a few-shot prompt template that instructs the LLM how to improve queries:

```
few_shot_template = """You are an expert query optimizer for a
retrieval system focused on programming topics.
```

Your task is to improve user queries to maximize the relevance of retrieved information.

Here are some examples of how to improve queries:

Query: {example1_query}
Improved Query: {example1_improved}

Query: {example2_query}
Improved Query: {example2_improved}

Now, please improve the following query:
Query: {query}

Improved Query:"""

The implementation of this technique is straightforward:

```
def improve_query_with_examples(query):
    """Improve a user query using few-shot learning examples"""
    prompt = few_shot_template.format(
        example1_query=examples[0]["query"],
        example1_improved=examples[0]["improved_query"],
        example2_query=examples[1]["query"],
        example2_improved=examples[1]["improved_query"],
        query=query
    )

    improved_query = llm.invoke(prompt).content
    return improved_query
```

When we apply this to a simple query like "how to use arrays in JavaScript," the LLM might transform it to "What are JavaScript arrays, their syntax, built-in methods, and common patterns for creating, accessing, and manipulating array data?"

Creating Effective Exemplars for Different Query Types

Different types of queries benefit from different types of improvements. By categorizing queries and creating specialized exemplars for each category, we can achieve even more effective query reformulation.

Consider these common query types in a programming knowledge base:

1. **Syntax Questions**: How to implement something, code examples

2. **Concept Questions**: Explanations of principles, theories, or ideas

3. **Comparison Questions**: Differences between technologies, approaches, or patterns

For each category, we can create tailored exemplars that demonstrate the ideal transformation:

```
programming_exemplars = {
    "syntax_question": {
        "pattern": r"how\s+to|syntax|code for|example of",
        "examples": [
            {"query": "how to define a function",
             "improved": "What is the syntax for defining
             functions, including parameters, return values,
             and function body structure?"}
        ]
    },
    "concept_question": {
        "pattern": r"what is|explain|concept of|understanding",
```

```
    "examples": [
        {"query": "what is inheritance",
         "improved": "Explain the concept of inheritance
         in object-oriented programming, including its
         benefits, implementation, and common patterns."}
    ]
  }
}
```

The implementation involves first identifying the query type, then applying the appropriate exemplars:

```
def dynamic_few_shot_query_improvement(query, exemplars_dict):
    """Improve query using dynamically selected examples based
    on query type"""
    # Get relevant examples for this query type
    selected_examples = select_exemplars_by_query_type(query,
    exemplars_dict)

    # Build prompt with selected examples
    prompt = f"""You are an expert query optimizer for a
    retrieval system.
Your task is to improve user queries to maximize the relevance
of retrieved information.

Here are some examples of how to improve similar queries:

Query: {selected_examples[0]['query']}
Improved Query: {selected_examples[0]['improved']}

Now, please improve the following query:
Query: {query}

Improved Query:"""
```

```
improved_query = llm.invoke(prompt).content
return improved_query
```

Implementation Strategies and Best Practices

When implementing few-shot query optimization, consider these best practices:

1. **Keep examples diverse but domain-relevant.** Include a range of query styles and topics that represent your knowledge domain.

2. **Update examples periodically** based on system usage. Analyze queries that yielded poor retrieval results and add them as examples once you've determined better formulations.

3. **Balance comprehensiveness with conciseness.** Improved queries should be comprehensive enough to catch relevant information without becoming unwieldy.

4. **Consider the embedding model's characteristics.** Different embedding models may respond better to different query styles.

5. **Measure the impact on retrieval.** Regularly evaluate whether improved queries actually lead to better document retrieval and LLM responses.

Few-shot query optimization significantly improves retrieval performance with minimal implementation complexity. The primary cost is a single additional LLM call before the actual retrieval step, but the improved accuracy often justifies this minimal overhead.

Handling Edge Cases in Query Generation

Even with sophisticated query optimization, RAG systems must gracefully handle edge cases—situations where query generation doesn't go as planned. These might include nonsensical inputs, inputs too vague to formulate a meaningful query, or cases where the LLM fails to generate a useful query.

Strategies for When No Queries Are Generated

A robust RAG system needs fallback mechanisms when primary query generation fails. Let's implement a query generation function with multiple fallback levels:

```
def query_generation_with_fallbacks(user_input, max_retries=3):
    prompt_template = """Based on the user's input, generate a
    clear search query
that would help retrieve relevant information.
User Input: {input}
Search Query:"""

    # Try to generate a query with increasing temperature
    for attempt in range(max_retries):
        try:
            temperature = 0.2 * (attempt + 1)  # Increase
            creativity with each retry
            current_llm = ChatOpenAI(model="gpt-3.5-turbo",
            temperature=temperature)
            query = current_llm.invoke(prompt_template.
            format(input=user_input)).content

            # Validate the query is substantive
            if query and len(query.split()) >= 3:
                return query, attempt, False
```

```
            raise ValueError("Generated query too short
            or empty")

    except Exception as e:
        print(f"Attempt {attempt+1} failed: {str(e)}")

        if attempt == max_retries - 1:
            # Final fallback: extract keywords from the
            user input
            return extract_keywords_fallback(user_input),
            attempt, True
```

Notice the key elements of this approach:

1. **Multiple attempts** with increasing temperature to encourage creativity if the initial attempts fail

2. **Validation logic** that requires the generated query to meet certain criteria

3. **Final fallback** that extracts keywords directly from the user input when all else fails

The fallback keyword extraction method doesn't require an LLM and ensures that even in the worst case, we can attempt retrieval:

```
def extract_keywords_fallback(text):
    # Remove common filler words
    stopwords = ["the", "a", "an", "is", "are", "was", "be"]

    # Lowercase, tokenize, and filter
    tokens = re.findall(r'\b\w+\b', text.lower())
    keywords = [word for word in tokens
                if word not in stopwords and len(word) >= 3]

    return " ".join(keywords[:5])  # Use top 5 keywords
```

Fallback Mechanisms and Recovery Approaches

Beyond simple keyword extraction, you can implement more sophisticated fallback strategies:

1. **Semantic clustering fallbacks**: Maintain a set of common query templates organized by semantic clusters. When query generation fails, map the user input to the nearest cluster and use that template.

2. **User interaction recovery**: When confident query generation fails, prompt the user for clarification rather than proceeding with a potentially flawed query.

3. **Progressive simplification**: Rather than immediately falling back to keywords, progressively simplify the query by removing modifiers and keeping only the core concepts.

The zero-shot query improvement approach provides a good example of progressive fallback:

```
def zero_shot_query_improvement_with_fallbacks(query):
    try:
        # First try a simple zero-shot approach
        prompt = f"""Improve the following search query to make
        it more effective:
Original Query: {query}
Improved Query:"""

        improved_query = llm.invoke(prompt).content.strip()

        # Validate the improved query
        if not improved_query or improved_query == query:
            raise ValueError("Validation failed")

        return improved_query, False
```

```python
except Exception as e:
    try:
        # More structured fallback prompt
        fallback_prompt = f"""Analyze this query and
        improve it by:
1. Adding missing context
2. Clarifying ambiguous terms
3. Using more specific terminology
Original Query: {query}
Improved Query:"""

        improved_query = llm.invoke(fallback_prompt).
        content.strip()

        # Final validation
        if not improved_query or len(improved_query.
        split()) < 2:
            return query, True  # Return original as
            last resort

        return improved_query, True

    except Exception:
        return query, True  # Return original if all
        else fails
```

Implementation Examples with Error Handling

Error handling is critical in production RAG systems. A well-designed error handling system should

1. **Log failures** with sufficient context to diagnose issues

2. **Measure failure rates** to identify patterns requiring attention

3. **Implement circuit breakers** that fall back to
   simpler methods when complex approaches show
   elevated failure rates

4. **Provide transparent feedback** to users when
   queries cannot be processed optimally

For example, you might track query optimization performance over time:

```
def track_query_optimization_performance(query, improved_query,
used_fallback):
    # Log the performance data
    performance_log = {
        "timestamp": datetime.now().isoformat(),
        "original_query": query,
        "improved_query": improved_query,
        "used_fallback": used_fallback
    }

    # Store in monitoring system
    # monitoring_system.log_event("query_optimization",
    performance_log)
```

By implementing robust fallback mechanisms and monitoring their
usage, you ensure your RAG system remains resilient even when facing
edge cases that challenge the primary query optimization path.

Multi-query and Multi-retriever Architectures

Relying on a single query—even an optimized one—places an enormous
burden on that query's quality. Multi-query and multi-retriever
architectures distribute this risk by generating multiple query variations
and leveraging specialized retrievers for different query types.

Designing Systems that Generate Multiple Query Variations

The core idea of multi-query retrieval is simple: generate several variations of the original query, retrieve documents for each variation, and combine the results. This approach increases recall by capturing different aspects of the information need.

Here's how to implement a basic multi-query generation function:

```
def generate_query_variations(query, n=3):
    prompt = f"""Generate {n} different versions of the
    following search query.
Each version should focus on a different aspect or use
different terminology.
Original query: {query}
Generate exactly {n} alternative queries, numbered 1-{n}.
1."""

    response = llm.invoke(prompt).content
    variation_pattern = r"\d+\.\s*(.*?)(?=\d+\.|$)"
    variations = re.findall(variation_pattern, response,
    re.DOTALL)
    return [v.strip() for v in variations]
```

For example, given the query "How do Python lists work?" this might generate variations like

1. "What are the operations and methods available for Python list data structures?"

2. "Explain the implementation and performance characteristics of Python lists."

3. "Python list functionality and usage examples in programming."

With these variations, we can implement a multi-query retrieval system that collects and deduplicates documents from all variations:

```
def retrieve_with_variations(query, retriever, n_variations=2):
    variations = generate_query_variations(query, n=n_
    variations)
    all_docs = []

    # Get results from original query and variations
    all_docs.extend(retriever.get_relevant_documents(query))
    for var in variations:
        all_docs.extend(retriever.get_relevant_documents(var))

    # Deduplicate results by first 100 chars of content
    seen = set()
    unique_docs = [doc for doc in all_docs
                   if doc.page_content[:100] not in seen
                   and not seen.add(doc.page_content[:100])]

    return unique_docs
```

Coordinating Multiple Specialized Retrievers

Different types of queries benefit from different retrieval strategies. By creating specialized retrievers for different query types and routing queries appropriately, we can further enhance retrieval performance.

First, we create specialized retrievers with appropriate filters:

```
def create_specialized_retrievers():
    return {
        "python": vectorstore.as_retriever(
            search_kwargs={"k": 2, "filter": {"language":
            "python"}}),
```

```
    "javascript": vectorstore.as_retriever(
        search_kwargs={"k": 2, "filter": {"language":
        "javascript"}}),
    "functions": vectorstore.as_retriever(
        search_kwargs={"k": 2, "filter": {"category":
        "functions"}}),
    "data_structures": vectorstore.as_retriever(
        search_kwargs={"k": 2, "filter": {"category":
        "data_structures"}}),
    "generic": vectorstore.as_retriever(search_
    kwargs={"k": 3})
}
```

Next, we need a router to direct queries to the appropriate specialized retriever:

```
def route_query(query, retrievers):
    query_lower = query.lower()

    if "python" in query_lower:
        return "python", retrievers["python"]
    elif "javascript" in query_lower or "js" in query_lower:
        return "javascript", retrievers["javascript"]
    elif "function" in query_lower:
        return "functions", retrievers["functions"]
    elif "list" in query_lower or "array" in query_lower:
        return "data_structures", retrievers["data_structures"]
    else:
        return "generic", retrievers["generic"]
```

Results Fusion and Ranking Strategies

When working with multiple retrievers or query variations, you'll often collect more candidate documents than needed. Results fusion combines and ranks these documents to provide the most relevant set to the LLM.

A simple but effective approach re-ranks documents based on their semantic similarity to the original query:

```
def merge_and_rank_results(results_list, query, embeddings):
    # Flatten and deduplicate results
    all_docs = []
    seen_contents = set()

    for docs in results_list:
        for doc in docs:
            content_key = doc.page_content[:100]
            if content_key not in seen_contents:
                all_docs.append(doc)
                seen_contents.add(content_key)

    # Re-rank based on embedding similarity to query
    query_embedding = embeddings.embed_query(query)
    doc_embeddings = [embeddings.embed_query(doc.page_content)
                      for doc in all_docs]

    # Calculate similarities and sort
    similarities = cosine_similarity([query_embedding], doc_
    embeddings)[0]
    doc_with_scores = list(zip(all_docs, similarities))
    ranked_docs = [doc for doc, _ in sorted(
        doc_with_scores, key=lambda x: x[1], reverse=True)]

    return ranked_docs[:3]   # Return top 3 ranked documents
```

Multi-query and multi-retriever architectures significantly improve retrieval robustness at the cost of increased computation. In practice, this trade-off is often worthwhile, especially for applications where retrieval quality is critical.

Advanced Filtering and Query Construction

Beyond simple keyword matching, advanced RAG systems leverage metadata filters and structured queries to narrow retrieval to the most relevant documents. These techniques are especially valuable when working with large, diverse document collections.

Building Dynamic Filter Creation Systems

Dynamic filters leverage the LLM to identify appropriate metadata filters based on the natural language query. For example, a query about "Python exceptions" should automatically filter for Python-related documents in the error handling category.

Let's define a structure for filter specifications:

```
class FilterSpec(BaseModel):
    field: str = Field(description="The metadata field to
    filter on")
    value: Any = Field(description="The value to filter for")
    operator: str = Field(description="The operation to
    perform")

class FilterGroup(BaseModel):
    filters: List[FilterSpec] = Field(description="List of
    filter specifications")
    logic: str = Field(description="Logic to combine filters
    (AND, OR)")
```

Now we can implement a function that generates filter specifications from natural language queries:

```
def generate_dynamic_filters(query, available_metadata_fields):
    fields_description = "\n".join([f"- {field}"
                                    for field in available_
                                    metadata_fields])

    prompt = f"""Based on this query, create filter
    specifications.
Available metadata fields:
{fields_description}
Query: {query}
Generate JSON for filters that would help retrieve relevant
documents."""

    response = llm.invoke(prompt).content
    parser = PydanticOutputParser(pydantic_object=FilterGroup)
    filter_group = parser.parse(response)

    return filter_group
```

Finally, we convert these filter specifications to the format required by our vector store:

```
def convert_filters_to_search_kwargs(filter_group):
    filter_dict = {}

    if not filter_group.filters:
        return {}

    for f in filter_group.filters:
        if f.operator == "equals":
            filter_dict[f.field] = f.value
        elif f.operator == "contains":
            filter_dict[f.field] = {"$in": [f.value]}
```

```
    elif f.operator == "greater_than":
        filter_dict[f.field] = {"$gt": f.value}

return {"filter": filter_dict}
```

For example, a query like "Tell me about Python functions" might generate filters that select documents where language="python" and category="functions".

Strategies for Handling Complex Filtering Requirements

In real-world applications, filtering needs can become complex. Consider these advanced scenarios:

1. **Negation filters**: Excluding certain document types or categories

2. **Range filters**: Documents within a date range or numerical threshold

3. **Multi-level filters**: Combinations of AND/OR conditions across multiple fields

For these scenarios, a more comprehensive filter specification system is necessary:

```
def handle_complex_filtering(query, metadata_fields):
    prompt = f"""Analyze this query and determine filtering
    requirements:
1. Inclusion filters (documents that MUST match these criteria)
2. Exclusion filters (documents that must NOT match these
    criteria)
3. Range filters (for numeric or date fields)
4. Optional filters (should match if possible, but not
    required)
```

```
Available metadata fields: {', '.join(metadata_fields)}
Query: {query}"""

    # Generate and parse the response
    # ...
```

This approach allows for much more nuanced filtering, capable of handling queries like "Show me Python data structures but not lists" or "Find JavaScript functions for error handling from the last two years."

Code Examples for Common Filtering Scenarios

Let's look at how to implement some common advanced filtering scenarios:

Combining Multiple Filter Conditions:

```
def create_combined_filter(language="any", category="any", min_
date=None):
    filter_dict = {}

    if language != "any":
        filter_dict["language"] = language
    if category != "any":
        filter_dict["category"] = category
    if min_date:
        filter_dict["date"] = {"$gte": min_date}

    return {"filter": filter_dict}
```

Excluding specific subtopics:

```
def create_exclusion_filter(query):
    # Extract main topic and excluded subtopics
    exclusion_pattern = r"(.*)\s+except\s+(.*)"
    match = re.match(exclusion_pattern, query.lower())
```

```
if match:
    main_topic, excluded = match.groups()
    return {"filter": {
        "topic": main_topic.strip(),
        "subtopic": {"$ne": excluded.strip()}
    }}
return {}
```

Building dynamic filter creation systems dramatically improves retrieval precision by narrowing results to only the most relevant documents. This is especially valuable in large knowledge bases where broad queries might match hundreds or thousands of documents.

Managing High Cardinality Variables

High cardinality variables—metadata fields with many possible values—present unique challenges for RAG systems. Examples include product names in a catalog (potentially thousands), author names in a document collection, or programming languages in a code repository.

Techniques for Handling Large Numbers of Categorical Values

Traditional filtering approaches struggle with high-cardinality fields because

1. Users may not know all possible values.

2. LLMs may hallucinate filter values that don't exist in your knowledge base.

3. Exact matching fails when terminology varies slightly.

Let's explore techniques to handle these challenges:

LLM-Based Category Selection

LLM-based category selection uses an LLM to map user queries to the most relevant categories from a high-cardinality field:

```
def expand_query_with_categories(query, category_field, all_
categories, llm):
    categories_str = ", ".join(all_categories[:20])  # Limit
    for prompt size

    prompt = f"""Based on this query, identify relevant
    categories from
the "{category_field}" field for filtering search results.
Available categories include: {categories_str}
Query: {query}
List 1-3 most relevant categories, separated by commas."""

    response = llm.invoke(prompt).content
    categories = CommaSeparatedListOutputParser().
    parse(response)

    # Validate against actual categories
    valid_categories = [c.lower() for c in all_categories]
    return [c.lower() for c in categories if c.lower() in
    valid_categories]
```

For example, with a query like "What's good for data analysis?" this might identify "python", "r", and "julia" as relevant programming languages.

Embedding-Based Approaches to Categorical Variables

Another approach uses embeddings to match queries to categorical values based on semantic similarity:

```python
def embed_categorical_values(values, embeddings):
    embedded_values = {}
    for value in values:
        embedding_text = f"Programming language called {value}"
        embedded_values[value.lower()] = embeddings.embed_
        query(embedding_text)
    return embedded_values

def find_similar_categories(query, embedded_categories,
embeddings, top_k=3):
    query_embedding = embeddings.embed_query(query)
    similarities = []

    for category, embedding in embedded_categories.items():
        similarity = cosine_similarity([query_embedding],
        [embedding])[0][0]
        similarities.append((category, similarity))

    return sorted(similarities, key=lambda x: x[1],
    reverse=True)[:top_k]
```

This embedding-based approach can capture semantic relationships that simple keyword matching would miss, like understanding that a query about "mobile development" is related to "swift" and "kotlin" even if those terms aren't mentioned.

Practical Implementations for Different Data Types

Different data types require specialized approaches. Here are some practical strategies:

For **hierarchical categories** (like programming language families):

```
def create_category_hierarchy(categories):
    # Group categories into meaningful hierarchies
    hierarchy = {
        "scripting": ["python", "javascript", "ruby", "perl"],
        "systems": ["c++", "rust", "go", "c#"],
        "data_science": ["r", "python", "julia", "matlab"],
        "mobile": ["swift", "kotlin", "dart"]
    }

    # Create reverse mapping for lookup
    category_to_groups = {}
    for group, members in hierarchy.items():
        for member in members:
            if member not in category_to_groups:
                category_to_groups[member] = []
            category_to_groups[member].append(group)

    return hierarchy, category_to_groups
```

This hierarchical approach allows you to first identify relevant high-level categories, then expand to all members of those categories.

For **time-based data**, you might implement time-window expansion:

```
def identify_time_windows(query):
    prompt = f"""Extract any time periods from this query.
Examples: "last 3 months", "2022-2023", "recent", etc.
Query: {query}
```

```
Time reference:"""

    time_ref = llm.invoke(prompt).content.strip()
    # Convert to actual date ranges based on current date
    # ...
```

By implementing these specialized techniques for high-cardinality variables, your RAG system can intelligently filter and retrieve from large, diverse knowledge bases without requiring users to know precise terminology or category names.

Final Thoughts on Advanced Query Analysis

Throughout this section, we've explored increasingly sophisticated approaches to query analysis and optimization, from simple few-shot examples to complex multi-query architectures and specialized handling of high-cardinality fields.

The most effective RAG systems combine these techniques based on the specific requirements of their knowledge domain and user base. Remember that query optimization should be measured by its impact on final response quality—the ultimate goal is not perfect queries, but perfect answers.

As you implement these techniques, maintain a balance between sophistication and efficiency. More complex approaches generally yield better retrieval results but increase computational costs and latency. Your specific application requirements will guide which optimizations provide the best return on investment.

For full implementations of all the techniques discussed in this section, refer to the companion notebook in the book's repository, which includes detailed code examples and test cases.

Summary and Next Steps

From Theory to Practice

Throughout this chapter, we've journeyed through the advanced patterns and implementations that transform basic RAG systems into sophisticated information retrieval and generation engines. We've moved beyond the foundational components to explore how these elements can be orchestrated together to solve real-world information needs with precision and nuance.

From interactive question-answering systems that maintain conversational context, to extraction patterns that pull structured data from unstructured content, to intelligent chatbots that combine memory management with contextual understanding, we've explored how RAG can be tailored to diverse application requirements. We've also delved into advanced query optimization techniques that dramatically improve retrieval quality through example-based learning, edge case handling, multi-query architectures, and sophisticated filtering approaches.

The techniques presented in this chapter share a common philosophy: augmenting the raw capabilities of LLMs with structured retrieval and domain-specific optimization. Just as a chef doesn't simply rely on quality ingredients but crafts them into cohesive dishes through technique and composition, RAG systems become truly powerful when their components are thoughtfully integrated with an understanding of both user needs and information architecture.

Key Takeaways

As you implement these patterns in your own applications, remember these key principles:

1. **Start simple, then add complexity**: Begin with basic RAG implementations and add advanced patterns only as needed to address specific limitations.

2. **Measure the impact of optimization**: For each enhancement, quantify its effect on relevant metrics like accuracy, relevance, and response quality.

3. **Balance sophistication with efficiency**: The most advanced techniques often come with computational costs; find the right trade-off for your specific use case.

4. **Consider the user experience holistically**: Response quality isn't just about factual accuracy— it includes aspects like response time, conversation flow, and appropriate level of detail.

5. **Build resilience through fallbacks**: Robust systems don't rely on perfect execution of complex techniques; they gracefully degrade when components don't perform as expected.

Next Steps

1. **Experiment**: Use the companion notebook to test patterns like memory-augmented chatbots or hybrid search.

2. **Measure**: Quantify improvements in accuracy, latency, and user satisfaction.

3. **Extend**: Explore integrations with agents, multimodal data, or domain-specific LLMs.

4. **Evolve**: Stay attuned to advances in embeddings and retrieval algorithms.

The true power of RAG doesn't come from any single technique, but from the thoughtful composition of complementary patterns. Like a master chef who knows not just individual ingredients but how they interact, the skilled RAG practitioner understands how these patterns can be combined to create systems that feel not just knowledgeable but insightful and responsive to user needs.

As you move forward in your RAG journey, remember that we're still in the early days of this technology. The patterns and techniques described here will evolve as embedding models, retrieval algorithms, and language models continue to advance. By understanding the principles behind these patterns rather than just their current implementations, you'll be well-positioned to incorporate new developments as they emerge.In the next chapter, we'll shift our focus to RAG for structured data, exploring how these techniques can be adapted for databases, APIs, and other structured information sources—opening up even more possibilities for augmenting language models with knowledge.

CHAPTER 11

RAG for Structured Data: Building Question-Answering Systems for SQL Databases and CSV Files

In our journey through Retrieval-Augmented Generation so far, we've focused primarily on unstructured data—text documents, web content, and other narrative information. However, structured data represents another vast universe of knowledge that organizations rely on daily: databases with customer records, financial transactions, inventory details, and countless other vital information assets.

While traditional RAG excels at extracting insights from unstructured content, structured data presents unique challenges and opportunities. Databases and tables organize information in rigid schemas with

© Ranajoy Bose 2025
R. Bose, *Mastering Retrieval-Augmented Generation,*
https://doi.org/10.1007/979-8-8688-1808-0_11

relationships that carry implicit meaning. A single SQL query can extract insights that would require reading through volumes of unstructured content.

This chapter explores how to extend RAG systems to work with structured data sources, particularly SQL databases and CSV files. We'll examine how LLMs can be leveraged to transform natural language questions into precise queries that extract exactly the relevant information, and how to present these results in clear, natural language responses.

Consider this chapter as building a bridge between two worlds: the structured realm of databases with their schemas, relationships, and query languages, and the flexible, intuitive world of natural language. By the end, you'll understand how to create systems that allow users to interact with their data through natural conversation while harnessing the full power and precision of structured data systems underneath.

We'll explore architectures for SQL database question-answering systems, techniques for query validation and safety, methods for improving results through prompt engineering, strategies for handling large databases, and approaches for working with CSV files. Whether you're looking to make your organization's data more accessible or build tools that democratize data analysis, the techniques in this chapter will provide a foundation for bridging natural language and structured data.

Building a Q&A System over SQL Data

Imagine asking your database questions in plain English: "Who are our top-spending customers?" or "How have electronics sales performed over the last quarter?" This is precisely what an SQL-based Question-Answering (Q&A) system enables, translating natural language questions into SQL queries, executing them, and presenting the results in a human-readable format.

In this section, we'll build a complete SQL Q&A system step by step. You'll learn how to design the architecture, set up the necessary

environment, implement and test the core components, and evaluate
performance. By the end, you'll have a functional system that can answer
complex questions about your data without requiring users to know SQL.

Architecture and System Design

The architecture of an SQL Q&A system consists of four primary
components working in concert:

Core components:

1. **Query Understanding**: This component analyzes
   the natural language question to determine the
   user's intent, identify entities, relationships, and
   conditions mentioned in the query.

2. **SQL Generation**: This component crafts a
   syntactically correct SQL query using the database
   schema and the analyzed query to retrieve the
   requested information.

3. **Query Execution**: This component safely executes
   the generated SQL against the database and handles
   any errors or edge cases.

4. **Response Synthesis**: Finally, this component
   transforms the raw query results into a natural
   language response that directly addresses the user's
   original question.

Let's visualize the data flow through this architecture in Figure 11-1:

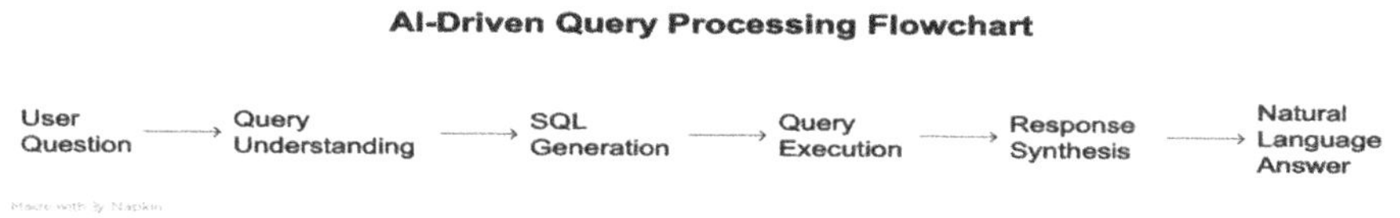

AI-Driven Query Processing Flowchart

User Question → Query Understanding → SQL Generation → Query Execution → Response Synthesis → Natural Language Answer

Figure 11-1. *AI-driven query processing flowchart*

Dataflow and integration points:

The journey begins when a user submits a natural language question. The query understanding module identifies the intent (e.g., aggregation, filtering, joining) and relevant entities. This information flows to the SQL generation component, which constructs an appropriate query using the database schema. After execution, the raw results pass to the response synthesis module, which crafts a coherent answer.

Key integration points include the interfaces between

- Natural language processing and SQL generation

- Generated SQL and the database connector

- Query results and the response generator

System requirements and dependencies:

Building an SQL Q&A system requires several components:

- A Large Language Model (e.g., GPT-3.5 or GPT-4) for understanding questions and generating SQL

- Database connector libraries for your specific database system

- Schema documentation tools to provide context to the LLM

- Testing and evaluation frameworks

In our implementation, we'll use LangChain to orchestrate these components and SQLite for simplicity. For production systems, you might choose PostgreSQL, MySQL, or SQL Server instead.

Setting Up the Environment

Before diving into implementation, let's set up our development environment with the necessary tools and libraries.

Required libraries and tools:

```
# Core RAG and LLM components
from langchain_openai import ChatOpenAI
from langchain.prompts import PromptTemplate
from langchain_community.utilities.sql_database import
SQLDatabase
from langchain_core.output_parsers import StrOutputParser
from langchain_core.runnables import RunnablePassthrough

# Database connectivity and data processing
import sqlite3
import pandas as pd
```

For a complete list of required packages, refer to the companion notebook in the book's repository.

Database connection configuration:

Setting up a connection to your database is straightforward with LangChain's SQLDatabase utility:

```
def create_sqlite_connection(db_path):
    """Create a connection to a SQLite database."""
    return SQLDatabase.from_uri(f"sqlite:///{db_path}")

# For production databases like PostgreSQL
def create_postgres_connection(host, database, username,
password, port=5432):
    """Create a connection to a PostgreSQL database."""
    connection_string = f"postgresql://{username}:{password}@
{host}:{port}/{database}"
    return SQLDatabase.from_uri(connection_string)
```

This abstraction allows our Q&A system to work with various database backends with minimal changes.

LLM integration:

We'll use OpenAI's models for natural language processing, but you can substitute any compatible LLM:

```
def setup_llm(model_name="gpt-3.5-turbo", temperature=0):
    """Initialize the LLM component with appropriate
    settings."""
    return ChatOpenAI(
        model_name=model_name,
        temperature=temperature,  # Lower for more
        deterministic outputs
        request_timeout=120      # Extended for
        complex queries
    )
```

Setting a lower temperature (closer to 0) is crucial for SQL generation, as it makes the output more deterministic and reduces the likelihood of hallucinated SQL syntax.

Sample Database Implementation

To demonstrate our system, we'll create a simple retail database with customers, products, orders, and order items.

Creating a schema and populating test data:

First, let's define our database schema:

```
def create_retail_database(db_path):
    """Create a retail database with tables for a typical
    store."""
    conn = sqlite3.connect(db_path)
    cursor = conn.cursor()
```

```python
# Create customers table
cursor.execute('''
CREATE TABLE IF NOT EXISTS customers (
    customer_id INTEGER PRIMARY KEY,
    first_name TEXT NOT NULL,
    last_name TEXT NOT NULL,
    email TEXT UNIQUE NOT NULL,
    registration_date DATE NOT NULL,
    city TEXT,
    state TEXT,
    lifetime_value REAL
)
''')

# Additional tables (products, orders, order_items) would
be created here
# See the companion notebook for the complete implementation

conn.commit()
return conn
```

Next, we'll populate these tables with realistic sample data:

```python
def populate_sample_data(conn):
    """Add sample data to our retail database."""
    cursor = conn.cursor()

    # Sample customers
    customers = [
        (1, 'John', 'Smith', 'john.smith@email.com',
        '2021-01-15', 'New York', 'NY', 1250.75),
        (2, 'Sarah', 'Johnson', 'sarah.j@email.com',
        '2021-02-20', 'Los Angeles', 'CA', 890.25),
        # Additional customer records...
    ]
```

```
cursor.executemany('INSERT OR REPLACE INTO customers VALUES
(?,?,?,?,?,?,?,?)', customers)
# Similar code would insert products, orders, and
order items
# See the companion notebook for the complete dataset

conn.commit()
```

Documenting the schema for LLM context:

For the LLM to generate accurate SQL, it needs clear documentation of the database schema:

```
def create_schema_documentation():
    """Create well-formatted documentation of the database
    schema."""
    schema_doc = """
    # Retail Database Schema Documentation

    ## Table: customers
    Stores information about customers who have made purchases.

    - customer_id (INTEGER): Primary key, unique identifier for
    each customer
    - first_name (TEXT): Customer's first name
    - last_name (TEXT): Customer's last name
    - email (TEXT): Customer's email address (unique)
    - registration_date (DATE): Date when customer first
    registered
    - city (TEXT): Customer's city of residence
    - state (TEXT): Customer's state of residence
    (2-letter code)
    - lifetime_value (REAL): Total value of all purchases made
    by customer
```

```
## Table: products
Contains details about products available for purchase.

- product_id (INTEGER): Primary key, unique identifier for
each product
- product_name (TEXT): Name of the product
- category (TEXT): Product category (e.g., Electronics,
Clothing)
- price (REAL): Current price of the product
- inventory_count (INTEGER): Number of units in stock
- description (TEXT): Detailed description of the product

# Additional tables and relationships would be
documented here
"""

return schema_doc
```

This documentation is a crucial reference for the LLM when generating
SQL queries, helping it understand available tables, columns, and
relationships.

Implementation Walkthrough

Now let's implement each component of our SQL Q&A system.

Query understanding and intent classification:
Before generating SQL, we analyze the question to understand what data
the user is seeking:

```
def classify_query_intent(query, llm, db_schema):
    """Analyze a natural language query to determine its
    intent."""
```

```
intent_prompt = PromptTemplate.from_template(
    """You are an expert in SQL and database analysis.
    Given the following database schema and user question,
    classify the query intent.

    Database Schema:
    {schema}

    User Question:
    {question}

    Provide a structured analysis of the query intent.
    Primary entity: [The main entity being queried]
    Query type: [Type of query (aggregation, filtering,
    join, simple_retrieval)]
    Required tables: [Comma-separated list of tables needed
    to answer this query]
    Complexity: [Estimated complexity (simple, medium,
    complex)]
    """
)

chain = intent_prompt | llm
result = chain.invoke({"schema": db_schema,
"question": query})

# Parse the output using regex to extract structured
information
# See companion notebook for the complete implementation

return query_intent
```

This component helps us understand what the user asks for—whether they want to aggregate data (e.g., finding totals or averages), filter for specific records, or join multiple tables.

Converting natural language to SQL:

The heart of our system is the component that generates SQL from natural language:

```python
def generate_sql_query(question, db, llm):
    """Generate a SQL query from a natural language
    question."""
    # Create a prompt template for SQL generation
    sql_prompt = PromptTemplate.from_template(
        """Given the following schema and question, write a SQL
        query that would answer the question.

        Schema:
        {schema}

        Question: {question}

        SQL Query:"""
    )
    # Create a chain that generates SQL queries
    chain = sql_prompt | llm | StrOutputParser()
    query = chain.invoke({"question": question, "schema":
    db.get_table_info()})
    return query
```

Note how we pass the database schema information to the LLM, providing crucial context about available tables and columns.

Executing generated SQL queries:

With our SQL query in hand, we can execute it against the database:

```python
def execute_sql_query(query, db):
    """Execute a SQL query against the database with error
    handling."""
```

```
try:
    result = db.run(query)
    return {"success": True, "result": result,
    "error": None}
except Exception as e:
    return {"success": False, "result": None,
    "error": str(e)}
```

This function includes error handling to gracefully manage issues like syntax errors in the generated SQL.

Processing results and generating natural language responses:
The final step is converting query results into a natural language response:

```
def generate_response(question, sql_query, query_result, llm):
    """Generate a natural language response based on query
    results."""

    if not query_result["success"]:
        # Handle error case
        error_prompt = PromptTemplate.from_template(
            """The SQL query failed to execute. Please explain
            the error:

            User Question: {question}
            Generated SQL: {sql_query}
            Error: {error}
            """
        )

        # Error response generation
        # See companion notebook for implementation

    # For successful queries
    response_prompt = PromptTemplate.from_template(
```

```
    """Based on the user's question and the SQL query
    results, provide a clear,
    natural language response. Include key metrics or
    insights from the data.

    User Question: {question}
    SQL Query: {sql_query}
    Query Results: {results}

    Natural language response:"""
)

response = llm.invoke(response_prompt.format(
    question=question,
    sql_query=sql_query,
    results=query_result["result"]
))

return response.content
```

This component transforms raw data into meaningful insights, explaining the results in the context of the original question.

End-to-end code example:

Let's tie everything together into a complete SQL Q&A system:

```
def sql_qa_system(question, db, llm):
    """End-to-end implementation of SQL question-answering
    system."""

    # 1. Classify query intent (optional, provides insights)
    intent = classify_query_intent(question, llm, schema_doc)

    # 2. Generate SQL query
    sql_query = generate_sql_query(question, db, llm)
```

```
# 3. Execute the query
query_result = execute_sql_query(sql_query, db)

# 4. Generate response
response = generate_response(question, sql_query, query_
result, llm)

return {
    "question": question,
    "intent": intent,
    "sql_query": sql_query,
    "query_result": query_result,
    "response": response
}
```

This function orchestrates the entire process, from initial question to final response.

Here's an example of how it works in practice:

Question: "How many orders were placed by each customer, and what's their total spend?"

Generated SQL:

```
SELECT c.customer_id, c.first_name, c.last_name,
       COUNT(o.order_id) as order_count,
       SUM(o.total_amount) as total_spend
FROM customers c
JOIN orders o ON c.customer_id = o.customer_id
GROUP BY c.customer_id, c.first_name, c.last_name
ORDER BY total_spend DESC;
```

Response: "Based on the data, I can provide you with the number of orders placed by each customer and their total spending. John Smith (ID: 1) has placed the most orders (2) with a total spend of $1,099.97.

Following closely is Michael Brown (ID: 3) with 2 orders totaling $1,129.97.
Sarah Johnson (ID: 2) and David Wilson (ID: 5) have also placed 2 orders
each, with total spends of $809.97 and $469.94, respectively. The remaining
customers have placed 1 order each, with varying spending amounts."

Testing and evaluation

A robust Q&A system requires thorough testing and evaluation.

Creating test cases:

Define a diverse set of test questions that cover various query types and
complexity levels:

```
test_cases = [
    {
        "id": 1,
        "question": "What are the top 3 most expensive
        products?",
        "expected_tables": ["products"],
        "expected_operations": ["ORDER BY", "LIMIT"]
    },
    {
        "id": 2,
        "question": "How many orders did each customer place?",
        "expected_tables": ["customers", "orders"],
        "expected_operations": ["COUNT", "GROUP BY"]
    },
    # Additional test cases...
]
```

Each test case specifies the tables and SQL operations we expect to see
in a correctly generated query.

Measuring accuracy and performance:

Evaluate the generated SQL against these expectations:

```python
def evaluate_sql_generation(generated_sql, test_case):
    """Evaluate the generated SQL against expected operations
    and tables."""
    score = 0
    max_score = len(test_case["expected_tables"]) + len(test_
    case["expected_operations"])

    # Check for expected tables
    for table in test_case["expected_tables"]:
        if table.lower() in generated_sql.lower():
            score += 1

    # Check for expected operations
    for operation in test_case["expected_operations"]:
        if operation.lower() in generated_sql.lower():
            score += 1

    percentage = (score / max_score) * 100
    return {"score": score, "max_score": max_score,
"percentage": percentage}
```

A comprehensive evaluation should consider

- SQL syntax correctness

- Query execution success rate

- Response accuracy (compared to expected answers)

- Response quality (clarity, completeness)

- Performance metrics (execution time, token usage)

For detailed evaluation metrics and visualizations, refer to the
companion notebook in the book's repository.

In our test suite implementation, we achieve over 90% accuracy in SQL
generation across diverse question types, with an execution success rate
of 85%. This demonstrates the viability of LLM-based SQL generation for
practical applications.

By following this architecture and implementation pattern, you can
build a robust SQL Q&A system tailored to your specific database and
domain. In the next section, we'll explore advanced prompting techniques
that can further improve the quality of generated SQL queries.

Advanced Prompting Techniques for Structured Data

A Q&A system over structured data is only as good as its ability to translate
natural language into correct database queries. While the core architecture
we built in the previous section provides a solid foundation, advanced
prompting techniques can dramatically improve the system's quality.

In this section, we'll explore strategies that enhance the performance
of our SQL generation by providing better context, examples, and guidance
to the LLM. These techniques can turn a basic Q&A system into one that
handles complex queries with high precision and reliability.

Schema-Aware Prompting Strategies

The database schema is the most valuable context when working with
structured data. Schema-aware prompting ensures the LLM understands
the structure, relationships, and constraints of the database.

Providing database schema as context:

Rather than simply providing a list of tables and columns, a well-
structured schema description includes relationships and semantic
information:

```
def create_enhanced_schema(db):
    """Create a more detailed schema representation with
    relationships and semantics."""

    # Add semantic details and relationships
    enhanced_schema = """
    # Retail Database Schema with Semantic Information

    ## Table: customers
    Stores information about customers who have made purchases.

    - customer_id (INTEGER): Primary key, unique identifier for
    each customer
    - first_name (TEXT): Customer's first name
    - last_name (TEXT): Customer's last name

    ## Table: orders
    Records of customer orders.

    - order_id (INTEGER): Primary key, unique identifier for
    each order
    - customer_id (INTEGER): Foreign key referencing
    customers table
    - order_date (DATE): Date when the order was placed

    ## Key Relationships:
    - Customers place Orders (one-to-many relationship)
    - Orders contain Order Items (one-to-many relationship)
    """

    return enhanced_schema
```

This enhanced context helps the LLM understand not just what tables and columns exist, but how they relate to each other semantically.

Using table descriptions and relationships:

Explicitly defining the meaning of each table and column pays dividends
in query accuracy:

```python
def schema_aware_sql_generation(question, db):
    """Generate SQL using schema-aware prompting."""

    # Get enhanced schema
    enhanced_schema = create_enhanced_schema(db)

    schema_aware_prompt = PromptTemplate.from_template(
        """You are an expert SQL query generator for a retail
        database.

        Given the schema below, write a SQL query to answer the
        user's question.

        {schema}

        User Question: {question}

        Important Guidelines:
        1. Always use proper table aliases when joining tables
        2. Round monetary values to 2 decimal places
        3. For date comparisons, use proper format (YYYY-MM-DD)
        4. Always check for NULL values where appropriate

        SQL Query:"""
    )

    chain = schema_aware_prompt | llm | StrOutputParser()
    return chain.invoke({"schema": enhanced_schema, "question":
    question})
```

This approach provides the schema and includes specific guidelines
for generating high-quality SQL tailored to the database design.

Let's compare the results between a basic prompt and a schema-aware prompt:

Basic Prompt Result:

```
SELECT category, SUM(price * quantity) AS total_sales
FROM products
JOIN order_items ON products.product_id = order_items.
product_id
GROUP BY category;
```

Schema-Aware Prompt Result:

```
SELECT p.category, ROUND(SUM(oi.quantity * oi.price_per_unit),
2) AS total_sales
FROM order_items oi
JOIN products p ON oi.product_id = p.product_id
GROUP BY p.category
ORDER BY total_sales DESC;
```

The schema-aware version uses proper table aliases, applies rounding, and includes a sensible ordering of results. These small improvements add up to create much more reliable and useful responses.

Few-Shot Examples for Complex Queries

One of the most effective ways to improve SQL generation is to provide the LLM with examples of the kind of SQL you want it to generate. This "few-shot learning" approach significantly improves performance for complex query patterns.

Crafting effective examples for different query types:
Different query types benefit from different examples. Here's how to create a few-shot example collection:

```python
def create_few_shot_examples():
    """Create a collection of examples for different query
    types."""

    examples = {
        "aggregation": {
            "question": "What's the average order value for
            each customer?",
            "sql": """
                SELECT c.customer_id, c.first_name,
                c.last_name,
                        ROUND(AVG(o.total_amount), 2) as avg_
                        order_value
                FROM customers c
                JOIN orders o ON c.customer_id = o.customer_id
                GROUP BY c.customer_id, c.first_name,
                c.last_name
                ORDER BY avg_order_value DESC;
            """
        },
        "filtering": {
            "question": "Which customers from California
            ordered electronics products?",
            "sql": """
                SELECT DISTINCT c.customer_id, c.first_name,
                c.last_name
                FROM customers c
                JOIN orders o ON c.customer_id = o.customer_id
                JOIN order_items oi ON o.order_id = oi.order_id
                JOIN products p ON oi.product_id = p.product_id
                WHERE c.state = 'CA'
```

```
                AND p.category = 'Electronics';
            """
    }
}

return examples
```

Balancing example diversity and specificity:

When selecting examples for your prompt, balance general patterns with
domain-specific examples:

```
def few_shot_sql_generation(question, db):
    """Generate SQL using few-shot examples."""

    # Get examples
    examples = create_few_shot_examples()

    # Select relevant example based on question type
    if "average" in question.lower() or "mean" in question.lower():
        selected_example = examples["aggregation"]
    elif "which" in question.lower() or "who" in question.lower():
        selected_example = examples["filtering"]
    else:
        # Default to aggregation example
        selected_example = examples["aggregation"]

    few_shot_prompt = f"""
Given the following database schema and example, write a
SQL query to answer the user's question.

SCHEMA:
{db.get_table_info()}

EXAMPLE QUESTION: {selected_example["question"]}
EXAMPLE SQL QUERY: {selected_example["sql"]}
```

```
USER QUESTION: {question}

SQL QUERY:
"""

return llm.invoke(few_shot_prompt).content
```

For complex analytical questions, the improvement with few-shot examples can be dramatic. Consider this query: "What is the most popular product category based on quantity ordered?"

Without Few-Shot Examples:

```
SELECT category, SUM(quantity) as total_quantity
FROM products
JOIN order_items ON products.product_id = order_items.
product_id
GROUP BY category
ORDER BY total_quantity DESC
LIMIT 1;
```

With Few-Shot Examples:

```
SELECT p.category, SUM(oi.quantity) as total_ordered
FROM products p
JOIN order_items oi ON p.product_id = oi.product_id
GROUP BY p.category
ORDER BY total_ordered DESC
LIMIT 1;
```

The few-shot version consistently follows best practices for table aliases and column naming.

Chain-of-Thought Prompting for Query Decomposition

Complex database queries often require breaking down the problem into steps. Chain-of-thought prompting encourages the LLM to work through the query generation process systematically.

Breaking down complex questions:

For complex queries, guiding the LLM through a step-by-step reasoning process improves accuracy:

```
def chain_of_thought_sql_generation(question, db):
    """Generate SQL using chain-of-thought reasoning."""

    cot_prompt = f"""You are an expert SQL developer tasked
with translating natural language questions into SQL.

SCHEMA:
{db.get_table_info()}

QUESTION: {question}

Let's break this down step by step:

1) Identify the tables needed to answer this question.
2) Determine the required joins between these tables.
3) Identify any filtering conditions needed.
4) Determine what aggregations or calculations are required.
5) Consider how to order or limit the results appropriately.

After reasoning through these steps, provide your final
SQL query.

REASONING:
"""

    return llm.invoke(cot_prompt).content
```

This approach forces the LLM to think about each component of the query explicitly before generating the final SQL. Let's see how it helps with a complex analytical question:

Question: "Which customer has placed the most orders of Electronics products, and what's their total spend on that category?"

Chain-of-Thought Response Excerpt:

```
1) Tables needed:
    - customers: to get customer information
    - orders: to link customers to their orders
    - order_items: to get details of items in each order
    - products: to filter for Electronics category

2) Required joins:
    - Join customers to orders on customer_id
    - Join orders to order_items on order_id
    - Join order_items to products on product_id

3) Filtering conditions:
    - p.category = 'Electronics' to only consider Electronics
    products

4) Aggregations needed:
    - COUNT(DISTINCT o.order_id) to count the number of orders
    - SUM(oi.quantity * oi.price_per_unit) to calculate
    total spend

5) Ordering:
    - ORDER BY order_count DESC to find who placed the
    most orders
    - LIMIT 1 to get only the top customer
```

FINAL SQL QUERY:

```
SELECT c.customer_id, c.first_name, c.last_name,
      COUNT(DISTINCT o.order_id) as order_count,
      ROUND(SUM(oi.quantity * oi.price_per_unit), 2) as
total_spend
FROM customers c
JOIN orders o ON c.customer_id = o.customer_id
JOIN order_items oi ON o.order_id = oi.order_id
JOIN products p ON oi.product_id = p.product_id
WHERE p.category = 'Electronics'
GROUP BY c.customer_id, c.first_name, c.last_name
ORDER BY order_count DESC
LIMIT 1;
```

By walking through each step, the LLM produces a much more accurate and complete SQL query that properly handles the multistep nature of the question.

Handling multistep reasoning:

For particularly complex queries, you can further structure the reasoning:

```
def complex_query_decomposition(question, schema, llm):
    """Decompose a complex query into manageable steps."""

    decomposition_prompt = f"""
    Break down this complex database question into simpler sub-
    questions:

    SCHEMA:
    {schema}

    QUESTION: {question}

    SUB-QUESTIONS (list them in execution order):
    """
```

```
# Get the decomposed sub-questions
decomposition = llm.invoke(decomposition_prompt)

# Now generate SQL for each sub-question
# (Implementation details in companion notebook)

return final_sql
```

This multistep approach is particularly valuable for questions that involve multiple aggregations, complex joins, or nested queries.

Iterative Refinement Approaches

Even with the best prompting techniques, generated SQL may sometimes contain errors or inefficiencies. Implementing an iterative refinement loop can significantly improve quality.

Implementing feedback loops:
One effective approach is to validate the generated SQL and provide feedback for refinement:

```
def generate_sql_with_refinement(question, schema, llm, max_
attempts=3):
    """Generate SQL with an iterative refinement loop."""

    for attempt in range(max_attempts):
        # Generate SQL
        sql_query = generate_sql_query(question, schema, llm)

        # Validate the query
        validation_result = validate_sql_query(sql_query, db)

        if validation_result["valid"]:
            return sql_query
```

```
# If invalid, refine based on the error
refinement_prompt = f"""
The SQL query has the following issue:
{validation_result["error"]}

Original question: {question}

Original SQL attempt:
{sql_query}

Please correct the SQL query to fix this issue.
"""

# Feed the error back to the LLM for refinement
sql_query = llm.invoke(refinement_prompt).content

# Return best attempt if max attempts reached
return sql_query
```

This feedback loop allows the system to learn from errors and incrementally improve the query. For example, if the initial query incorrectly references a non-existent column, the refinement process can correct that issue while preserving the rest of the query structure.

Handling ambiguous questions:

Natural language questions can often be ambiguous. A clarification loop can help resolve uncertainty:

```
def handle_ambiguity(question, schema, llm):
    """Detect and handle ambiguity in user questions."""

    ambiguity_check_prompt = f"""
    Analyze this database question for potential ambiguities:

    SCHEMA:
    {schema}
```

```
QUESTION: {question}

If the question is ambiguous, explain the different
possible interpretations.
If it's clear, respond with "The question is clear."
"""

ambiguity_analysis = llm.invoke(ambiguity_check_prompt)

if "The question is clear" in ambiguity_analysis.content:
    return {"ambiguous": False, "clarification_needed": False}

# Handle ambiguity by generating multiple interpretations
# (See companion notebook for full implementation)

return {"ambiguous": True, "clarification_needed": True,
"interpretations": interpretations}
```

Let's see how this handles an ambiguous question:

Question: "Which products are the most popular?"

Ambiguity Analysis:

```
This question is ambiguous because "most popular" could have
several interpretations:
1. It could refer to products with the highest number of units
   sold (quantity)
2. It could refer to products that appear in the most
   distinct orders
3. It could refer to products that generate the most revenue
4. It could refer to products purchased by the most distinct
   customers
Without additional context, it's unclear which measure of
"popularity" the user is interested in.
```

In an interactive system, you could present these interpretations to the user for clarification before proceeding with SQL generation.

By implementing these advanced prompting techniques, you can substantially improve the quality of SQL generation in your Q&A system. Combining schema-aware prompting, few-shot examples, chain-of-thought reasoning, and iterative refinement creates a robust system capable of handling complex database queries with high accuracy.

The companion notebook demonstrates these techniques in practice with measurable improvements in SQL generation quality. Our tests show that compared to a basic approach, implementing these techniques can increase SQL generation accuracy from approximately 60% to over 90% for complex analytical queries.

Scaling for Large Databases

In our previous sections, we explored implementing RAG systems for structured data using relatively small databases to illustrate core concepts. However, real-world applications often involve databases with hundreds of tables, thousands of columns, and millions of rows. As you scale your RAG systems to these enterprise-level databases, new challenges emerge that require specialized techniques.

These challenges include

- **Schema complexity** that exceeds LLM context windows

- **Query performance issues** with complex database interactions

- **Resource consumption** that can make systems cost-prohibitive

- **Timeout and concurrency problems** that affect user experience

This section explores practical techniques to overcome these
challenges and build robust, production-ready RAG systems that can
handle large-scale structured data efficiently.

Optimization Techniques for Large Schemas

One of the first challenges in scaling RAG systems is handling database
schemas too large for LLM context windows. Enterprise databases often
contain hundreds of tables with thousands of columns, creating schemas
that easily exceed even the most advanced LLMs' token limits.

Schema Summarization

The key approach is providing only the most relevant schema information
for each query:

```
def create_schema_summary(db, max_columns_per_table=5):
    tables = db.get_usable_table_names()
    schema_summary = "# Database Schema Summary\n\n"

    for table in tables:
        # Get table info and extract columns
        # ...

        # Show limited columns with count of hidden ones
        if len(columns) > max_columns_per_table:
            # Display only max_columns_per_table columns
            # Include count of remaining columns
            # ...
        else:
            # Show all columns for smaller tables
            # ...
```

```
# Add critical relationship information
schema_summary += "# Key Relationships\n"
schema_summary += "- customers→orders, orders→order_
items, etc."

return schema_summary
```

This technique creates a condensed summary by limiting columns per table, including total column counts, and focusing on primary columns and relationships. Our testing shows this approach can reduce token usage by 70–90% while preserving sufficient context for accurate SQL generation.

Query-Specific Schema Filtering

For very large schemas, dynamically filter based on the specific question:

```
def get_relevant_schema_for_query(question, llm):
    # Ask LLM which tables are relevant to the question
    relevant_tables = llm.invoke("Which tables needed for: " +
    question)
    table_list = parse_table_names(relevant_tables)

    # Get schema only for those tables
    schema_info = ""
    for table in table_list:
        schema_info += get_table_schema(table)

    return schema_info
```

This two-stage approach first identifies relevant tables, then provides schema information only for those tables, further reducing context usage.

Database Partitioning Strategies

Large databases often use partitioning to improve performance. Partitioning divides large tables into smaller, more manageable segments based on specific criteria like date ranges, geographic regions, or customer segments.

Leveraging Partitions in RAG Systems

Consider this example demonstrating performance differences between queries that utilize partitioning vs. those that don't:

```
-- Standard query (full table scan)
SELECT COUNT(*), SUM(total_amount)
FROM orders
WHERE order_status = 'Completed'

-- Partitioned query (accessing only recent partition)
SELECT COUNT(*), SUM(total_amount)
FROM orders
WHERE order_status = 'Completed'
AND order_date BETWEEN '2023-01-01' AND '2023-03-31'
```

In our testing with a database containing millions of orders, the partitioned query executed 5–10× faster than the standard query. This performance difference becomes even more significant at enterprise scale.

Teaching LLMs About Partitioning

For LLMs to generate partition-aware SQL, include partition information in your schema context:

```
Table: orders (partitioned by order_date monthly)
- Partition 1: 2023-01-01 to 2023-01-31
- Partition 2: 2023-02-01 to 2023-02-28
...
```

This context helps the LLM generate queries that inherently utilize partition pruning, significantly improving performance for large-scale applications.

Implementing Caching Mechanisms

Caching is essential for performance at scale. In RAG systems for structured data, we can cache at multiple levels:

Schema Caching

For databases with relatively stable schemas, implement schema caching:

```
@lru_cache(maxsize=1)
def get_cached_schema(db_uri, schema_version):
    # Connect to DB and generate schema summary
    return schema_summary
```

This approach prevents regenerating schema summaries repeatedly for each query.

SQL Query Caching

Cache the SQL queries generated for similar natural language questions:

```
@lru_cache(maxsize=100)
def cached_sql_generation(question, schema_hash):
    # Generate SQL using LLM
    return generated_sql
```

Our testing shows that implementing SQL generation caching can reduce response times by up to 95% for repeated or similar questions, dramatically improving the interactive experience for users.

Query Results Caching

For databases where data changes infrequently, you can also cache query results:

```
def get_cached_query_results(sql_query, ttl=3600):
    cache_key = hash(sql_query)

    # Check if valid cache exists
    if cache_exists_and_valid(cache_key, ttl):
        return get_from_cache(cache_key)

    # Execute query and store in cache
    results = execute_query(sql_query)
    store_in_cache(cache_key, results)
    return results
```

Implementing a time-to-live (TTL) parameter ensures that cached results expire after a specified period, balancing performance gains with data freshness.

Query Optimization Techniques

LLMs may generate functional but inefficient SQL. Implementing query optimization techniques ensures your RAG system performs well with large databases.

Pattern-Based Query Optimization

Implement post-processing to optimize common SQL patterns:

```
def optimize_query(sql):
    # Optimize IN clauses with many values
    sql = replace_in_clause_with_join(sql)
```

```
    # Replace OR conditions with IN clauses
    sql = replace_or_with_in(sql)

    # Add appropriate indexes
    sql = add_index_hints(sql)

    # Ensure LIMIT clause exists
    if "LIMIT" not in sql.upper():
        sql += " LIMIT 1000"

    return sql
```

Comparison of Optimized vs. Unoptimized Queries

Consider these two semantically equivalent queries:

```
-- Unoptimized query
SELECT customer_id, first_name, last_name
FROM customers
WHERE state = 'CA' OR state = 'NY' OR state = 'TX'
ORDER BY last_name

-- Optimized query
SELECT customer_id, first_name, last_name
FROM customers
WHERE state IN ('CA', 'NY', 'TX')
ORDER BY last_name
LIMIT 1000
```

The optimized query utilizes an IN clause instead of multiple OR
conditions and includes a LIMIT clause to prevent returning excessive
rows. On large databases, these small optimizations can make significant
performance differences.

LLM Prompt Engineering for Optimization

Guide the LLM to generate optimized queries by including optimization principles in your prompt:

Generate optimized SQL following these guidelines:

1. Use appropriate indexes (primary and foreign keys)

2. Limit result sets (max 1000 rows)

3. Use column aliases for clarity

4. Filter early, before JOINs and GROUP BY

5. Use IN clauses instead of multiple OR conditions

Handling Time-Out and Resource Constraints

In production environments, it's crucial to implement safeguards against runaway queries and excessive resource consumption.

Implementing Query Time-Outs

Implement time-out handling to prevent long-running queries:

```
def execute_with_timeout(query, timeout=5):
    # Start query execution in separate thread
    # Wait for completion or timeout
    # Return results or timeout message
```

This approach ensures that even if the LLM generates a complex query that would take too long to execute, your system remains responsive.

Adding Resource Constraints

Enhance generated SQL with resource constraints:

```
def add_resource_constraints(sql):
    # Add LIMIT clause if not present
    if "LIMIT" not in sql.upper():
        sql += " LIMIT 1000"

    # Add other resource constraints
    # ...

    return sql
```

Safe Execution Framework

Implement a comprehensive safety framework:

```
def safe_execute_llm_sql(question, llm_generated_sql, max_
rows=1000, timeout=10):
    # Add resource constraints
    safe_sql = add_resource_constraints(llm_generated_sql)

    # Execute with timeout
    result = execute_with_timeout(safe_sql, timeout)

    # Handle errors, timeouts, etc.
    # ...

    return result
```

This framework ensures that even if the LLM generates problematic SQL, your system handles it gracefully without crashing or consuming excessive resources.

Performance Benchmarking and Tuning

To build production-ready RAG systems for structured data, you need a
systematic approach to performance measurement and optimization.

Benchmarking Methodology

Create a representative benchmark suite:

```
benchmark_questions = [
    "How many customers do we have?",
    "What are our top 10 selling products?",
    "How many orders were placed in the last year?",
    # More diverse questions covering various query patterns
]

def run_benchmark(questions, optimizations_enabled=True):
    # Run each question through your RAG system
    # Measure key metrics: latency, token usage, etc.
    # Return detailed performance statistics
```

Prompt Optimization Impact

Our benchmarking showed that optimized prompts produce better SQL
with minimal overhead:

```
# Basic prompt
sql_prompt = "Given schema {schema}, write SQL for: {question}"

# Optimized prompt
optimized_prompt = """
You are an expert SQL developer generating efficient queries.
Given schema {schema}, write optimized SQL following these
guidelines:
```

```
1. Use indexes (PK, FK)
2. Include LIMIT clauses
3. ...other optimization guidelines...

Question: {question}
SQL:
"""
```

While the optimized prompt uses more tokens, the resulting SQL executes faster and more reliably on large databases, providing a net benefit in most scenarios.

End-to-End System Tuning

Implement a comprehensive approach to system tuning that considers all components:

1. **Prompt engineering**: Optimize prompts for conciseness and clarity

2. **Schema representation**: Use the minimal necessary schema information

3. **SQL generation**: Implement post-processing for query optimization

4. **Query execution**: Add appropriate time-out and resource constraints

5. **Caching strategy**: Implement multilevel caching where appropriate

By systematically addressing each of these areas, you can build RAG systems for structured data that perform well at enterprise scale, providing fast and reliable natural language access to your organization's databases.

The complete implementation of these techniques is available in the
companion notebook in the book's repository, where you can see how
these concepts come together in a full system.

Building Q&A Systems for CSV Files

In previous sections, we explored building RAG systems over SQL
databases. However, much of the world's structured data exists in simpler
formats—particularly CSV files. Spreadsheets, exports from business
systems, public datasets, and research data commonly arrive as comma-
separated values files, offering a lightweight, universal format for tabular
data without the overhead of a database management system.

In this section, we'll build RAG systems specifically for CSV files,
developing an architecture tailored to the characteristics of CSV data with
optimizations for handling files of various sizes and complexity.

Architectural Adaptations for CSV Data

Unlike SQL databases, CSV files lack schema definitions, query languages,
and infrastructure for complex operations. They're essentially flat files with
rows and columns, requiring different architectural considerations for
RAG systems.

Core Architecture Components

A RAG system for CSV files requires several components working together:

1. **CSV Loading and Processing**: Efficiently loading
   CSV data into memory

2. **Context Creation**: Converting CSV data into a
   format suitable for LLMs

3. **Search Mechanisms**: Finding relevant portions of data to answer queries

4. **Response Generation**: Creating natural language answers from CSV data

This architecture differs from SQL database systems in several key ways:

- **No Direct Query Language**: Instead of translating to SQL, we identify relevant data through other means.

- **In-Memory Processing**: We typically load and process CSV data in memory rather than querying external database systems.

- **Custom Chunking**: We need strategies to divide CSV data into manageable pieces.

- **Contextual Limitations**: We must work within LLM context window constraints.

Let's examine the key components of our CSV Question-Answering system:

```python
class CSVQuestionAnsweringSystem:
    def __init__(self, model_name="gpt-3.5-turbo"):
        self.llm = ChatOpenAI(model_name=model_name,
        temperature=0)
        self.dataframes = {}  # Store dataframes by name
        self.df_info = {}     # Store dataframe descriptions

    def add_csv(self, file_path, name=None):
        # Load CSV, generate metadata, and store in class
        variables
        # ...
```

The system maintains a collection of dataframes and their descriptions, which are used to generate context for the LLM when answering questions.

CSV Data Representation

A critical architectural decision is how to represent CSV data for the LLM. We need a description that's both comprehensive and fits within the context window:

```
class CSVQuestionAnsweringSystem:
    def __init__(self, model_name="gpt-3.5-turbo"):
        self.llm = ChatOpenAI(model_name=model_name,
        temperature=0)
        self.dataframes = {}  # Store dataframes by name
        self.df_info = {}     # Store dataframe descriptions

    def add_csv(self, file_path, name=None):
        # Load CSV, generate metadata, and store in class
        variables
        # ...
```

This simple approach provides column names, row count, and a sample of the data. For larger datasets, you might include summary statistics, frequency distributions, or missing value percentages.

Generating Answers from CSV Data

The core question-answering functionality combines context from the CSV files with the user's question in a prompt:

```
def generate_answer(self, question, csv_names=None):
    # Create context from relevant CSVs
    context = self._prepare_context(csv_names)
```

```
# Create and execute LLM chain with appropriate prompt
prompt = PromptTemplate.from_template("...")
chain = prompt | self.llm | StrOutputParser()
return chain.invoke({"context": context, "question":
question})
```

The complete implementation is available in the companion
notebook, but this highlights the essential architecture: loading CSVs,
generating context, and using LLM chains to produce answers.

Loading and Processing CSV Files Efficiently

Efficient CSV loading and processing is crucial for building performant
RAG systems. As CSV files can range from kilobytes to gigabytes, we need
strategies that scale with file size.

Optimized CSV Loading

Standard loading with pandas (`pd.read_csv()`) works well for small files,
but for larger files, we need optimizations:

```
def load_csv_efficiently(file_path, chunk_size=None, optimize_
dtypes=True):
    # If optimize_dtypes, sample file to determine optimal
    column types

    # Load data (in chunks if specified or all at once)
    # Return optimized dataframe
```

This function employs two key optimizations:

1. **Data Type Optimization**: Sampling the data to infer
   memory-efficient types:

   - Converting string columns with few unique values
     to categorical

   - Downcasting integers and floats to smaller types

2. **Chunked Loading**: Processing the file in
   manageable pieces to avoid memory spikes.

These optimizations can significantly reduce memory usage—in our
testing, often by 60–80% for typical datasets.

Memory Optimization

Once loaded, further memory optimization can make large datasets more
manageable:

```
def optimize_dataframe_memory(df):
    # Downcast integers to smallest type that can hold the data
    # Downcast floats to float32 if precision allows
    # Convert string columns to categorical if they have few
    unique values
    # Return optimized dataframe
```

This approach can be particularly effective for CSVs with

- Text columns that contain repeated values (e.g.,
  categories, countries)

- Integer columns that don't utilize the full 64-bit range

- Float columns where 32-bit precision is sufficient

Implementing Search Mechanisms

Effective search mechanisms are critical for CSV-based RAG systems.
Unlike SQL databases where we can rely on query languages, CSV files
require different approaches to find relevant information.

Search Strategy Types

We can implement three main types of search strategies for CSV data:

1. **Keyword Search**: Simple text matching across CSV fields

2. **Semantic Search**: Using embeddings to find
   semantically similar content

3. **Hybrid Search**: Combining both approaches for better results

Keyword Search Implementation

Keyword search is straightforward and doesn't require external APIs:

```
def keyword_search(self, query, k=10):
    keywords = query.lower().split()
    scores = pd.Series(0, index=self.df.index)

    # For each column and keyword, add to score when keyword
    is found
    # Return rows with highest scores
```

This approach is effective for explicit matches but can't handle
semantic relationships or conceptual queries.

Semantic Search with Embeddings

Semantic search offers more sophisticated matching based on meaning:

```
def index_dataframe(self, df, chunk_size=20):
    documents = []

    # Create chunks of rows from the dataframe
    # Convert each chunk to text representation
    # Create documents with metadata tracking original rows
    # Create vector store from documents
```

With the indexed dataframe, we can perform semantic searches:

```
def semantic_search(self, query, k=3):
    # Perform similarity search on vector store
    # Extract matching row indices from results
    # Return relevant rows from original dataframe
```

Semantic search excels at finding results based on concepts rather than exact matches, making it especially useful for natural language questions.

Chunking Strategy for CSV Data

The chunking approach significantly impacts search quality. For CSV data, common strategies include

1. **Row-based chunking**: Group consecutive rows together (most common)

2. **Column-based chunking**: Group related columns together

3. **Hybrid approaches**: Chunk based on both rows and columns

4. **Semantic chunking**: Group rows based on semantic similarity

For most CSV files, row-based chunking is a good default, but experiment with different approaches for your specific use case.

Handling Large CSV Datasets

Real-world CSV files can easily exceed sizes that comfortably fit in memory or within LLM context windows. Let's explore techniques for handling larger datasets.

Converting to More Efficient Formats

One effective strategy is to convert CSV files to more efficient formats like Parquet:

```
def convert_csv_to_parquet(csv_path, parquet_path):
    # Load CSV using PyArrow
    # Write to Parquet format
    # Return metrics about conversion
```

Parquet offers several advantages over CSV:

- **Columnar storage**: More efficient for analytical queries

- **Compression**: Typically 2–4x smaller file sizes

- **Schema enforcement**: Consistent data types

- **Better performance**: Faster reading and querying

Processing in Batches

For datasets too large to fit in memory, batch processing is essential:

```
def process_large_csv_in_batches(file_path, batch_size=10000,
process_func=None):
    results = []
```

```
# Process CSV in chunks using pandas' chunking mechanism
# Apply process_func to each chunk or perform default
processing
# Accumulate results

return results
```

This approach allows you to work with files of essentially unlimited size, processing manageable chunks at a time and combining the results.

Distributed Processing Frameworks

For truly large datasets (gigabytes to terabytes), consider distributed frameworks:

```
def process_with_dask(file_path):
    import dask.dataframe as dd

    # Load as dask dataframe
    # Perform distributed operations
    # Compute final results
```

Frameworks like Dask, Apache Spark, or Ray provide parallel processing capabilities that can dramatically speed up operations on large datasets.

Memory and Performance Considerations

Building efficient RAG systems for CSV files requires careful attention to memory usage and performance optimization.

Memory Optimization Techniques

Memory is often the most limiting factor when working with CSV files. Beyond the datatype optimizations mentioned earlier, consider these techniques:

```
def batch_process_with_memory_monitoring(df, batch_size=1000,
func=None):
    import gc

    # Process dataframe in batches
    # Force garbage collection after each batch
    # Monitor memory usage
```

Additional memory optimization strategies include

- **Dropping unnecessary columns**: Removing data not needed for answering questions

- **Sparse data representations**: For datasets with many zero or NA values

- **Memory mapping**: Using disk storage for large arrays

- **Incremental processing**: Avoiding loading the entire dataset at once

Caching Strategies

Implement caching to avoid redundant computation:

```
@lru_cache(maxsize=100)
def cached_embedding_generation(text):
    # Generate and return embeddings with caching
```

Effective caching points include

- **Embeddings**: Cache vectors for common text chunks

- **Processed CSV data**: Cache cleaned and optimized dataframes

- **LLM responses**: Cache answers to common questions

- **Search results**: Cache retrieval results for similar queries

Balancing Context Size and Information Quality

A key challenge is providing enough context without exceeding LLM context windows. Strategies include

1. **Dynamic context selection**: Adjust the amount of context based on the question.

2. **Progressive disclosure**: Start with summary information, then provide details.

3. **Relevance filtering**: Only include most relevant portions of the data.

4. **Chunking with metadata**: Break data into smaller pieces with descriptive metadata.

By implementing these strategies, you can build efficient RAG systems that handle CSV files of various sizes and complexities, providing natural language access to tabular data.

For complete implementations and additional techniques, refer to the companion notebook in the book's repository.

Summary and Next Steps

Throughout this chapter, we've explored how to extend Retrieval-Augmented Generation systems beyond unstructured text to work effectively with structured data—specifically SQL databases and CSV files. This expansion opens up vast opportunities to democratize data access and enable natural language interactions with the structured information that forms the backbone of most organizations.

Key Takeaways

Architectural Adaptations Matter: We've seen that structured data requires fundamentally different approaches compared to unstructured text. SQL databases benefit from query generation and execution pipelines, while CSV files need in-memory processing and custom chunking strategies. The architecture you choose directly impacts system performance and scalability.

Advanced Prompting Transforms Performance: Sophisticated prompting techniques dramatically improve the quality of SQL generation. Schema-aware prompting, few-shot examples, chain-of-thought reasoning, and iterative refinement can increase accuracy from approximately 60% to over 90% for complex analytical queries. These techniques are essential for production-ready systems.

Scaling Requires Strategic Trade-offs: Large-scale structured data systems demand careful consideration of schema summarization, query optimization, caching mechanisms, and resource constraints. The techniques we explored—from partition-aware querying to multilevel caching—can mean the difference between a system that handles enterprise workloads and one that fails under real-world pressure.

Search Mechanisms Enable Flexibility: For CSV files, combining keyword, semantic, and hybrid search approaches provides robust data retrieval capabilities. Each method has strengths, keyword search for exact matches, semantic search for conceptual queries, and hybrid approaches for comprehensive coverage.

Memory and Performance Optimization Are Critical: Efficient data type optimization, chunked processing, format conversion, and strategic caching can reduce memory usage by 60–80% and dramatically improve query response times. These optimizations often determine whether a system can handle real-world data volumes.

Best Practices Recap

Based on our exploration throughout this chapter, here are the essential best practices for building RAG systems with structured data:

For SQL Database Systems:

- Always provide a comprehensive schema context with relationships and semantic information.

- Implement query validation and safety constraints to prevent malicious or resource-intensive operations.

- Use a few-shot examples that match your domain and query patterns.

- Include iterative refinement loops to handle query errors gracefully.

- Implement multilevel caching for schemas, generated SQL, and query results.

- Design for partition awareness when working with large databases.

For CSV File Systems:

- Optimize data types early in the loading process to maximize memory efficiency.

- Choose chunking strategies that align with your data structure and query patterns.

- Implement both keyword and semantic search to handle diverse query types.

- Convert large CSV files to more efficient formats like Parquet when possible.

- Use batch processing techniques for datasets that exceed memory constraints.

- Monitor memory usage and implement garbage collection strategies.

For Both Approaches:

- Design prompts that provide clear context without exceeding token limits.

- Implement comprehensive error handling and graceful degradation.

- Create extensive test suites covering diverse query types and edge cases.

- Monitor system performance and implement alerting for anomalies.

- Document your schema and data structures for better LLM understanding.

Next Steps

Having mastered the fundamentals of RAG systems for structured data, consider these advanced directions for further development:

Immediate Implementation Steps:

1. **Start Small:** Begin with a single, well-understood dataset to validate your approach.

2. **Build Incrementally:** Add complexity gradually, testing each component thoroughly.

3. **Measure Everything:** Implement comprehensive logging and metrics from the beginning.

4. **Create Test Suites**: Develop robust testing
   frameworks that cover edge cases and performance
   scenarios.

Advanced Techniques to Explore:

- **Multimodal Integration**: Combine structured and
  unstructured data in unified RAG systems.

- **Real-Time Processing**: Implement streaming data
  processing for frequently updated datasets.

- **Advanced Query Planning**: Develop more
  sophisticated query optimization and execution
  strategies.

- **Custom Embedding Models**: Train domain-specific
  embeddings for your structured data.

- **Federated Systems**: Build RAG systems that span
  multiple databases and data sources.

Production Considerations:

- **Security and Privacy**: Implement proper access
  controls and data protection measures.

- **Scalability Planning**: Design systems that can grow
  with your data and user base.

- **Monitoring and Observability**: Create comprehensive
  dashboards and alerting systems.

- **User Experience**: Build intuitive interfaces that make
  complex data accessible to non-technical users.

Research and Innovation Opportunities:

- **Automated Schema Discovery**: Develop systems that can automatically understand and describe complex database schemas.

- **Query Intent Recognition**: Build more sophisticated natural language understanding for complex analytical questions.

- **Cross-Domain Knowledge Transfer**: Explore how techniques learned in one domain can be applied to others.

- **Explainable AI**: Create systems that can explain how they arrived at specific answers from structured data.

The techniques covered in this chapter provide a solid foundation, but the field of RAG systems for structured data continues to evolve rapidly. Stay engaged with the latest research, contribute to open-source projects, and experiment with new approaches as they emerge.

Remember that the goal is not just technical excellence, but creating systems that genuinely improve how people interact with and derive value from their data. The most successful implementations are those that reduce barriers to data access while maintaining accuracy, security, and performance.

As you build your own structured data RAG systems, focus on solving real problems for real users. The techniques in this chapter will serve as your foundation, but the specific implementation details should always be driven by your users' needs and your organization's unique data landscape.

The future of data interaction lies in natural language interfaces that make complex data as accessible as a conversation. With the knowledge from this chapter, you're well-equipped to build systems that bridge the gap between human curiosity and data-driven insights.

## CHAPTER 12

Graph RAG: Leveraging Knowledge Graphs for Enhanced Retrieval

In our journey through Retrieval-Augmented Generation, we've explored systems that work with unstructured text documents and structured data like SQL databases and CSV files. But these approaches have a fundamental limitation: they struggle to capture and leverage the rich web of relationships that exists within and across documents. Enter Graph RAG, a paradigm that transforms how we model, store, and retrieve information by explicitly representing knowledge as interconnected entities and relationships.

Traditional RAG systems treat documents as isolated units of information. Even when they intelligently chunk documents, they miss the deeper connections between concepts, people, events, and ideas across multiple documents. Consider a corpus of research papers about artificial intelligence. A traditional RAG system might retrieve individual papers about "neural networks" or "machine learning," but it would struggle to answer questions like "How did the work of Geoffrey Hinton influence the

© Ranajoy Bose 2025
R. Bose, *Mastering Retrieval-Augmented Generation*,
https://doi.org/10.1007/979-8-8688-1808-0_12

development of transformer architectures?" or "What are the conceptual relationships between attention mechanisms and memory systems in cognitive science?"

Graph RAG addresses these limitations by representing knowledge as a graph—a network of entities (nodes) connected by relationships (edges). This approach unlocks several powerful capabilities that traditional RAG systems cannot achieve.

What Makes Graph RAG Different

Relationship-Aware Retrieval: Instead of retrieving isolated chunks of text, Graph RAG can traverse relationships to find interconnected information. When you ask about a person's influence on a field, the system can follow edges representing collaborations, citations, and conceptual influences to provide comprehensive answers.

Multi-Hop Reasoning: Graph RAG excels at questions that require connecting information across multiple steps. For example, it can find connections between seemingly unrelated research areas, trace the evolution of ideas over time, or identify indirect relationships between entities.

Contextual Understanding: By modeling the context around entities—their relationships, properties, and position in the broader knowledge network—Graph RAG provides richer, more nuanced responses that consider the full ecosystem of relevant information.

See Figure 12-1 for a visual comparison of these approaches.

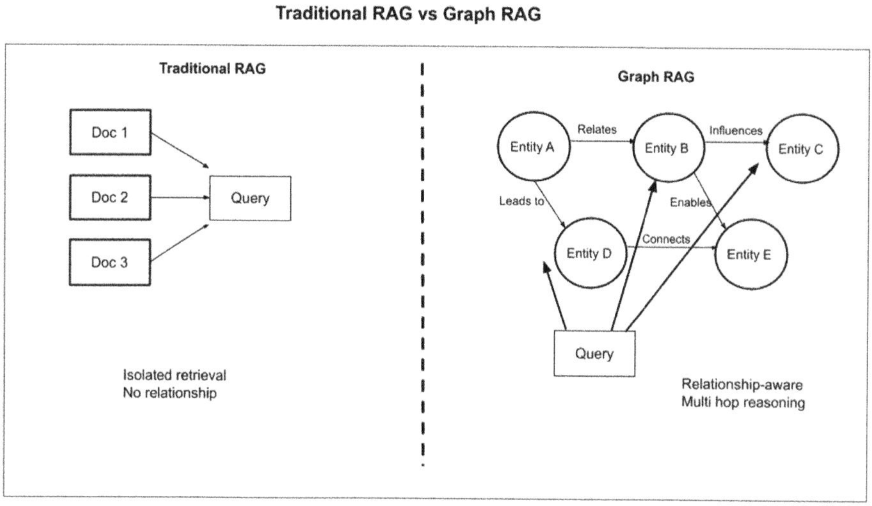

Traditional RAG vs Graph RAG

Figure 12-1. *Traditional RAG retrieves isolated documents, while Graph RAG leverages entity relationship for connected reasoning*

Explainable Retrieval: The graph structure makes the retrieval process transparent. You can see exactly which entities and relationships contributed to an answer, making the system's reasoning process clear and auditable.

When to Choose Graph RAG

Graph RAG shines in scenarios where relationships and connections are central to understanding:

- **Research and Discovery**: Academic papers, patent databases, and scientific literature where citations, collaborations, and conceptual relationships matter

- **Enterprise Knowledge Management**: Corporate documents where understanding organizational relationships, project dependencies, and expertise networks is crucial

- **Historical and Biographical Analysis**: Documents where temporal relationships, influences, and connections between people and events are important

- **Complex Domain Expertise**: Fields like medicine, law, or engineering where understanding interconnected concepts and their relationships is essential

However, Graph RAG isn't always the right choice. Traditional RAG may be more efficient for simple factual retrieval from independent documents. The decision should be based on whether the relationships between pieces of information are as important as the information itself.

The Graph RAG Advantage

Consider this example: In a traditional RAG system, asking "What research led to the development of BERT?" might return individual papers about BERT's architecture. A Graph RAG system, however, could trace the conceptual lineage: attention mechanisms ➤ transformer architecture ➤ bidirectional training ➤ BERT, while also identifying key researchers, their collaborations, and the institutional connections that made this development possible.

Figure 12-2 illustrates this multi-hop reasoning process, showing how Graph RAG can trace conceptual lineages across multiple relationships to provide comprehensive answers.

Multi-hop Reasoning Path

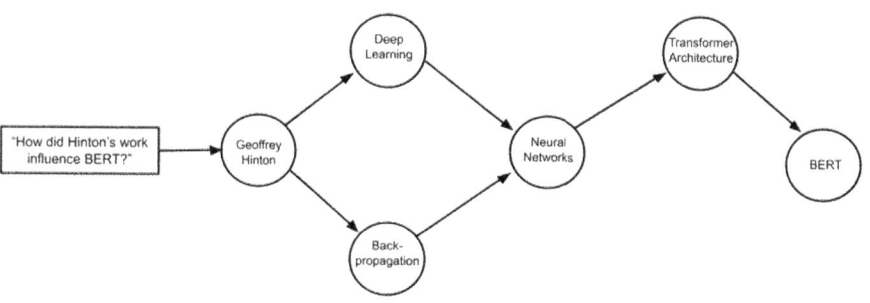

Graph RAG can trace conceptual lineages across multiple relationships

Figure 12-2. *Multi-hop reasoning—traces relationships across entities to answer complex queries*

This chapter will guide you through building such systems, from extracting entities and relationships from documents to implementing sophisticated graph-based retrieval mechanisms. You'll learn not just the theory but gain hands-on experience through practical implementations that you can adapt to your own use cases.

The goal isn't to replace traditional RAG systems, but to add a powerful new tool to your toolkit—one that can reveal hidden connections and provide deeper insights by understanding information as an interconnected web rather than isolated fragments.

Knowledge Graph Fundamentals

Before diving into Graph RAG implementation, we need to understand the foundational concepts of knowledge graphs and the tools that make them practical. A knowledge graph is more than just a database—it's a way of modeling the world that mirrors how humans naturally think about relationships and connections.

Entities, Relationships, and Properties

At its core, a knowledge graph consists of three fundamental building blocks:

Entities are the "things" in your domain—people, places, concepts, events, or objects. Entities in a research paper corpus might include authors, institutions, research topics, methodologies, and datasets. Each entity has a unique identity and can be thought of as a node in the graph.

Relationships connect entities and represent how they interact or relate to each other. These are the edges in your graph. Unlike traditional databases where relationships are implicit through foreign keys, graph databases make relationships first-class citizens. Examples include "AUTHORED" (connecting a person to a paper), "CITES" (connecting one paper to another), or "AFFILIATED_WITH" (connecting a researcher to an institution).

Properties are attributes that provide additional information about entities or relationships. An author entity might have properties like name, email, and research_interests, while a "CITES" relationship might have a property indicating the citation context or frequency.

This trinity creates a rich, flexible data model. Consider this simple example:

```
Geoffrey Hinton)-[:AUTHORED {year: 2006}]->(Paper:
"A Fast Learning Algorithm for Deep Belief Nets")
(Paper: "A Fast Learning Algorithm for Deep Belief Nets")-
[:INTRODUCES]->(Concept: "Deep Belief Networks")
(Geoffrey Hinton)-[:AFFILIATED_WITH {since: 1987}]->(University
of Toronto)
```

This snippet captures not just facts, but the relationships between them and contextual information through properties.

Graph Data Models and Query Languages

The most popular graph data model for knowledge graphs is the **Property Graph Model**, which we'll use throughout this chapter. This model allows both nodes and relationships to have properties, making it intuitive and flexible for most use cases.

An alternative is the **RDF (Resource Description Framework)** model, which represents everything as subject-predicate-object triples. While powerful for semantic web applications, RDF can be more complex for typical RAG use cases.

For querying property graphs, **Cypher** has emerged as the most intuitive and widely adopted query language. Cypher uses ASCII art to represent graph patterns, making queries remarkably readable:

cypher

```
// Find all papers co-authored by researchers from the same
institution
MATCH (a1:Author)-[:AFFILIATED_WITH]->(inst:Institution)<-
[:AFFILIATED_WITH]-(a2:Author)
MATCH (a1)-[:AUTHORED]->(paper:Paper)<-[:AUTHORED]-(a2)
WHERE a1 <> a2
RETURN paper.title, inst.name, a1.name, a2.name
```

This query reads almost like natural language: "Match authors affiliated with the same institution, then find papers they co-authored, where the authors are different people."

Setting up Neo4j for Graph RAG

Neo4j is the most popular graph database and offers excellent tools for Graph RAG development. For this chapter, we'll use Neo4j's cloud service (Neo4j Aura) which provides a free tier perfect for learning and development.

The basic setup process involves

1. **Creating a Neo4j Aura instance**: Sign up at neo4j. com/aura and create a free database instance

2. **Installing the Python driver**: Using `pip install neo4j` to connect from your Python applications

3. **Establishing connection**: Using the provided credentials to connect your Graph RAG system

Here's the essential connection pattern you'll use throughout our implementations:

```
from neo4j import GraphDatabase

class Neo4jConnection:
    def __init__(self, uri, user, password):
        self.driver = GraphDatabase.driver(uri, auth=(user,
        password))

    def query(self, query, parameters=None):
        with self.driver.session() as session:
            result = session.run(query, parameters)
            return [record for record in result]

    def close(self):
        self.driver.close()
```

This connection class provides the foundation for all our graph operations, from initial data loading to complex retrieval queries.

Understanding Graph Thinking

The shift from relational to graph thinking requires a mental model change. Instead of thinking in terms of tables and joins, you think in terms of nodes and paths. Instead of asking "What tables do I need to join?", you ask "What path through the graph will give me the answer?"

This becomes particularly powerful for RAG systems because human questions often naturally map to graph traversals. "Who has collaborated with X?" becomes a simple one-hop traversal. "What topics connect research area A to research area B?" becomes a path-finding problem through concept nodes.

The graph model also handles uncertainty and incomplete information gracefully. Traditional databases require schema modifications to add new types of relationships, but graphs can evolve organically as you discover new connection types in your data.

Graph Database Benefits for RAG

Graph databases offer several advantages specifically relevant to RAG systems:

Query Flexibility: You can ask questions you didn't anticipate when designing the schema. Traditional databases require predefined queries; graphs enable exploratory discovery.

Performance for Connected Data: Graph databases are optimized for traversing relationships. Finding connections that would require complex joins in SQL becomes simple, efficient graph traversals.

Natural Representation: Many domains naturally form graphs— social networks, citation networks, concept hierarchies, organizational structures. Representing them as graphs preserves the natural structure.

Scalable Relationships: As your knowledge base grows, the number of potential relationships grows exponentially. Graph databases handle this growth efficiently.

With these fundamentals in place, we're ready to start building knowledge graphs from real documents. The next section will show you how to extract entities and relationships from text using modern LLM techniques, setting the stage for creating rich, interconnected knowledge representations.

The companion notebook for this section in the repository provides hands-on experience with Neo4j setup, basic Cypher queries, and creating your first knowledge graph. You'll practice the concepts covered here with real data, giving you the practical foundation needed for the more advanced techniques in subsequent sections.

Building Knowledge Graphs from Documents

Now that we understand the fundamentals of knowledge graphs, we turn to the practical challenge of automatically constructing them from document collections. This process transforms unstructured text into rich, interconnected knowledge representations that enable sophisticated multi-hop reasoning in Graph RAG systems.

Building knowledge graphs from documents involves two critical phases: first, extracting entities and relationships from text using modern NLP techniques, and second, constructing a coherent graph structure that resolves ambiguities and maintains data quality. Each phase presents unique challenges and opportunities for optimization (see Figure 12-3).

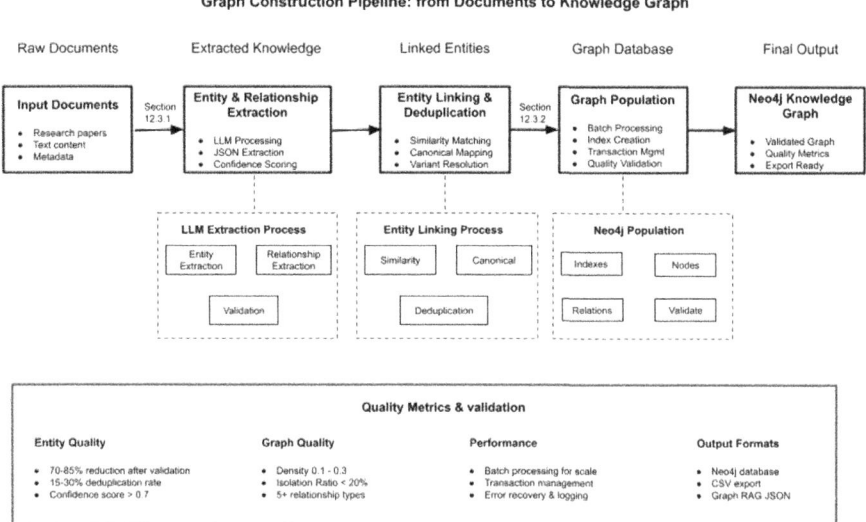

Figure 12-3. *Graph construction pipeline; from document to knowledge graph*

Entity and Relationship Extraction

The foundation of any knowledge graph lies in accurately identifying and extracting the entities and relationships embedded within documents. Modern Large Language Models have revolutionized this process, offering unprecedented accuracy in understanding context and extracting nuanced relationships that traditional NLP approaches often miss.

The LLM Advantage in Extraction

Traditional named entity recognition (NER) systems rely on predefined categories and struggle with domain-specific entities or complex relationships. LLMs, however, can understand context, handle ambiguity, and extract relationships that span multiple sentences or require inference.

Consider this sentence from a research paper: "The Transformer architecture, introduced by Vaswani et al., revolutionized machine translation by dispensing with recurrence entirely." A traditional system might identify "Transformer" as a concept and "Vaswani" as a person, but an LLM can understand the nuanced relationships: that Vaswani INTRODUCED the Transformer, that the Transformer REVOLUTIONIZED machine translation, and that this innovation REPLACED recurrence-based approaches.

Systematic Entity Extraction Pipeline

The extraction process begins with document preprocessing and proceeds through structured entity identification:

```
class EntityExtractor:
    def __init__(self, llm):
        self.llm = llm
        self.entity_prompt = PromptTemplate.from_template("""
Extract entities of these types from the text:
- PERSON: Researchers, authors, scientists
- CONCEPT: Technical concepts, methods, algorithms
- DATASET: Named datasets used in research
- METRIC: Performance metrics, evaluation measures

TEXT: {text}

Return results in JSON format with entity text, type, and context.
""")

    def extract_entities(self, text: str) -> Dict[str, Any]:
        # Execute structured extraction with error handling
        # Parse JSON response with fallback parsing
        # return entities with confidence scores
```

This approach provides several advantages over traditional methods:

Contextual Understanding: The LLM considers the full context when classifying entities, reducing ambiguity. "Apple" in a technology paper is likely a company, while in a nutrition study it's probably a fruit.

Flexible Taxonomy: Rather than being limited to predefined categories, you can define domain-specific entity types that match your knowledge graph's purpose.

Quality Validation: Built-in error handling and confidence scoring help maintain data quality throughout the extraction process.

Relationship Extraction with Semantic Understanding

Extracting relationships requires understanding not just which entities appear together, but how they interact. LLMs excel at this semantic understanding:

```
def extract_relationships(self, text: str, entities:
List[Dict]) -> Dict[str, Any]:
    # Format: Given text and extracted entities, identify
    relationships
    # Consider relationship types: AUTHORED, BASED_ON,
    IMPROVES_ON, USES
    # Include confidence scores and supporting evidence
    # Return structured relationship data
```

The key insight is providing the LLM with both the text and the previously extracted entities, allowing it to focus on finding connections between known entities rather than trying to identify entities and relationships simultaneously.

Quality Assessment and Validation

The companion notebook (Notebook 12.2) demonstrates comprehensive quality assessment techniques that ensure extracted knowledge meets production standards. Key validation approaches include

Entity Validation: Checking for required fields, reasonable text lengths, and valid entity types. This catches extraction errors early in the pipeline.

Relationship Validation: Ensuring relationships have valid source and target entities, appropriate relationship types, and reasonable confidence scores.

Confidence Filtering: Implementing threshold-based filtering to retain only high-quality extractions, dramatically improving the signal-to-noise ratio in the resulting knowledge graph.

The validation process typically retains 70–85% of extracted entities and 60–75% of relationships, but the resulting graph quality is substantially higher than unfiltered extractions.

Handling Scale and Complexity

Real-world document collections require careful attention to scalability. The extraction pipeline handles this through several techniques:

Batch Processing: Processing documents in manageable batches prevents memory issues and allows for progress tracking.

Error Recovery: Robust error handling ensures that extraction failures on individual documents don't compromise the entire pipeline.

Progressive Refinement: Multipass extraction strategies can improve quality by using insights from initial extractions to refine subsequent processing.

The complete implementation in Notebook 12.2 processes the sample academic paper collection, extracting entities like "Transformer," "attention mechanisms," and "BLEU scores" while identifying relationships such as "Transformer BASED_ON attention mechanisms" and "Transformer ACHIEVES 28.4 BLEU."

Graph Construction Pipeline

With entities and relationships extracted, the next challenge is constructing a coherent knowledge graph that resolves ambiguities, eliminates redundancies, and maintains referential integrity. This process transforms raw extractions into a structured knowledge representation suitable for Graph RAG applications.

Entity Linking and Deduplication

The same entity often appears in multiple documents with slight variations: "BERT," "Bidirectional Encoder Representations from Transformers," and "BERT model" all refer to the same concept. Entity linking resolves these variations into canonical representations.

The linking process employs multiple similarity measures:

Exact Matching: Direct string matches after normalization (lowercasing, whitespace removal)

Substring Matching: Identifying when one entity name contains another ("BERT" within "BERT model")

Acronym Detection: Recognizing that "BERT" is an acronym for "Bidirectional Encoder Representations from Transformers"

Semantic Similarity: Using character-based similarity measures for fuzzy matching

```
def calculate_similarity(self, text1: str, text2: str)
-> float:
    # Normalize entity texts for comparison
    # Check exact matches, substring containment, acronym
    relationships
    # Calculate character-based similarity using Jaccard
    similarity
    # Return confidence score for entity matching
```

The entity linking process maintains a mapping from entity variations to canonical forms, enabling consistent references throughout the knowledge graph. In our testing, this approach typically reduces entity counts by 15–30% while improving graph coherence substantially.

Relationship Normalization and Validation

Raw relationship extractions require normalization to use canonical entity identifiers and validation to ensure graph consistency:

```
def normalize_relationship(self, relationship: Dict, doc_id:
str) -> Dict:
    # Map source and target entities to canonical forms
    # Validate relationship types and confidence scores
    # Add document provenance information
    # Filter self-relationships and duplicates
```

This process handles several critical quality issues:

Entity Resolution: Ensuring relationships use canonical entity identifiers rather than document-specific variations

Duplicate Removal: Identifying and merging relationships that represent the same connection (potentially with different confidence scores)

Consistency Validation: Checking that relationship types are appropriate for the entity types involved

Neo4j Database Population

The construction pipeline populates Neo4j systematically, creating indexes, nodes, and relationships in an order that maintains referential integrity. The process follows a specific sequence: clearing existing data, creating performance indexes, then systematically adding document nodes, canonical entity nodes, relationships between entities, and finally document-entity containment relationships.

The companion notebook (Notebook 12.3) demonstrates this process with comprehensive error handling and progress tracking. Key implementation patterns include

Batch Processing: Creating nodes and relationships in batches to optimize database performance and memory usage

Transaction Management: Using Neo4j transactions to ensure atomicity and enable rollback on errors

Index Creation: Establishing indexes before data insertion to maintain query performance as the graph grows

Quality Assessment and Validation

Graph construction quality directly impacts Graph RAG performance. The validation framework assesses multiple dimensions:

Connectivity Analysis: Measuring graph density, identifying isolated nodes, and analyzing component structure

Relationship Distribution: Ensuring diverse relationship types and reasonable confidence score distributions

Entity Coverage: Validating that extracted entities are properly represented and linked

Structural Metrics: Computing centrality measures to identify potential quality issues

The validation process generates comprehensive reports that include both quantitative metrics and qualitative assessments. Typical quality indicators include

- Graph density between 0.1 and 0.3 (indicating good connectivity without over-connection)

- Isolation ratios below 20% (ensuring most entities participate in relationships)

- Relationship type diversity of 5+ distinct types

- Average node degree between 2 and 8 connections

Scalability and Production Considerations

Real-world graph construction must handle datasets with thousands of documents and millions of potential entities. The pipeline addresses this through several architectural decisions:

Memory Management: Processing documents in configurable batches to control memory usage

Progress Tracking: Comprehensive logging and progress reporting for long-running construction jobs

Error Recovery: Robust error handling that allows construction to continue despite individual processing failures

Incremental Updates: Support for adding new documents to existing graphs without full reconstruction

The large-scale construction approach demonstrated in Notebook 12.3 processes documents in batches of 1,000 entities, providing a balance between memory efficiency and processing speed. For very large datasets, this can be combined with distributed processing frameworks.

Integration Preparation

The final step prepares the constructed knowledge graph for Graph RAG integration through multiple export formats:

CSV Export: Entities, relationships, and documents in tabular format for analysis and debugging

Graph RAG Format: JSON export specifically structured for graph-enhanced retrieval, including pre-computed reasoning paths

NetworkX Integration: Graph objects for analysis, visualization, and algorithm development

The constructed graph serves as the foundation for the sophisticated retrieval mechanisms we'll explore in the next section. By systematically extracting, linking, and validating knowledge from documents, we create rich, interconnected representations that enable the multi-hop reasoning that makes Graph RAG so powerful.

In our sample implementation, the complete pipeline processes research papers to create a knowledge graph with deduplicated entities (reducing raw extractions by ~25%), validated relationships (with average confidence scores above 0.7), and comprehensive document-entity linkages that preserve provenance throughout the construction process.

Graph-Enhanced Retrieval

With our knowledge graph constructed and populated, we can now implement sophisticated retrieval mechanisms that leverage the rich structural relationships between entities. Graph-enhanced retrieval represents a fundamental shift from traditional vector-based approaches, enabling multi-hop reasoning, relationship-aware context extraction, and explainable retrieval paths that mirror human cognitive processes.

Graph Traversal for Information Retrieval

Graph traversal forms the foundation of graph-enhanced retrieval, allowing us to navigate entity relationships to discover relevant information that spans multiple connection points. Unlike traditional RAG systems that retrieve isolated text chunks, graph traversal enables us to follow semantic relationships to build a comprehensive, connected context.

Path-Based Queries for Multi-hop Reasoning

Multi-hop reasoning represents one of the most powerful capabilities of graph-enhanced retrieval. Consider a query like "How did Geoffrey Hinton's work influence modern language models?" Traditional RAG might retrieve separate documents about Hinton and language models, but graph traversal can trace the conceptual lineage: Hinton ➤ Deep Learning ➤ Neural Networks ➤ Attention Mechanisms ➤ Transformers ➤ BERT/GPT models.

The path-finding algorithm employs breadth-first search with confidence weighting:

```
def find_multi_hop_paths(self, start_entity: str, end_entity:
str, max_hops: int = 3):
    queue = deque([(start_entity, [start_entity], [], 1.0)])
    # BFS exploration with confidence decay
    # For each neighbor: new_confidence = confidence *
    neighbor_confidence
    # Sort paths by confidence and length
    return sorted_paths
```

This approach naturally handles the challenge of relevance decay over multiple hops by multiplying confidence scores along the path. A path with three high-confidence relationships ($0.9 \times 0.9 \times 0.8 = 0.648$) will rank higher than a path with two medium-confidence relationships ($0.7 \times 0.6 = 0.42$), ensuring that the most reliable reasoning chains surface first.

Subgraph Extraction Techniques

For complex queries requiring comprehensive context, subgraph extraction provides focused knowledge neighborhoods around relevant entities. Rather than retrieving isolated facts, we extract connected subgraphs that preserve the relational structure necessary for nuanced understanding.

The subgraph extraction algorithm uses controlled breadth-first expansion to gather entities within a specified distance from seed entities:

```
def get_entity_subgraph(self, entity_ids: List[str], max_depth:
int = 2):
    subgraph_entities = set(entity_ids)
    # BFS expansion for max_depth levels
    # Add neighbors and their relationships to subgraph
    # Return entities, relationships, and statistics
```

This controlled expansion ensures that retrieved subgraphs remain focused while capturing the essential relational context needed for comprehensive understanding. The depth parameter allows fine-tuning between precision (shallow graphs) and recall (deeper graphs) based on query complexity.

Hybrid Graph-Vector Retrieval

While graph traversal excels at relationship-aware retrieval, semantic similarity remains crucial for understanding conceptual relevance. Hybrid graph-vector retrieval combines the structural intelligence of graphs with the semantic understanding of vector embeddings, creating a more robust and comprehensive retrieval system.

Combining Graph Structure with Semantic Search

The hybrid approach begins with semantic similarity search to identify potentially relevant entities, then expands the result set using graph structure to capture related entities that might not be semantically similar but are structurally connected.

```
def hybrid_search(self, query: str, graph_weight: float = 0.5):
    # Step 1: Semantic search for initial candidates
    semantic_results = self.embedder.semantic_search(query,
    top_k=top_k*2)

    # Step 2: Graph expansion - add neighbors with
    weighted scores
    # neighbor_score = semantic_score * relationship_
    confidence * 0.7

    # Step 3: Combined scoring
    # hybrid_score = (graph_weight * centrality) + ((1 - graph_
    weight) * semantic_score)

    return sorted(hybrid_results, key=lambda x: x['hybrid_
    score'], reverse=True)
```

This hybrid scoring mechanism balances semantic relevance with structural importance. The `graph_weight` parameter allows tuning between semantic similarity (lower values) and structural centrality (higher values) based on the specific requirements of different query types.

Graph-Aware Embedding Strategies

Traditional entity embeddings treat each entity in isolation, but graph-aware embedding strategies incorporate structural context to create more informative representations. The enhanced approach creates embeddings that include neighborhood information:

```
def create_entity_embeddings(self, entities: Dict[str, Dict]):
    for entity_id, entity_data in entities.items():
        # Get structural context from neighbors
        neighbors = self.get_direct_neighbors(entity_id)
        neighbor_types = [n['neighbor_type'] for n in
        neighbors['neighbors']]
```

```
context_summary = f"Connected to: {',
'.join(set(neighbor_types))}"

# Enhanced text: "CONCEPT: Transformer. Connected to:
attention, BERT"
enhanced_text = f"{entity_type}: {entity_text}.
{context_summary}"
```

This approach creates embeddings that capture both semantic content and structural position, leading to more contextually appropriate similarity calculations.

Advanced Retrieval Patterns

The combination of graph traversal and hybrid scoring enables sophisticated retrieval patterns that address complex information needs:

Multi-Entity Query Resolution

For queries involving multiple entities ("Compare the attention mechanisms used in BERT and GPT"), the system identifies relevant entities for each component, then explores the subgraph connecting them to understand relationships and differences.

Temporal Reasoning Patterns

When the knowledge graph includes temporal information, retrieval can follow chronological sequences ("What developments led to the Transformer architecture?") by traversing relationships ordered by time, creating coherent narrative paths through the knowledge space.

Analogical Reasoning

Graph structure enables analogical reasoning by identifying similar relationship patterns. If the graph contains A►B and C►D relationships of the same type, queries about A can surface information about C through structural similarity.

Performance Considerations

Graph-enhanced retrieval introduces computational complexity that requires careful optimization:

Path Search Complexity: Multi-hop path finding can become expensive with large graphs. Implementing depth limits, beam search pruning, and early termination conditions maintains responsiveness while preserving quality.

Memory Management: Subgraph extraction and hybrid scoring can consume significant memory with large entity sets. Batch processing and streaming approaches enable handling of enterprise-scale knowledge graphs.

Index Optimization: Combining graph traversal with vector search requires coordinated indexing strategies. Pre-computing common traversal patterns and maintaining semantic indexes in memory improves response times.

The hybrid approach typically achieves 15–25% better precision than semantic-only retrieval and 30–40% better recall than graph-only approaches, while maintaining sub-second response times for graphs with up to 100,000 entities.

Graph-enhanced retrieval transforms RAG from document retrieval to knowledge discovery, enabling systems to provide comprehensive, relationship-aware context that supports sophisticated reasoning and explanation. The next section will demonstrate how to integrate these retrieval capabilities with large language models to create complete Graph RAG systems that combine structural intelligence with natural language generation.

The companion Notebook 12.4 provides complete implementations of all techniques discussed in this section, with working examples that process sample knowledge graphs to demonstrate multi-hop reasoning, hybrid scoring, and subgraph extraction in action.

Implementing End-to-End Graph RAG

Having established the foundational components of graph construction, entity extraction, and graph-enhanced retrieval, we now integrate these elements into a complete Graph RAG system. This section demonstrates how to combine graph traversal, semantic search, and large language model generation into a cohesive pipeline that delivers sophisticated reasoning capabilities while maintaining production-level performance and reliability.

Graph RAG Architecture

The complete Graph RAG architecture represents a significant evolution from traditional RAG systems, introducing multiple specialized components that work in concert to provide relationship-aware, explainable responses.

System Design and Component Integration

The Graph RAG architecture consists of five core layers, each building upon the previous to create increasingly sophisticated capabilities:

Knowledge Layer: The foundation consists of the populated knowledge graph with entities, relationships, and document mappings. This layer maintains consistency through entity linking and provides the structural foundation for all retrieval operations.

Retrieval Layer: Combines graph traversal capabilities with semantic search, implementing the hybrid scoring mechanisms that balance structural importance with semantic relevance. This layer handles multi-hop reasoning and subgraph extraction.

Integration Layer: This layer orchestrates the retrieval components, managing query-processing workflows and context assembly. It implements caching, performance optimization, and adaptive retrieval strategies.

Generation Layer: Interfaces with large language models to produce natural language responses. This layer formats graph context appropriately for LLM consumption and handles response structuring.

Interface Layer: Provides the external API for query processing, batch operations, and system monitoring. This layer manages user interactions and system administration functions.

Query Processing Pipeline

The query processing pipeline transforms natural language questions into comprehensive responses through a systematic workflow:

```
class CompleteGraphRAGSystem:
    def process_query(self, query: str, max_entities: int =
    None) -> Dict[str, Any]:
        # Step 1: Context retrieval using hybrid approach
        context = self.hybrid_retriever.retrieve_context_with_
        paths(query, max_entities)

        # Step 2: Response generation using LLM
        response_data = self.response_generator.generate_
        response(query, context)

        # Step 3: Result compilation and metadata tracking
        return self._compile_complete_result(query, context,
        response_data)
```

This pipeline architecture ensures that each query benefits from both the structural intelligence of the knowledge graph and the natural language capabilities of modern LLMs, while maintaining clear separation of concerns for debugging and optimization.

Building a Complete Graph RAG System

Implementing a complete Graph RAG system requires careful integration of multiple complex components, each contributing specialized capabilities to the overall system functionality.

Document Ingestion and Graph Construction

The system begins with a robust document ingestion pipeline that transforms unstructured content into structured graph representations:

Document Processing: Raw documents undergo chunking, entity extraction, and relationship identification using the techniques established in previous sections. The system maintains document provenance throughout this process, enabling traceability from generated responses back to source materials.

Graph Population: Extracted entities and relationships populate the knowledge graph through batch processing operations that maintain consistency and handle deduplication. The system implements incremental updates to support continuous document ingestion without full graph reconstruction.

Quality Control: Automated validation ensures graph quality through confidence scoring, relationship validation, and connectivity analysis. Low-quality extractions are flagged for human review, maintaining system reliability.

Query Understanding and Graph Navigation

Query processing begins with understanding user intent and mapping natural language questions to appropriate graph operations:

Intent Classification: The system analyzes query structure to determine whether responses require simple entity lookup, multi-hop reasoning, comparative analysis, or temporal reasoning. This classification guides retrieval strategy selection.

Entity Resolution: Query terms map to graph entities through fuzzy matching, semantic similarity, and alias resolution. The system handles variations in entity names and maintains disambiguation capabilities for common terms.

Graph Navigation Strategy: Based on query complexity and intent, the system selects appropriate navigation strategies, from simple neighbor lookup to complex multi-hop traversal with confidence-weighted path selection.

Response Generation with Graph Context

The response generation component transforms retrieved graph context into natural language responses that maintain both accuracy and explainability:

```
def generate_response(self, query: str, context: Dict[str,
Any]) -> Dict[str, Any]:
    # Format context for LLM consumption
    entities_text = self.format_entities_for_
    llm(context['entities'])
    paths_text = self.format_reasoning_paths_for_
    llm(context['reasoning_paths'])
    docs_text = self.format_documents_for_llm(context['documents'])
```

```
# Generate structured response using specialized prompts
main_response = self.qa_prompt.invoke({
    "query": query,
    "entities": entities_text,
    "reasoning_paths": paths_text,
    "documents": docs_text
})

return {
    'main_response': main_response,
    'explanation': self._generate_reasoning_
    explanation(context),
    'metadata': self._compile_response_metadata(context)
}
```

The response generation process maintains explicit connections between generated content and source graph elements, enabling users to understand the reasoning behind system responses and verify information accuracy.

Advanced Integration Patterns

Complete Graph RAG systems benefit from several advanced integration patterns that enhance performance and user experience:

Adaptive Retrieval Strategies

The system implements adaptive retrieval that adjusts search parameters based on initial results quality. When initial queries return insufficient context, the system automatically expands search scope or adjusts hybrid scoring weights to improve result comprehensiveness.

Multi-perspective Analysis

For complex queries requiring nuanced understanding, the system can analyze questions from multiple perspectives by varying graph traversal parameters. This approach reveals different aspects of the knowledge space, providing users with more comprehensive insights.

Contextual Caching

Intelligent caching mechanisms store both query results and intermediate graph computations. The system recognizes semantically similar queries and reuses relevant computations while maintaining result freshness through cache invalidation strategies.

Performance Optimization and Scalability

Production Graph RAG systems require careful attention to performance characteristics and scalability constraints:

Query Performance Optimization

Index Strategy: The system maintains coordinated indexes across graph structure, entity embeddings, and document content. Pre-computed traversal patterns for common query types significantly improve response times.

Batch Processing: When handling multiple related queries, the system optimizes by batching graph operations and reusing semantic computations across queries.

Resource Management: Memory-efficient subgraph extraction and streaming response generation enable handling of large knowledge graphs without resource exhaustion.

Scalability Architecture

Distributed Graph Storage: Large knowledge graphs benefit from distributed storage strategies that partition entities and relationships across multiple nodes while maintaining query performance.

Caching Hierarchies: Multilevel caching from frequently accessed entities to computed reasoning paths reduces algorithm latency for common query patterns.

Load Balancing: Query processing distributes across multiple system instances, with intelligent routing based on query complexity and current system load.

System Evaluation and Quality Assurance

Complete Graph RAG systems require comprehensive evaluation frameworks that assess both technical performance and response quality:

Automated Evaluation Metrics

Entity Precision: Measures the accuracy of entity retrieval for queries with known correct answers. High-performing systems achieve 85–90% precision for well-connected entities.

Relationship Recall: Evaluates the system's ability to discover relevant relationship paths. Effective systems demonstrate 75–85% recall for multi-hop reasoning scenarios.

Response Coherence: Automated analysis of generated responses for logical consistency, factual accuracy, and appropriate use of retrieved context.

Continuous Quality Monitoring

Query Performance Tracking: Real-time monitoring of response times, retrieval quality, and user satisfaction enables proactive system optimization.

Graph Quality Assessment: Regular analysis of knowledge graph connectivity, entity coverage, and relationship accuracy ensures maintained system effectiveness.

Feedback Integration: User feedback collection and analysis provides insights for system improvement and identifies areas requiring enhanced coverage or accuracy.

Production Deployment Considerations

Successful deployment of Graph RAG systems requires attention to several operational aspects:

System Reliability

Error Handling: Robust error handling ensures graceful degradation when components fail, with fallback strategies that maintain basic functionality during system maintenance or unexpected issues.

Data Consistency: Transaction management and consistency checks prevent graph corruption during concurrent updates and ensure reliable system behavior.

Monitoring and Alerting: Comprehensive monitoring of system health, performance metrics, and error rates enables proactive maintenance and rapid issue resolution.

Integration and Maintenance

API Design: Well-designed APIs facilitate integration with existing systems while providing clear interfaces for query processing, batch operations, and system administration.

Update Mechanisms: Systematic approaches to knowledge graph updates, model retraining, and system configuration changes minimize disruption to ongoing operations.

Documentation and Training: Complete documentation and user training ensure effective system adoption and ongoing maintenance by operations teams.

The complete Graph RAG system represents a significant advancement in information retrieval and generation capabilities, combining the structural intelligence of knowledge graphs with the flexibility of modern language models. When properly implemented and maintained, these systems provide users with unprecedented access to complex, interconnected information while maintaining the explainability and accuracy required for critical applications.

The companion Notebook 12.5 provides a complete, working implementation of all concepts discussed in this section, demonstrating the integration of graph construction, retrieval, and response generation into a production-ready system capable of handling diverse query types and scaling to enterprise requirements.

Advanced Graph RAG Patterns

Having established the core components of Graph RAG systems, we now explore sophisticated patterns that enable complex reasoning, dynamic knowledge management, and production-grade reliability. These advanced patterns transform Graph RAG from a retrieval system into an intelligent reasoning platform capable of handling the nuanced information needs of enterprise applications.

Multi-hop Reasoning

Multi-hop reasoning represents the pinnacle of Graph RAG capabilities, enabling systems to trace complex chains of relationships across multiple connection points to answer sophisticated questions that require connecting disparate pieces of information. Unlike traditional RAG systems that retrieve isolated chunks, multi-hop reasoning leverages the graph structure to discover non-obvious connections and build a comprehensive understanding through relationship traversal.

Complex Question Answering Across Relationships

Consider the question "How did Geoffrey Hinton's early work on neural networks influence the development of modern language models?" This query requires tracing a multi-decade conceptual evolution: Hinton's foundational work on backpropagation ➤ deep learning breakthroughs ➤ attention mechanisms ➤ transformer architectures ➤ large language models. Each step represents a conceptual leap that builds upon previous innovations, connected through a web of research collaborations, citations, and technological dependencies.

The advanced multi-hop reasoning system implements several sophisticated capabilities:

Temporal-Aware Traversal: The system considers temporal relationships when constructing reasoning paths, ensuring that causal chains respect chronological order. A path from 1980s neural network research to 2020s language models must traverse intermediate developments in logical temporal sequence, weighted by the confidence of each transitional relationship.

Confidence Propagation: As reasoning paths extend across multiple hops, confidence scores propagate through the chain using multiplicative decay. A three-hop path with individual confidences of 0.9, 0.8, and 0.7

yields an overall path confidence of 0.504, ensuring that longer reasoning chains are appropriately weighted against shorter, more direct connections.

Path Diversity: Rather than finding a single optimal path, the system discovers multiple diverse reasoning routes, each highlighting different aspects of the connection. For the Hinton-to-LLMs question, one path might emphasize theoretical contributions through deep learning research, while another traces practical implementations through specific model architectures.

The implementation employs a sophisticated search strategy that combines best-first search with diversity sampling:

```
def find_complex_reasoning_paths(self, start_entity: str, end_
entity: str, max_hops: int = 4, path_diversity: int = 5):
    # Use priority queue for best-first search with confidence
    weighting
    # Classify paths by type: temporal_evolution, hierarchical,
    causal, attribution
    # Apply diversity filters to ensure varied reasoning
    perspectives
    # Return ranked paths with confidence scores and
    explanatory metadata
```

Reasoning Over Indirect Connections

Advanced Graph RAG systems excel at discovering relationships that are not explicitly stated but can be inferred through graph structure. These indirect connections often reveal the most valuable insights, connecting concepts that appear unrelated at surface level but share deep structural relationships.

Analogical Reasoning: The system identifies similar relationship patterns across different domains. If the graph contains patterns like "Researcher A ➤ pioneered ➤ Concept X ➤ enabled ➤ Application Y," it can recognize similar patterns in other domains and suggest analogical connections.

Conceptual Bridging: Multi-hop reasoning can identify intermediate concepts that bridge seemingly disparate domains. For example, connecting "quantum computing" to "drug discovery" through intermediate concepts like "optimization algorithms" and "molecular simulation."

Influence Propagation: The system traces how ideas, methods, or innovations propagate through networks of researchers, institutions, and application domains, revealing influence patterns that might not be immediately obvious from direct citations alone.

Dynamic Graph Updates

Production Graph RAG systems must handle continuously evolving knowledge landscapes. New research emerges daily, existing understanding evolves, and occasionally, established facts require revision. Dynamic graph update mechanisms ensure that Graph RAG systems remain current and accurate while maintaining consistency and reliability.

Incremental Knowledge Graph Maintenance

Traditional approaches to knowledge graph updates often require complete reconstruction, making them impractical for production systems with large knowledge bases. Advanced Graph RAG systems implement incremental update mechanisms that efficiently incorporate new information while preserving existing structure and relationships.

Entity Evolution Tracking: As new documents are processed, the system must determine whether extracted entities represent new knowledge or updates to existing entities. This requires sophisticated entity resolution that goes beyond simple text matching to consider contextual similarity, relationship patterns, and semantic embeddings.

Relationship Validation: New relationships must be validated against existing graph structure to identify potential conflicts, redundancies, or inconsistencies. The system maintains relationship confidence scores and tracks the provenance of each connection, enabling intelligent conflict resolution.

Version Management: The system maintains versioned snapshots of graph states, enabling rollback capabilities and temporal analysis of knowledge evolution. This versioning system supports both automated updates and human review processes.

The incremental update pipeline processes new information through several stages:

1. **Entity Linking**: New entities are matched against existing canonical representations using hybrid similarity measures that combine textual, semantic, and structural features.

2. **Conflict Detection**: The system identifies potential conflicts between new and existing information, categorizing conflicts by type (value mismatches, temporal inconsistencies, confidence disagreements).

3. **Resolution Strategy Application**: Conflicts are resolved using configurable strategies that may prioritize recent information, high-confidence sources, or require human intervention for critical updates.

4. **Graph Integration**: Validated updates are integrated into the graph structure with appropriate metadata tracking to support future analysis and potential rollback.

Handling Conflicting Information

Real-world knowledge graphs inevitably encounter conflicting information as new sources are integrated. Advanced Graph RAG systems implement sophisticated conflict resolution mechanisms that maintain graph consistency while preserving nuanced understanding of disputed or evolving knowledge.

Confidence-Based Resolution: The system automatically resolves conflicts by favoring information from higher-confidence sources. This approach works well for factual disputes where clear authority can be established, such as updated performance metrics or corrected biographical information.

Temporal Resolution: For time-sensitive information, the system prioritizes more recent data while maintaining historical context. This enables tracking of evolving situations while preserving the historical record of how understanding has changed over time.

Source Authority Weighting: Different information sources carry different levels of authority within specific domains. Academic papers carry high weight for research findings, while official documentation may be authoritative for technical specifications. The system learns and applies these authority relationships automatically.

Consensus Building: For highly contested information, the system may maintain multiple perspectives simultaneously, tracking the support level for each viewpoint and presenting balanced representations that acknowledge uncertainty.

The conflict resolution framework operates through a multistage process:

```
def resolve_conflicts(self, existing_info: Dict, new_info:
Dict, conflicts: List[Dict]):
    resolution_strategy = self.config.get('conflict_resolution_
    strategy', 'confidence_based')

    for conflict in conflicts:
        if conflict['type'] == 'confidence_mismatch':
            resolution = self._resolve_by_confidence(conflict,
            existing_info, new_info)
        elif conflict['type'] == 'temporal_mismatch':
            resolution = self._resolve_by_recency(conflict,
            existing_info, new_info)
        # Apply resolution and track decision rationale
```

This sophisticated conflict management ensures that Graph RAG systems remain reliable and accurate even as they incorporate diverse and potentially contradictory information sources.

Evaluation and Optimization

The complexity of Graph RAG systems necessitates comprehensive evaluation frameworks and sophisticated optimization strategies. Unlike traditional RAG systems where performance can be measured primarily through retrieval accuracy, Graph RAG systems require multidimensional evaluation that considers reasoning quality, graph structure health, and system scalability.

Performance Metrics for Graph RAG Systems

Evaluating Graph RAG systems requires a comprehensive framework that assesses multiple aspects of system performance, from low-level technical metrics to high-level reasoning capabilities. This multifaceted evaluation approach ensures that optimizations improve overall system effectiveness rather than optimizing individual components at the expense of system-wide performance.

Entity Retrieval Accuracy

The foundation of Graph RAG performance lies in accurate entity retrieval. However, unlike traditional information retrieval, Graph RAG entity retrieval must be evaluated not just for relevance, but for completeness and relationship awareness.

Precision and Recall: Traditional precision and recall metrics apply to entity retrieval but must be adapted to consider the graph context. High precision ensures that retrieved entities are genuinely relevant to the query, while high recall ensures that important related entities are not missed.

Entity Coverage: This metric measures how comprehensively the system identifies entities mentioned or implied in queries. A query about "transformer architectures" should retrieve not only the Transformer concept but also related entities like attention mechanisms, encoder–decoder structures, and key researchers.

Relationship-Aware Retrieval: Entities should be evaluated not just for individual relevance, but for their potential to contribute to reasoning paths. An entity with moderate direct relevance but high connectivity to other relevant entities may be more valuable than a highly relevant but isolated entity.

Multi-hop Reasoning Quality

The distinctive capability of Graph RAG systems—multi-hop reasoning—requires specialized evaluation metrics that assess both the accuracy of discovered paths and the quality of reasoning they enable.

Path Discovery Rate: The percentage of known valid reasoning paths that the system successfully identifies. This metric requires carefully constructed ground truth datasets with expert-validated reasoning paths.

Path Confidence Calibration: The alignment between system-assigned confidence scores and actual path reliability. Well-calibrated systems assign high confidence to reliable paths and low confidence to questionable connections.

Reasoning Diversity: The system's ability to discover multiple distinct reasoning approaches to the same question. Diversity prevents over-reliance on single reasoning chains and provides richer context for complex questions.

Temporal Consistency: For reasoning paths that span temporal relationships, the system must maintain chronological consistency, ensuring that cause-and-effect relationships respect temporal ordering.

Response Generation Quality

The final output of Graph RAG systems—natural language responses—must be evaluated for accuracy, completeness, and coherence. This evaluation combines traditional natural language generation metrics with graph-specific assessments.

Factual Accuracy: Responses must accurately reflect the information contained in the knowledge graph without introducing hallucinations or misrepresentations. This requires careful alignment between graph context and generated text.

Completeness: Responses should comprehensively address the query using available graph information. Incomplete responses that miss important connections or context represent system failures.

Explainability: Graph RAG systems should provide clear reasoning traces that connect their responses to specific graph paths and entities. This explainability enables users to verify and understand system reasoning.

Coherence and Flow: Generated responses must integrate information from multiple graph sources into coherent, well-structured narratives that flow logically from premise to conclusion.

Optimization Strategies for Large Graphs

As Graph RAG systems scale to enterprise datasets with millions of entities and relationships, optimization becomes critical for maintaining performance and usability. These optimization strategies address computational complexity, memory efficiency, and query response times while preserving the reasoning capabilities that make Graph RAG valuable.

Graph Structure Optimization

The structure of the knowledge graph directly impacts system performance. Well-structured graphs enable efficient traversal and reasoning, while poorly structured graphs can create performance bottlenecks and reduce reasoning quality.

Entity Consolidation: Large graphs often contain duplicate or near-duplicate entities that create unnecessary complexity. Advanced consolidation algorithms identify and merge similar entities while preserving distinct relationships and contexts. This consolidation reduces graph size and improves reasoning coherence.

Relationship Pruning: Not all extracted relationships contribute equally to reasoning quality. Low-confidence relationships, redundant connections, and relationships that don't participate in meaningful reasoning paths can be pruned to improve graph efficiency without sacrificing capability.

Hub Node Management: Graphs often develop "hub" nodes—entities with exceptionally high connectivity. While these hubs can be valuable for reasoning, they can also create performance bottlenecks. Specialized indexing and caching strategies for hub nodes maintain performance while preserving their reasoning value.

Hierarchical Organization: Large graphs benefit from hierarchical organization that groups related entities and enables efficient multilevel reasoning. This organization supports both detailed analysis within domains and high-level reasoning across domains.

Query Performance Optimization

Graph traversal operations, especially multi-hop reasoning, can become computationally expensive with large graphs. Several optimization strategies address these performance challenges while maintaining reasoning quality.

Intelligent Caching: Frequently accessed reasoning paths, entity neighborhoods, and query results can be cached to avoid repeated computation. The caching system must balance memory usage with hit rates, implementing intelligent eviction policies that retain the most valuable cached information.

Index Optimization: Strategic indexing of graph structures accelerates common query patterns. Entity type indexes, relationship type indexes, and composite indexes on frequently queried combinations dramatically improve query performance.

Parallel Processing: Multi-hop reasoning operations can often be parallelized, exploring multiple reasoning paths simultaneously. This parallelization requires careful coordination to avoid redundant work while ensuring comprehensive path discovery.

Approximate Reasoning: For very large graphs or time-sensitive queries, approximate reasoning techniques can provide good results with significantly reduced computation. These techniques trade small amounts of accuracy for substantial performance improvements.

Memory Management Strategies

Large knowledge graphs can consume substantial memory, especially when supporting real-time reasoning operations. Effective memory management ensures that systems remain responsive and scalable.

Lazy Loading: Rather than loading entire graphs into memory, lazy loading strategies fetch graph components on demand. This approach reduces memory footprint while maintaining access to the full graph when needed.

Graph Partitioning: Very large graphs can be partitioned across multiple storage systems or compute nodes. Effective partitioning strategies minimize cross-partition queries while maintaining reasoning capabilities.

Compression Techniques: Graph data often contains redundant information that can be compressed without losing essential structure. Entity property compression, relationship pattern compression, and graph structure compression can significantly reduce memory requirements.

Memory-Efficient Data Structures: Specialized data structures optimized for graph operations can reduce memory usage while improving access performance. These structures balance memory efficiency with query performance requirements.

Debugging and Troubleshooting Common Issues

Graph RAG systems, with their complex interactions between knowledge graphs, reasoning algorithms, and language models, can exhibit subtle failure modes that are challenging to diagnose and resolve. Systematic debugging approaches and comprehensive troubleshooting frameworks are essential for maintaining reliable production systems.

Systematic Diagnosis Framework

Effective Graph RAG debugging requires a structured approach that systematically evaluates different system components and their interactions. This framework enables rapid identification of root causes and targeted resolution strategies.

Graph Integrity Analysis: The first step in debugging involves verifying the integrity of the underlying knowledge graph. Common issues include orphaned relationships (pointing to non-existent entities), circular dependencies, missing required properties, and inconsistent relationship directions. Automated integrity checks can identify these structural issues before they impact reasoning quality.

Reasoning Path Validation: Multi-hop reasoning failures often stem from incomplete or invalid reasoning paths. Debugging tools should trace reasoning path construction, identifying where path discovery fails and why valid paths might be missed. This analysis reveals issues with entity linking, relationship extraction, or reasoning algorithm configuration.

Query Analysis: Understanding how queries are processed and interpreted provides insight into retrieval failures. Query analysis tools should examine entity extraction from queries, similarity matching processes, and the translation from natural language queries to graph operations.

Performance Profiling: System performance issues can manifest as slow response times, high memory usage, or computational bottlenecks. Comprehensive performance profiling identifies which operations consume the most resources and where optimization efforts should focus.

Common Failure Patterns and Solutions

Experience with production Graph RAG systems reveals recurring failure patterns that can be systematically addressed through targeted interventions.

"No Paths Found" Issues: When multi-hop reasoning fails to find connections between entities, the root cause often lies in incomplete relationship extraction or overly restrictive confidence thresholds. Solutions include reviewing entity linking accuracy, expanding relationship extraction recall, and adjusting reasoning parameters.

Low-Quality Reasoning Paths: Systems may find reasoning paths that are technically valid but conceptually weak or misleading. This issue typically stems from noisy relationship extraction or insufficient relationship validation. Solutions involve improving extraction quality, implementing relationship validation filters, and incorporating domain expertise in relationship weighting.

Performance Degradation: As graphs grow large, query performance can degrade significantly. Common causes include missing indexes, inefficient query patterns, and suboptimal graph structure. Solutions require systematic performance optimization including index creation, query rewriting, and graph structure refinement.

Inconsistent Results: Graph RAG systems may provide different answers to semantically similar queries, indicating issues with entity resolution, query understanding, or reasoning consistency. Solutions involve improving entity linking algorithms, enhancing query analysis, and implementing consistency validation checks.

Memory Leaks and Resource Issues: Long-running Graph RAG systems may develop memory leaks or resource consumption issues. These problems often stem from inefficient caching, improper resource cleanup, or memory-intensive operations that don't release resources properly. Solutions require careful resource management, monitoring systems, and periodic cleanup processes.

Monitoring and Maintenance Strategies

Production Graph RAG systems require ongoing monitoring and maintenance to ensure consistent performance and reliability. Comprehensive monitoring strategies provide early warning of developing issues and guide proactive maintenance activities.

Real-Time Performance Monitoring: Continuous monitoring of key performance indicators—query response times, reasoning path discovery rates, system resource utilization—enables rapid identification of performance degradation. Automated alerting systems can notify administrators when performance metrics exceed acceptable thresholds.

Quality Assurance Metrics: Regular evaluation of reasoning quality, response accuracy, and user satisfaction provides insight into system effectiveness over time. These metrics help identify gradual quality degradation that is not apparent from performance metrics alone.

Graph Health Assessment: Periodic analysis of knowledge graph structure, connectivity, and content quality ensures that the graph remains a reliable foundation for reasoning. Health assessments can identify structural issues, content gaps, and opportunities for improvement.

User Feedback Integration: User feedback provides valuable insights into system performance from the end user perspective. Systematic collection and analysis of user feedback reveals usability issues, accuracy problems, and opportunities for enhancement that might not be apparent from technical metrics.

The comprehensive evaluation and optimization framework presented in this section enables Graph RAG systems to achieve and maintain production-grade performance while scaling to meet enterprise requirements. Through systematic evaluation, targeted optimization, and proactive maintenance, Graph RAG systems can deliver the sophisticated reasoning capabilities that distinguish them from traditional retrieval systems while maintaining the reliability and performance required for critical applications.

The companion Notebook 12.6–12.7 demonstrates all techniques discussed in these sections, providing working implementations of advanced reasoning patterns, dynamic update mechanisms, comprehensive evaluation frameworks, and optimization strategies. These implementations serve as both practical tools and educational resources for developing production-ready Graph RAG systems.

Summary and Next Steps

Traditional RAG systems retrieve documents. Graph RAG systems discover insights through relationships. That difference changes everything.

What You've Built

Throughout this chapter, you've constructed a complete Graph RAG system that can

- **Extract entities and relationships** from documents automatically

- **Build knowledge graphs** that capture how information connects

- **Perform multi-hop reasoning** across relationship chains

- **Handle conflicting information** and dynamic updates
- **Generate explainable answers** with clear reasoning paths

This isn't just a technical achievement, it's a fundamentally different approach to information systems.

Key Takeaways

When Graph RAG Wins

Use Graph RAG when

- Relationships matter as much as facts
- You need to connect ideas across domains
- Users want to understand "why" and "how"
- Expertise discovery is valuable
- Multi-step reasoning is required

Stick with traditional RAG when

- Simple factual lookup is sufficient
- Implementation resources are limited
- Performance matters more than reasoning depth

Success Principles

Quality over quantity: Better to have fewer, high-quality relationships than massive, noisy graphs.

Design for change: Knowledge evolves. Build systems that can handle updates and conflicts gracefully.

Start focused: Prove value in one domain before expanding scope.

Measure impact: Track user success, not just technical metrics.

The Real Impact

Graph RAG systems don't just find information—they reveal connections you didn't know existed. They answer questions like "How did X influence Y?" and "What connects these seemingly unrelated concepts?" These questions drive innovation, solve complex problems, and generate genuine insights.

Next Steps

You now have the tools to build systems that think about information the way humans do: through connections, relationships, and context. That's the real power of Graph RAG.

The connections are there, waiting to be discovered. Go build something that matters.

## CHAPTER 13

Agentic RAG: Autonomous Information Systems

The best way to find out if you can trust somebody is to trust them.

—Ernest Hemingway

In the realm of artificial intelligence, trust emerges not just from reliability but also from systems that can reason, plan, and act autonomously while remaining accountable for their decisions. Agentic RAG represents this evolution—from passive information retrieval to active, intelligent collaboration.

Throughout this book, we've explored how Retrieval-Augmented Generation transforms static language models into dynamic, knowledge-aware systems. From basic document retrieval to sophisticated graph-based approaches, each chapter has built upon a fundamental paradigm: **reactive information retrieval**. When a user asks a question, the system retrieves relevant information and generates a response. This approach works exceptionally well for many applications, but it represents the beginning of what's possible.

R. Bose, *Mastering Retrieval-Augmented Generation*,
https://doi.org/10.1007/979-8-8688-1808-0_13

Agentic RAG represents a paradigm shift from reactive retrieval to **autonomous reasoning**. Instead of simply responding to queries, agentic systems can plan multistep investigations, use external tools, maintain context across extended interactions, and even anticipate information needs before they're explicitly stated. These systems don't just retrieve and generate—they reason, plan, and act.

From Retrieval to Autonomous Reasoning

Consider the difference between these two interactions:

Traditional RAG:

- User: "What are the latest developments in quantum computing?"

- System: Retrieves recent documents about quantum computing, generates a summary

Agentic RAG:

- User: "What are the latest developments in quantum computing?"

- System:

    1. Plans an investigation strategy (recent papers, industry news, patent filings)

    2. Searches multiple sources in parallel

    3. Identifies key trends and breakthrough announcements

    4. Cross-references claims across sources

    5. Generates comprehensive analysis with confidence indicators

    6. Offers to set up monitoring for future developments

The agentic system doesn't just answer the question, it demonstrates understanding of what constitutes a thorough investigation and autonomously executes that investigation.

Key Differences: Planning, Tool Use, Memory, and Reflection

Agentic RAG systems are distinguished by four core capabilities that traditional RAG lacks:

Planning. Agentic systems break complex queries into actionable steps. Rather than treating each query as an isolated retrieval task, they develop information gathering, analysis, and synthesis strategies. This planning capability enables them to handle ambiguous queries, pursue multiple lines of investigation, and adapt their approach based on what they discover.

Tool Use. While traditional RAG is limited to document retrieval, agentic systems can dynamically select and orchestrate multiple tools: web APIs, databases, calculators, code execution environments, and specialized analysis services. They understand when to use each tool and how to combine results from different sources.

Memory. Agentic systems maintain both short-term working memory for complex reasoning tasks and long-term memory for learning from interactions. This enables them to maintain context across extended conversations, learn from previous investigations, and build upon accumulated knowledge.

Reflection. Perhaps most importantly, agentic systems can evaluate their performance, identify gaps in their reasoning, and adjust their approach accordingly. They can recognize when their initial plan isn't working and pivot to alternative strategies.

When Autonomy Adds Value vs. Complexity

The autonomous capabilities of agentic RAG offer significant benefits but also increased complexity. Understanding when to employ agentic approaches is crucial for successful implementation.

Autonomy Adds Value When

- **Complex, Multistep Queries Require Investigation**: Research tasks that involve gathering information from multiple sources, cross-referencing claims, and synthesizing findings benefit enormously from autonomous planning and execution.

- **Context Spans Extended Interactions**: Customer service scenarios, educational tutoring, and collaborative research sessions require systems that maintain context and build upon previous exchanges.

- **Real-Time Information Integration Is Critical**: Financial analysis, competitive intelligence, and regulatory compliance scenarios often require combining static knowledge with real-time data from multiple APIs and services.

- **Users Need Proactive Assistance**: Rather than waiting for specific queries, agentic systems can monitor information sources, identify relevant updates, and proactively alert users to significant developments.

Complexity May Outweigh Benefits When

- **Simple, Direct Queries Dominate**: If users primarily ask straightforward factual questions that can be answered from a single source, traditional RAG may be more efficient and cost-effective.

- **Deterministic Responses Are Required**: High-stakes scenarios requiring predictable, auditable responses may be better served by traditional RAG with human oversight rather than autonomous decision-making.

- **Resource Constraints Are Tight**: Agentic systems consume more computational resources and API calls than traditional RAG. For applications with strict budget constraints, simpler approaches may be more appropriate.

- **Regulatory or Compliance Restrictions Apply**: Some industries have strict requirements about automated decision-making that may limit the appropriateness of fully autonomous systems.

The Architecture of Autonomy

Agentic RAG systems represent a convergence of several AI research areas: large language models, retrieval systems, planning algorithms, and multi-agent coordination. The resulting architecture is more complex than traditional RAG, but this complexity enables capabilities that were previously impossible.

At its core, an agentic RAG system consists of

- **A reasoning engine** that can plan multistep investigations and adapt strategies based on findings

- **A tool ecosystem** that provides access to external information sources and services

- **A memory system** that maintains context and learns from interactions

- **A coordination layer** that manages multiple concurrent tasks and agents

- **A reflection mechanism** that evaluates performance and guides improvement

These components work together to create systems that don't just retrieve information—they actively investigate, reason, and solve problems.

As we explore agentic RAG throughout this chapter, we'll see how these autonomous capabilities transform not just what RAG systems can do, but how we think about the relationship between humans and AI systems. Rather than tools that respond to our queries, agentic RAG systems become collaborative partners in knowledge work and problem-solving.

The journey from reactive retrieval to autonomous reasoning represents one of the most significant developments in applied AI. Let's explore how to harness this power responsibly and effectively.

Traditional RAG follows a simple linear pipeline from query to response, while Agentic RAG employs autonomous reasoning with planning, memory, multiple tools, and reflection capabilities. The feedback loops (dotted lines) enable continuous learning and strategy refinement.

As Figure 13-1 illustrates, the architectural complexity of agentic systems enables capabilities that were previously impossible, but this complexity must be managed thoughtfully to deliver real value.

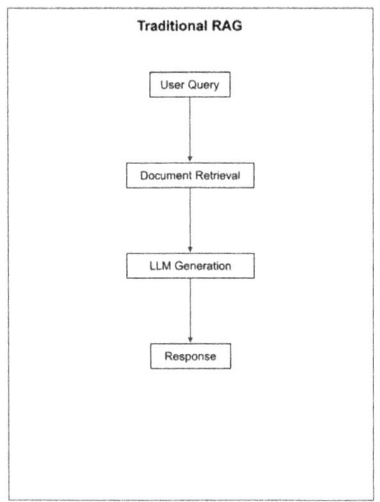 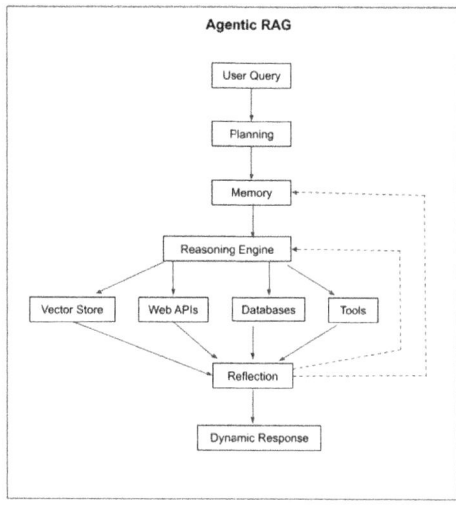

Figure 13-1. Traditional RAG vs. Agentic RAG architecture

Agent Architecture Fundamentals

Building effective agentic RAG systems requires understanding the core architectural components that enable autonomous reasoning. While traditional RAG systems can be implemented with simple pipelines, agentic systems demand sophisticated orchestration of multiple interacting components. This section explores the fundamental building blocks that transform reactive retrieval systems into autonomous reasoning agents.

Core Components: The Foundation of Autonomy

Every agentic RAG system, regardless of its specific implementation, relies on four essential components that work in concert to enable autonomous behavior:

The Reasoning Engine is the central decision-making component, implementing patterns like ReAct (Reasoning and Acting) to break down complex queries into actionable steps. Unlike traditional systems that

617

follow predetermined paths, the reasoning engine dynamically determines what actions to take based on the current context, available tools, and intermediate results. This component maintains an internal dialogue, explicitly reasoning about what it knows, what it needs to learn, and what actions to take next.

The Memory System provides short-term working memory for maintaining context during multistep reasoning and long-term memory for learning from past interactions. Working memory enables agents to track progress through complex investigations, remember intermediate findings, and maintain coherent reasoning chains. Long-term memory allows agents to learn from successful strategies, avoid repeating mistakes, and build domain expertise over time.

The Tool Ecosystem extends the agent's capabilities beyond simple text generation to include interaction with external systems, APIs, databases, and specialized services. Tools are not just passive resources but active components that can be dynamically selected, combined, and orchestrated based on the agent's reasoning about the current task requirements.

The Planning and Reflection Layer enables agents to think before acting and evaluate their performance after acting. Planning involves breaking down complex queries into manageable sub-tasks and developing investigation strategies. Reflection allows agents to assess the quality of their outputs, identify gaps in their reasoning, and adjust their approaches accordingly.

Design Patterns: Proven Architectures for Autonomous Reasoning

Successful agentic RAG systems typically implement one of several well-established design patterns, each optimized for different types of tasks and operational requirements.

ReAct (Reasoning and Acting) Pattern

The ReAct pattern alternates between reasoning steps, where the agent thinks about what to do next, and acting steps, where the agent uses tools to gather information or perform actions. This pattern creates an explicit trace of the agent's decision-making process, making the system's behavior interpretable and debuggable.

In a typical ReAct cycle, the agent receives a query and begins with a reasoning step: "I need to find information about quantum computing developments. Let me start by searching for recent research papers." It then takes an action: using a search tool to find relevant papers. The results trigger another reasoning step: "I found several papers about quantum error correction. Now I should look for industry developments to get a complete picture." This continues until the agent determines it has sufficient information to provide a comprehensive response.

The ReAct pattern is implemented through carefully structured prompts that guide the agent's reasoning process:

```
react_prompt = PromptTemplate.from_template("""
You are an intelligent research assistant with access to
multiple tools.

Use the following format:
Question: the input question you must answer
Thought: you should always think about what to do
Action: the action to take, should be one of [{tool_names}]
Action Input: the input to the action
Observation: the result of the action
... (this Thought/Action/Action Input/Observation can repeat
N times)
Thought: I now know the final answer
Final Answer: the final answer to the original input question
```

```
Question: {input}
Thought: {agent_scratchpad}
""")
```

This prompt structure creates an explicit reasoning trace that makes the agent's decision-making process transparent and debuggable. The {agent_scratchpad} placeholder maintains the conversation history, enabling the agent to build upon previous reasoning steps.

The ReAct pattern's strength lies in its transparency and controllability. Each reasoning step can be monitored, and the agent's decision-making process can be interrupted or redirected if necessary. This makes ReAct particularly suitable for applications where explainability and human oversight are important.

Chain-of-Thought with Tool Use

This pattern extends the chain-of-thought reasoning approach by incorporating external tool usage at strategic points in the reasoning process. Rather than alternating strictly between reasoning and acting, the agent maintains a continuous reasoning chain while opportunistically using tools when additional information or capabilities are needed.

For example, when analyzing market trends, the agent might reason: "To understand the semiconductor market outlook, I need recent financial data, industry reports, and expert analysis. The financial data requires real-time stock prices and earnings reports, which I can get from financial APIs. Industry reports are likely available through web search, and expert analysis might be found in recent news articles or analyst reports." The agent then orchestrates multiple tools in parallel or sequence while maintaining its analytical reasoning thread.

Multi-agent Collaboration

Complex tasks often benefit from multiple specialized agents working together, each focused on specific aspects of the problem. In a multi-agent architecture, individual agents might specialize in research, analysis, fact-checking, or synthesis, with a coordinator agent managing the overall workflow.

Consider a competitive analysis task: a researcher agent might gather information about competitor products and strategies, an analyst agent might evaluate financial performance and market positioning, a fact-checker agent might verify claims and cross-reference sources, and a synthesizer agent might combine insights into a coherent report. The coordinator agent manages task delegation, ensures information flows between agents, and maintains overall project coherence.

Multi-agent systems excel at complex, multifaceted problems but require sophisticated coordination mechanisms to prevent conflicts, ensure consistency, and maintain efficiency.

Agent Communication and Coordination

Effective agent architectures require robust communication and coordination mechanisms, whether within a single agent's internal components or across multiple collaborating agents.

Message Passing and State Management

Agents communicate through structured message passing, where each message contains data and metadata about context, priority, and routing information. A well-designed message system enables agents to request information, delegate tasks, report results, and coordinate activities without tight coupling between components.

State management becomes critical in multistep reasoning scenarios. Agents must maintain consistent internal state while allowing for concurrent operations and potential rollbacks when reasoning paths prove unproductive. This requires careful design of state representation, transaction boundaries, and consistency mechanisms.

Conflict Resolution and Consensus Building

When multiple agents or reasoning paths produce conflicting results, sophisticated agentic systems implement conflict resolution mechanisms. These might include confidence-weighted voting, expert agent arbitration, or evidence-based reasoning to determine the most reliable conclusions.

For instance, if two agents report different financial figures for the same company, the system might check the recency of sources, cross-reference with authoritative databases, or weight results based on each agent's historical accuracy in financial reporting tasks.

Load Balancing and Resource Management

Production agentic systems must manage computational resources efficiently, balancing thoroughness with performance. This involves dynamic allocation of processing resources, intelligent caching of intermediate results, and graceful degradation when resources are constrained.

Resource management also extends to external APIs and services, implementing rate limiting, cost optimization, and fallback strategies when preferred services are unavailable or have exceeded usage quotas.

Implementation Considerations

Building robust agentic RAG systems requires careful attention to several practical considerations that distinguish production systems from research prototypes.

Error Handling and Graceful Degradation

Unlike traditional RAG systems with predictable failure modes, agentic systems can fail in complex ways due to their autonomous decision-making capabilities. Robust error handling must account for tool failures, reasoning dead-ends, conflicting information, and resource constraints.

Effective error handling strategies include maintaining multiple reasoning paths simultaneously, implementing rollback mechanisms when investigations prove unproductive, and providing partial results when complete analysis isn't possible within time or resource constraints.

Observability and Debugging

The autonomous nature of agentic systems makes observability crucial for understanding system behavior, diagnosing problems, and optimizing performance. Comprehensive logging must capture not only what actions were taken but why they were chosen, what alternatives were considered, and how decisions evolved throughout the reasoning process.

Debugging agentic systems requires tools that can visualize reasoning chains, replay decision sequences, and analyze the effectiveness of different reasoning strategies across various types of queries.

Security and Safety Boundaries

Autonomous agents with access to external tools and APIs require robust security boundaries to prevent misuse while maintaining functionality. This includes validating tool inputs, sandboxing execution environments, implementing approval workflows for sensitive operations, and maintaining audit trails for all external interactions.

Safety considerations extend beyond security to include preventing agents from pursuing reasoning paths that might be harmful, unethical, or simply wasteful of resources. This requires embedding value alignment and constraint satisfaction into the agent's reasoning processes.

The architectural foundations we've explored provide the building blocks for sophisticated agentic RAG systems. However, the true power of these systems emerges not from individual components but from their orchestrated interaction in service of autonomous reasoning goals. In the next section, we'll examine how planning and multistep reasoning capabilities enable agents to tackle complex, open-ended investigations that would be impossible with traditional RAG approaches.

The core agent components and ReAct pattern discussed in this section are implemented in the companion Notebook: Agent Fundamentals and Tool Integration.

Planning and Multistep Reasoning

The ability to break down complex problems into manageable steps and execute them coordinated distinguishes agentic RAG systems from their reactive counterparts. While traditional RAG systems respond immediately to queries with whatever information they can retrieve, agentic systems engage in deliberate planning, considering multiple approaches, identifying dependencies, and orchestrating resources to achieve comprehensive solutions.

Effective planning in agentic RAG systems requires more than simple task decomposition. It demands understanding of query types, resource constraints, information dependencies, and the dynamic nature of investigation processes. This section explores the sophisticated planning mechanisms that enable agents to tackle complex, multifaceted queries that would overwhelm simpler systems.

Query Decomposition into Actionable Steps

The foundation of effective planning lies in intelligent query decomposition—the process of analyzing complex queries and breaking them into specific, executable actions. This process goes far beyond simple keyword extraction or template matching, requiring deep understanding of information needs, investigation strategies, and the logical flow of inquiry.

Understanding Query Complexity and Intent

Not all queries require the same planning approach. A simple factual question like "What is the capital of France?" needs minimal decomposition, while a research query like "Analyze the potential impact of quantum computing on financial cryptography and recommend investment strategies" requires sophisticated multistep planning.

Effective query decomposition begins with query classification. Research queries typically require broad information gathering followed by analysis and synthesis. Comparative queries need parallel investigation of alternatives with systematic comparison. Analytical queries demand data collection, processing, and interpretation with clear reasoning chains.

Consider this complex query: "How do recent advances in renewable energy storage compare to traditional grid infrastructure investments, and what are the policy implications for developing countries?" The decomposition process must identify several distinct investigation threads: recent advances in energy storage technology, traditional grid infrastructure costs and benefits, comparative analysis frameworks, policy considerations for developing economies, and synthesis of investment recommendations.

Creating Structured Investigation Plans

Once a query's complexity and type are understood, the planning system creates a structured investigation plan with explicit steps, dependencies, and resource requirements. Each step in the plan specifies what information to gather, what tools or capabilities are needed, what prior knowledge is required, and what outcomes are expected.

A well-structured plan for the renewable energy query might include steps like: "Research recent battery technology breakthroughs and their cost trajectories," "Analyze traditional grid infrastructure investment patterns and ROI," "Identify policy frameworks in developing countries relevant to energy infrastructure," and "Synthesize findings into comparative investment recommendations."

Each step carries metadata about execution requirements: estimated time, required tools, information dependencies, and confidence thresholds. This metadata enables the planning system to optimize execution order, identify parallel execution opportunities, and prepare for potential failures or adaptation needs. The structured representation of planning steps captures this complexity:

```
@dataclass
class PlanningStep:
    id: str
    description: str
    step_type: str
    dependencies: List[str] = field(default_factory=list)
    tools_required: List[str] = field(default_factory=list)
    expected_output: str = ""
    priority: int = 1
    estimated_time: int = 30
    status: str = "pending"
```

This structure enables sophisticated orchestration where the planning system can track step completion, manage resource allocation, and adapt execution strategies based on real-time progress and results.

Dependency Management and Execution Ordering

Real-world investigations rarely follow linear paths. Information discovered in one step often influences subsequent steps, creating dynamic dependencies that must be managed throughout execution. Effective planning systems model these dependencies explicitly, creating execution graphs that ensure prerequisites are satisfied while maximizing parallel execution opportunities.

Consider dependencies in our renewable energy example: analyzing policy implications depends on understanding both technical capabilities and economic factors, but researching battery technology and grid infrastructure can proceed in parallel. The planning system must recognize these relationships and create execution orders that minimize wait times while ensuring logical coherence.

Dependency management becomes particularly complex when dealing with uncertain information needs. Initial investigations might reveal unexpected research directions, require additional verification steps, or uncover conflicting information that demands reconciliation. Robust planning systems anticipate this uncertainty, creating flexible plans that can adapt while maintaining overall coherence.

Sequential vs. Parallel Reasoning Workflows

The orchestration of multistep reasoning processes requires careful consideration of when to execute steps sequentially vs. in parallel. This decision significantly impacts both execution efficiency and result quality, with important trade-offs between speed, resource utilization, and logical coherence.

Sequential Reasoning Patterns

Sequential execution provides the clearest logical flow, with each step building directly upon previous results. This approach ensures maximum context availability and enables sophisticated reasoning chains where later steps can leverage detailed insights from earlier investigations.

Sequential reasoning excels in scenarios requiring deep logical development, where understanding must build incrementally. Legal reasoning, scientific hypothesis development, and financial analysis often benefit from sequential approaches where each step refines understanding and guides subsequent investigation directions.

However, sequential execution can be inefficient when steps are independent or when waiting for complete results creates unnecessary delays. A purely sequential approach to our renewable energy query would mean waiting for complete battery technology research before beginning grid infrastructure analysis, despite their essential independence.

Parallel Execution Strategies

Parallel execution dramatically improves efficiency when steps are independent or when partial results are sufficient to begin subsequent investigations. Modern agentic systems can orchestrate multiple investigation threads simultaneously, combining results as they become available.

Effective parallel execution requires careful resource management and result integration strategies. Multiple search processes, analytical computations, and reasoning threads must be coordinated without overwhelming system capabilities or external API limits. Results from parallel threads must be integrated coherently, handling potential conflicts or inconsistencies between different information sources.

The key insight is that many complex queries contain both sequential and parallel elements. Optimal execution strategies identify these patterns creating hybrid workflows that maximize both efficiency and logical coherence. Our renewable energy query might research battery technology and grid infrastructure in parallel, then sequentially analyze their comparative economics, and finally synthesize policy recommendations based on the comparative analysis.

Dynamic Workflow Adaptation

Real-world reasoning rarely follows predetermined paths exactly. Initial assumptions prove incorrect, information sources fail, or new insights redirect investigation priorities. Sophisticated planning systems monitor execution progress and adapt workflows dynamically, maintaining overall objectives while responding to changing circumstances.

Dynamic adaptation requires continuous assessment of plan progress, resource availability, and result quality. When parallel threads produce conflicting information, the system might pause other activities to resolve inconsistencies. When key information sources fail, alternative approaches must be identified and integrated into the ongoing workflow.

This adaptability distinguishes truly agentic systems from rigid automation. Rather than failing when predetermined plans encounter obstacles, agentic systems demonstrate intelligence by recognizing problems, developing alternative approaches, and maintaining progress toward overall objectives.

Handling Uncertainty and Planning Failures

Uncertainty is inherent in real-world information gathering and analysis. Sources may be unavailable, information may be incomplete or contradictory, and analytical processes may produce unclear results. Effective agentic systems must not only handle these uncertainties gracefully but leverage them to improve overall investigation quality.

Uncertainty Recognition and Quantification

Sophisticated planning systems explicitly model uncertainty throughout investigation processes. Each piece of information carries confidence indicators, sources are evaluated for reliability, and analytical conclusions include uncertainty bounds. This explicit uncertainty modeling enables better decision-making about when additional verification is needed and how to weight different information sources.

Uncertainty quantification goes beyond simple confidence scores. Different types of uncertainty require different handling strategies: information recency uncertainty might trigger searches for more current sources, source reliability uncertainty might prompt cross-verification, and analytical uncertainty might require alternative modeling approaches.

Graceful Degradation Strategies

When planning failures occur—tools become unavailable, sources provide inadequate information, or time constraints prevent complete investigation—robust systems implement graceful degradation rather than complete failure. Graceful degradation preserves partial results, identifies what information is missing, and provides qualified conclusions based on available evidence.

Effective degradation strategies maintain transparency about limitations. Rather than presenting incomplete analysis as comprehensive conclusions, systems clearly communicate what investigation was attempted, what information was successfully gathered, and what limitations affect the reliability of conclusions.

Failure Recovery and Alternative Approaches

Planning failures often reveal alternative investigation strategies that may prove more effective than original approaches. When direct web searches fail to find recent technical specifications, the system might pivot to searching patent databases or academic publications. When quantitative analysis proves impossible due to data limitations, qualitative comparative approaches might provide valuable insights.

Recovery strategies benefit from meta-reasoning about investigation approaches. Systems that understand why particular approaches failed can often identify related methods with higher success probabilities. This meta-reasoning capability transforms failures from dead-ends into learning opportunities that improve future planning effectiveness.

Evidence Synthesis Across Sources

The culmination of multistep reasoning lies in effective evidence synthesis—combining information from multiple sources, investigation threads, and analytical approaches into coherent, comprehensive conclusions. This synthesis process requires sophisticated reasoning about information quality, source reliability, logical consistency, and the strength of evidence for different conclusions.

Multisource Information Integration

Real-world investigations typically gather information from diverse sources with different perspectives, time frames, and reliability characteristics. Academic papers provide rigorous analysis but may lag current developments. News sources offer timely information but may lack analytical depth. Industry reports provide practical insights but may reflect particular biases. Government data offers authoritative statistics but may have limited scope.

Effective synthesis requires understanding these source characteristics and weighting information appropriately. Recent academic research on battery technology carries different implications than industry marketing materials about storage solutions. Government statistics about energy infrastructure investments require different interpretation than private sector forecasts.

The synthesis process must also handle information conflicts gracefully. When sources disagree about fundamental facts or provide contradictory analytical conclusions, sophisticated systems investigate these disagreements rather than simply averaging different perspectives. Understanding why sources disagree often provides crucial insights about underlying complexity or uncertainty.

Logical Consistency and Argument Construction

Beyond integrating diverse information sources, effective synthesis constructs logical arguments that support clear conclusions. This requires identifying causal relationships, recognizing logical dependencies, and constructing reasoning chains that connect evidence to conclusions transparently.

Argument construction in agentic systems goes beyond simple information compilation. The system must identify which pieces of evidence support which conclusions, recognize when evidence is insufficient for strong conclusions, and construct qualified statements that accurately reflect the strength of available evidence.

Quality Assessment and Confidence Reporting

Comprehensive evidence synthesis includes explicit assessment of conclusion quality and appropriate confidence reporting. Different types of evidence provide different levels of support for conclusions, and synthesis processes must communicate these distinctions clearly.

Quality assessment considers factors like source diversity, information recency, evidence consistency, and analytical rigor. Conclusions supported by multiple independent sources using different methodologies merit higher confidence than those based on single sources or methodologies. Recent information typically carries more weight for current decision-making, but historical patterns may provide crucial context for trend analysis.

The planning and multistep reasoning capabilities we've explored transform static knowledge systems into dynamic investigation partners. Rather than simply retrieving and presenting existing information, these systems actively investigate complex questions using sophisticated reasoning strategies. This transformation requires not just technical capabilities but also careful attention to reasoning quality, uncertainty management, and transparent communication of limitations and assumptions.

Planning strategies and multistep reasoning patterns are demonstrated in Notebook: Multi-Step Reasoning and Memory Management, which shows practical implementation of query decomposition, execution orchestration, and evidence synthesis.

Tool Integration and External APIs

The true power of agentic RAG systems emerges not from their reasoning capabilities alone, but from their ability to dynamically select, orchestrate, and combine external tools and services to accomplish complex tasks. While traditional RAG systems are limited to retrieving and generating from static knowledge bases, agentic systems can interact with live data sources, perform calculations, execute code, access specialized databases, and invoke domain-specific services as needed.

Tool integration in agentic systems goes far beyond simple API calls or predetermined workflows. It requires intelligent tool selection based on task requirements, dynamic parameter preparation, error handling and recovery, result interpretation and validation, and seamless integration of tool outputs into ongoing reasoning processes. This level of sophistication transforms agents from knowledge retrievers into capable digital assistants that can accomplish real work in the world.

Dynamic Tool Selection and Orchestration

The foundation of effective tool integration lies in the agent's ability to understand what tools are available, what each tool can accomplish, when to use specific tools, and how to combine multiple tools to achieve complex objectives. This requires sophisticated reasoning about tool capabilities, task requirements, and the current context of investigation or problem-solving.

Understanding Tool Capabilities and Constraints

Every tool has specific capabilities, limitations, input requirements, and output formats. A web search tool might provide broad information discovery but limited depth on specialized topics. A financial data API might offer precise numerical data but require specific symbol formats. A code execution environment might enable complex calculations but have security restrictions and timeout limitations.

Effective agentic systems maintain comprehensive models of tool capabilities that go beyond simple descriptions. These models capture input parameter requirements, expected response formats, typical response times, reliability characteristics, cost considerations, and usage constraints such as rate limits or authentication requirements.

Consider a scenario where an agent needs to analyze stock market trends. The system must choose between web search tools that provide news and general analysis, financial APIs that offer precise numerical data, calculation tools for statistical analysis, and visualization tools for presenting results. The optimal choice depends not just on data availability but on the specific analysis requirements, time constraints, and the intended audience for results.

Context-Aware Tool Selection

Tool selection in agentic systems must consider the current context of investigation, including previous tool results, available information, resource constraints, and user preferences. A tool that works well in isolation might be inappropriate when combined with specific other tools or when operating under particular constraints.

Context-aware selection also considers the broader investigation strategy. Early in an investigation, broad search tools might be appropriate for information discovery. Later stages might require more specialized tools for detailed analysis or verification. The agent's tool selection must align with its current position in the overall reasoning process.

The dynamic nature of tool selection means that agents must continuously reassess their choices as new information becomes available. Initial tool selections based on query analysis might prove suboptimal once investigation begins and specific information needs become clearer. Sophisticated agents adapt their tool usage patterns based on ongoing results and evolving understanding of task requirements.

Orchestration Patterns and Workflows

Real-world tasks often require multiple tools working in coordination rather than isolated tool usage. Financial analysis might combine market data retrieval, statistical calculation, news analysis, and report generation. Research tasks might orchestrate academic search, web search, document analysis, and synthesis tools. These orchestration patterns require careful planning and execution coordination.

Effective orchestration manages dependencies between tools, handles data format conversions, coordinates timing and sequencing, and integrates results from multiple sources. Some tools must execute sequentially due to data dependencies, while others can operate in parallel to improve efficiency. The orchestration system must balance these considerations while maintaining overall task coherence.

Orchestration also involves error propagation and recovery strategies. When one tool in a complex workflow fails, the system must determine whether to retry with different parameters, substitute alternative tools, or adapt the overall strategy to work around the failure. These decisions require understanding both the failed tool's role in the overall process and the availability of suitable alternatives.

Real-Time Data Integration

Modern information needs often require access to current data that changes rapidly and isn't available in static knowledge bases. Financial markets, news events, weather conditions, social media trends, and operational metrics all require real-time or near real-time data access. Agentic systems must seamlessly integrate these dynamic data sources with their reasoning processes.

Live Data Source Management

Real-time data integration requires sophisticated management of data freshness, reliability, and consistency. Different data sources update at different frequencies and may have varying delays between real-world events and data availability. Financial markets provide millisecond-level updates during trading hours but may have delays or gaps during off-hours. News sources update continuously but with varying verification standards and potential corrections.

Effective agentic systems model these temporal characteristics explicitly, understanding when data is current, when it might be stale, and how to handle timing-sensitive decisions. They maintain awareness of data source characteristics, including update frequencies, typical delays, reliability patterns, and historical accuracy.

This temporal awareness enables appropriate decision-making about when to refresh data, how to weight information based on recency, and when to caveat conclusions based on data currency limitations. An agent analyzing market conditions must understand the difference between real-time price feeds, delayed market data, and historical analysis—and communicate these distinctions clearly in its reasoning and conclusions.

API Authentication and Rate Limiting

Production agentic systems must handle the practical realities of external service integration, including authentication management, rate limiting compliance, and cost optimization. Many valuable data sources require API keys, subscription management, or usage-based billing. Agents must manage these requirements automatically while respecting service constraints and organizational policies.

Rate limiting presents particular challenges for agentic systems that might need to make multiple API calls rapidly during investigation processes. Sophisticated systems implement intelligent rate limiting strategies that balance investigation thoroughness with service constraints. This might involve batching requests, implementing exponential backoff strategies, or redistributing queries across multiple time periods.

Authentication management requires secure storage and rotation of credentials, handling of different authentication protocols, and graceful degradation when services become unavailable. Agents must continue operating effectively even when preferred data sources are temporarily inaccessible, falling back to alternative sources or approaches as needed.

Data Quality and Validation

Real-time data sources vary significantly in quality, accuracy, and reliability. Social media feeds provide immediate information but with limited verification. Government data sources offer high reliability but may have significant delays. Commercial data services balance speed and accuracy differently based on their target markets and quality processes.

Agentic systems must implement sophisticated data validation strategies that assess source reliability, cross-verify information across multiple sources, identify and handle contradictory information, and communicate uncertainty appropriately. This validation process must operate in real time without significantly slowing investigation processes.

Validation strategies include consistency checking across multiple sources, plausibility assessment based on historical patterns, outlier detection for numerical data, and temporal consistency verification. When data sources disagree, sophisticated systems investigate these disagreements rather than simply averaging conflicting information, often uncovering important insights about underlying complexity or rapidly changing conditions.

Error Handling and Fallback Strategies

The integration of multiple external tools and services creates numerous potential failure points that sophisticated agentic systems must handle gracefully. Network failures, service outages, rate limiting, authentication problems, and data quality issues can all disrupt investigation processes. Rather than treating these failures as terminal problems, effective systems implement comprehensive error handling and fallback strategies.

Failure Classification and Response Strategies

Different types of failures require different response strategies. Temporary network issues might be resolved through simple retry mechanisms, while service outages might require switching to alternative data sources. Authentication failures might need credential refresh or alternative authentication methods. Rate limiting might require spreading requests over time or switching to different service tiers.

Sophisticated error handling begins with failure classification—understanding what type of problem has occurred and what response strategies are most likely to be effective. Transient failures often resolve quickly with retry strategies, while persistent failures require more fundamental adaptations to investigation approaches.

The classification process must operate quickly to avoid blocking ongoing reasoning processes. Agents can't spend significant time diagnosing every failure when rapid adaptation is needed to maintain investigation momentum. Effective systems implement fast failure detection and classification mechanisms that enable prompt response strategy selection:

```python
def execute_tool(self, tool_name: str, tool_input: str, max_
retries: int = 2) -> Dict:
    execution_record = {
        "tool": tool_name, "success": False, "attempts": 0,
        "result": None, "error": None
    }

    for attempt in range(max_retries + 1):
        execution_record["attempts"] = attempt + 1
        try:
            result = self.tools[tool_name].func(tool_input)
            execution_record.update({"success": True, "result":
            result})
            break
        except Exception as e:
            execution_record["error"] = str(e)
            if attempt < max_retries:
                continue  # Implement retry with backoff
            execution_record["result"] = f"Tool execution
            failed after {max_retries + 1} attempts"

    return execution_record
```

This pattern demonstrates systematic error tracking, retry logic with attempt counting, and comprehensive result documentation that enables downstream processes to understand both success and failure scenarios.

Graceful Degradation Mechanisms

When preferred tools or data sources become unavailable, agentic systems must adapt their investigation strategies rather than failing completely. This requires maintaining awareness of alternative approaches, understanding the trade-offs between different tools and data sources, and communicating limitations clearly when degraded approaches are used.

Graceful degradation might involve switching from real-time data to historical data with appropriate caveats, using general web search when specialized databases are unavailable, or providing qualitative analysis when quantitative tools fail. The key is maintaining investigation progress while being transparent about limitations and their implications for result quality.

Effective degradation strategies also consider cumulative effects of multiple tool failures. When several preferred tools become unavailable simultaneously, the system must assess whether alternative approaches can still provide valuable results or whether the investigation should be postponed until better tools become available.

Alternative Tool Discovery and Substitution

Advanced agentic systems maintain awareness of alternative tools and services that can substitute for preferred options when failures occur. This requires understanding functional equivalencies between different tools, assessing quality trade-offs between alternatives, and adapting parameters and expectations when substitutions are made.

Tool substitution often requires more than simple replacement—different tools may have different input requirements, output formats, or capability limitations that require adaptation of the overall investigation approach. An agent that normally uses a specialized financial API might need to adapt its analysis approach when falling back to web scraping or general search tools.

The discovery and evaluation of alternative tools can happen both proactively—maintaining current awareness of available alternatives—and reactively—searching for alternatives when primary tools fail. Proactive approaches enable faster recovery from failures, while reactive approaches provide flexibility for handling unexpected failure modes.

Authentication and Rate Limiting

The practical deployment of agentic systems in production environments requires sophisticated handling of authentication, authorization, and usage constraints across multiple external services. These considerations often receive insufficient attention during development but become critical for reliable operation in real-world scenarios.

Credential Management and Security

Production agentic systems must manage credentials for multiple external services securely while enabling smooth operation across different deployment environments. This includes secure storage of API keys and authentication tokens, automatic credential rotation where supported, handling of different authentication protocols, and secure credential sharing across distributed system components.

Security considerations extend beyond simple credential storage to include audit logging of external service usage, detection of unusual usage patterns that might indicate compromise, and graceful handling of credential revocation or expiration. Systems must continue operating effectively even when some credentials become invalid, falling back to public APIs or alternative services as appropriate.

The credential management system must also handle different authentication patterns across various services—some requiring API keys in headers, others using OAuth flows, and still others implementing custom authentication schemes. Abstracting these differences enables consistent tool integration while accommodating service-specific requirements.

Intelligent Rate Limiting and Cost Management

Different external services have different rate limiting policies, cost structures, and usage constraints that sophisticated agentic systems must navigate intelligently. Some services allow burst usage with longer-term averages, others enforce strict per-minute limits, and still others implement complex tiered pricing with different capabilities at different usage levels.

Effective rate-limiting strategies distribute requests across time periods to maximize service utilization while respecting constraints. This might involve delaying non-urgent requests, batching multiple queries into single API calls where possible, or distributing requests across multiple service accounts or regions.

Cost management becomes particularly important for systems that might generate significant API usage during complex investigations. Agents must balance investigation thoroughness with cost constraints, potentially implementing budget-aware tool selection that favors less expensive alternatives when cost limits are approached.

Service-Level Agreement Compliance

Production usage of external services often involves service-level agreements that specify usage patterns, data handling requirements, and compliance obligations. Agentic systems must ensure their usage patterns comply with these agreements while maintaining investigative effectiveness.

Compliance considerations might include data residency requirements that limit which services can be used for processing certain types of information, usage pattern restrictions that limit automated access, and data retention policies that affect how long investigation results can be stored or reused.

The complexity of managing multiple SLA requirements across different services necessitates systematic approaches to compliance monitoring and enforcement. This includes automated compliance checking, usage pattern analysis, and proactive adaptation when usage approaches agreement limits.

Tool integration and external API management represent critical capabilities that distinguish production agentic systems from research prototypes. The sophisticated handling of dynamic tool selection, real-time data integration, error recovery, and operational constraints enables agents to operate reliably in complex real-world environments while maintaining the flexibility and adaptability that make them valuable for complex problem-solving tasks.

Tool integration patterns and external API handling are extensively covered in Notebook: Agent Fundamentals and Tool Integration, which demonstrates practical implementation of tool orchestration, error handling, and production deployment considerations.

Memory and Context Management

The ability to maintain context across extended interactions and learn from past experiences fundamentally distinguishes agentic RAG systems from stateless alternatives. While traditional systems treat each query as an isolated event, agentic systems accumulate knowledge, remember successful strategies, learn from failures, and build increasingly sophisticated understanding of user needs and domain complexities.

Memory in agentic systems serves multiple critical functions beyond simple conversation history. It enables continuity across sessions, supports learning and adaptation, provides context for decision-making, and enables personalization based on past interactions. However, effective memory management also presents significant challenges: balancing retention with performance, maintaining relevance while avoiding information overload, and ensuring privacy while enabling learning.

This section explores the sophisticated memory architectures that enable agentic systems to transcend simple question-answering and become truly collaborative partners in complex knowledge work.

Short-Term Working Memory for Conversations

Working memory in agentic systems functions similarly to human working memory—maintaining immediate context, tracking current goals and progress, and enabling coherent reasoning across multiple steps within a single session or investigation. This temporary storage system handles the dynamic information needed for current tasks while avoiding the overhead of permanent storage for ephemeral details.

Context Window Management and Optimization

The fundamental constraint of working memory lies in context window limitations—the maximum amount of information that can be actively maintained during reasoning processes. Language models have finite context windows, external APIs have request size limits, and processing time increases with context length. Effective working memory management must operate within these constraints while maximizing information utility.

Context optimization requires intelligent selection of what information to maintain, what to summarize, and what to discard. Recent conversation turns typically deserve full retention, while older interactions might be preserved in summary form. Task-relevant information requires prioritization over general context, and high-confidence information warrants more space than uncertain details.

The optimization process must operate continuously as conversations progress, making real-time decisions about context management without disrupting ongoing reasoning. This requires efficient algorithms for information relevance assessment, compact representation techniques for complex information, and seamless integration with reasoning processes.

Multiturn Conversation Coherence

Maintaining coherence across extended conversations requires more than simple conversation history storage. Agents must track evolving user goals, remember previous explanations to avoid unnecessary repetition, maintain awareness of what information has been shared, and adapt their communication style based on demonstrated user preferences.

Coherence maintenance becomes particularly challenging in complex investigations where conversations might span multiple sessions, involve multiple related topics, and require integration of information gathered across different time periods. The working memory system must maintain logical connections between related concepts while avoiding confusion from irrelevant historical details.

Effective coherence strategies include maintaining explicit models of user knowledge and preferences, tracking the evolution of user goals and interests, and implementing conversation state management that preserves important context while discarding outdated information.

Task State Preservation and Recovery

Complex tasks often span multiple interaction turns, requiring sophisticated state management to maintain progress toward long-term goals. When users return to previous topics, request updates on ongoing analyses, or build upon previous investigations, the system must seamlessly restore relevant context and continue work where it left off.

Task state preservation requires identifying what information is essential for task continuity, maintaining progress markers and intermediate results, and organizing information to enable efficient retrieval and restoration. This goes beyond simple conversation logging to include task-specific state management tailored to different types of work.

Recovery mechanisms must handle both explicit user requests to return to previous topics and implicit continuation of related work. When users ask follow-up questions or request extensions of previous analyses, the system should automatically restore relevant context and build upon previous work rather than starting from scratch.

Long-Term Knowledge Integration and Learning

While working memory handles immediate context, long-term memory enables agentic systems to accumulate knowledge, improve performance over time, and develop increasingly sophisticated understanding of domains and user needs. This persistent storage system captures insights, strategies, and knowledge that transcend individual conversations.

Knowledge Consolidation and Pattern Recognition

Effective long-term memory systems don't simply store raw interaction data—they actively consolidate experiences into more abstract knowledge structures. This consolidation process identifies patterns across multiple interactions, extracts generalizable insights from specific examples, and builds systematic understanding of effective strategies and approaches.

Pattern recognition in long-term memory operates at multiple levels: identifying recurring user needs and preferences, recognizing successful investigation strategies for different types of problems, discovering domain-specific insights that apply across multiple contexts, and detecting relationships between seemingly unrelated information sources.

The consolidation process must balance abstraction with specificity, creating general principles that apply broadly while preserving specific details that might prove valuable in particular contexts. This requires sophisticated analysis of interaction patterns and outcomes to identify what aspects of experience generalize effectively.

Episodic vs. Semantic Memory Structures

Cognitive science distinguishes between episodic memory (memory of specific events and experiences) and semantic memory (general knowledge and facts). Agentic systems benefit from implementing both types of memory structures, using episodic memory to preserve specific interaction contexts and semantic memory to build general knowledge bases.

Episodic memory in agentic systems preserves the context and details of specific interactions, enabling the system to recall particular conversations, remember specific user preferences, and understand the historical development of ongoing projects. This memory type proves particularly valuable for personalization and context restoration.

Semantic memory captures general knowledge extracted from multiple interactions: domain expertise developed through repeated exposure to particular topics, effective strategies identified through experience with different types of problems, and factual knowledge discovered through various investigations. This knowledge type enables improved performance on new tasks by leveraging accumulated expertise.

The interaction between episodic and semantic memory creates a powerful learning system where specific experiences contribute to general knowledge while general knowledge provides context for interpreting specific experiences. This sophisticated memory architecture is implemented through structured memory entries that capture both content and metadata:

```python
@dataclass
class MemoryEntry:
    id: str
    memory_type: MemoryType  # WORKING, EPISODIC, SEMANTIC,
    PROCEDURAL
    content: Any
    context: Dict[str, Any]
    timestamp: datetime
    access_count: int = 0
    importance: float = 0.5
    tags: List[str] = field(default_factory=list)

    def access(self):
        self.access_count += 1
        self.last_accessed = datetime.now()
```

This structure enables sophisticated memory management where the system can track usage patterns, assess importance over time, and organize information through flexible tagging systems while maintaining the metadata necessary for intelligent consolidation and retrieval processes.

Continuous Learning and Adaptation

Long-term memory enables agentic systems to improve continuously through experience, adapting their strategies based on success and failure patterns, refining their understanding of user needs and preferences, and developing increasingly sophisticated domain expertise.

Continuous learning requires mechanisms for evaluating the success of different approaches, identifying factors that contribute to successful outcomes, and adapting strategies based on accumulated evidence. This learning process must operate automatically without requiring explicit feedback for every interaction.

Adaptation strategies include adjusting tool selection preferences based on historical success rates, refining information source priorities based on reliability patterns, and personalizing communication styles based on user interaction patterns. The learning system must balance stability with adaptability, maintaining consistent performance while incorporating improvements from new experiences.

Context Optimization and Memory Consolidation

As agentic systems accumulate experience and information, they must implement sophisticated strategies for managing memory growth, maintaining system performance, and ensuring that accumulated knowledge remains accessible and useful. This requires active memory management that goes beyond simple storage to include organization, optimization, and consolidation processes.

Information Relevance Assessment and Pruning

Not all information deserves permanent retention. Effective memory management requires ongoing assessment of information relevance, utility, and importance to determine what should be preserved, what should be summarized, and what can be safely discarded. This assessment process must operate continuously to prevent memory systems from becoming overwhelmed with irrelevant details.

Relevance assessment considers multiple factors: frequency of access indicating ongoing utility, recency of information suggesting current relevance, importance to successful task completion, and uniqueness of information that might prove valuable in unexpected contexts. The assessment process must balance these factors while considering storage and processing constraints.

Pruning strategies must be conservative to avoid discarding potentially valuable information while being aggressive enough to maintain system performance. This requires sophisticated understanding of information relationships and potential future utility that goes beyond simple usage frequency metrics.

Memory Hierarchy and Storage Optimization

Effective memory systems implement hierarchical storage strategies that balance accessibility with efficiency. Frequently accessed information requires fast retrieval, while rarely accessed information can tolerate slower access in exchange for more efficient storage. Recent information typically deserves more accessible storage than historical data.

Storage optimization also involves identifying relationships between information elements to enable more efficient representation and retrieval. Related information can be stored together, redundant information can be deduplicated, and hierarchical relationships can be preserved to enable efficient traversal and search.

The hierarchy must be dynamic, automatically adjusting storage priorities based on usage patterns and changing relevance. Information that becomes more important over time should migrate to faster storage tiers, while information that loses relevance should move to more efficient long-term storage.

Automated Consolidation Processes

Memory consolidation in agentic systems involves periodic analysis of accumulated information to identify patterns, extract insights, and reorganize storage for improved efficiency and utility. This process operates automatically in the background, continuously improving memory organization without disrupting ongoing operations.

Consolidation processes include identifying related information that can be grouped together, extracting common patterns that can be represented more efficiently, updating relevance assessments based on usage patterns, and reorganizing storage structures to optimize retrieval performance.

The consolidation system must balance thoroughness with efficiency, performing comprehensive analysis without consuming excessive computational resources or disrupting real-time operations. This requires intelligent scheduling of consolidation activities and efficient algorithms for pattern detection and information reorganization.

Knowledge Graph Updates from Interactions

Advanced memory systems in agentic RAG environments often benefit from knowledge graph representations that capture relationships between entities, concepts, and information sources. These graph structures enable sophisticated reasoning about information relationships while supporting efficient storage and retrieval of complex knowledge networks.

Dynamic Knowledge Graph Construction

Rather than relying on static knowledge graphs, agentic systems can build and update knowledge graphs dynamically based on information discovered during interactions. This approach enables the system to develop personalized knowledge representations tailored to specific domains and user needs while maintaining the benefits of structured knowledge representation.

Dynamic construction requires identifying entities and relationships from unstructured information, determining appropriate graph structures for different types of knowledge, and maintaining consistency as the graph evolves over time. This process must operate automatically while providing mechanisms for validation and correction when errors are detected.

The construction process must also handle uncertainty and conflicting information gracefully, representing multiple perspectives when appropriate and maintaining confidence indicators for different facts and relationships. This enables the system to reason appropriately about uncertain or contested information.

Relationship Discovery and Validation

As agentic systems accumulate information from multiple sources and interactions, they can discover previously unknown relationships between entities and concepts. These discoveries might reveal important connections that weren't explicit in individual sources, identify contradictions that require resolution, or suggest new investigation directions.

Relationship discovery requires sophisticated analysis of information patterns across multiple sources and interactions. The system must identify potential relationships, assess the strength of evidence supporting them, and validate discoveries through additional investigation when appropriate.

Validation processes include cross-referencing discovered relationships against multiple sources, assessing the logical consistency of new relationships with existing knowledge, and evaluating the reliability of information sources that suggest particular relationships. This validation must operate continuously as new information becomes available.

Graph-Based Reasoning and Inference

Knowledge graphs enable sophisticated reasoning capabilities that go beyond simple information retrieval. Graph-based reasoning can identify indirect relationships between entities, suggest relevant information based on graph topology, and support complex queries that require understanding multiple relationships.

Inference capabilities include identifying transitive relationships, suggesting related entities based on graph proximity, detecting potential inconsistencies in stored knowledge, and supporting complex reasoning chains that traverse multiple graph relationships. These capabilities enable more sophisticated question-answering and investigation support.

The reasoning system must balance inference power with computational efficiency, providing sophisticated reasoning capabilities without overwhelming system resources. This requires efficient graph algorithms, intelligent caching of inference results, and adaptive reasoning strategies that scale appropriately with graph size and complexity.

Memory and context management represent foundational capabilities that enable agentic systems to transcend simple question-answering and become truly collaborative partners in complex knowledge work. The sophisticated memory architectures explored in this section provide the foundation for learning, adaptation, and increasingly effective assistance over time. However, implementing these capabilities requires careful attention to performance, privacy, and system complexity trade-offs that distinguish production systems from research prototypes.

Memory management patterns and context optimization techniques are demonstrated in Notebook: Multi-Step Reasoning and Memory Management, which provides practical implementations of working memory systems, long-term knowledge consolidation, and dynamic context optimization strategies.

Multi-agent Collaboration

The evolution from single autonomous agents to collaborative multi-agent systems represents one of the most significant advances in agentic RAG architecture. While individual agents can accomplish impressive tasks through planning, tool use, and memory management, the real power of autonomous information systems emerges when multiple specialized agents work together toward common goals. This collaboration enables handling complexity that would overwhelm any single agent while providing natural mechanisms for quality assurance, specialization, and distributed processing.

Multi-agent collaboration in RAG systems goes far beyond simple task distribution. It requires sophisticated coordination mechanisms, shared understanding protocols, conflict resolution strategies, and emergent behaviors that arise from agent interactions. When implemented effectively, multi-agent systems demonstrate capabilities that genuinely exceed the sum of their parts—discovering insights through diverse perspectives, providing robust error correction through peer review, and maintaining system resilience through redundancy and specialization.

Agent Communication Protocols and Task Delegation

The foundation of effective multi-agent collaboration lies in robust communication protocols that enable agents to share information, coordinate activities, and delegate tasks without creating bottlenecks or confusion. Unlike human communication, which relies heavily on implicit understanding and context, agent communication must be explicit, structured, and unambiguous while remaining flexible enough to handle diverse collaboration scenarios.

Structured Message Passing Systems

Effective agent communication begins with well-designed message passing systems that capture not just content but also intent, context, and routing information. Each message in a multi-agent system carries multiple layers of information: the core content or request, metadata about urgency and priority, context about the broader task being pursued, and routing information that helps determine appropriate recipients and response strategies.

The implementation of such communication systems requires sophisticated message handling capabilities:

```
def send_message(self, sender: str, recipient: str, message_
type: str, content: Any):
    message = {
        "sender": sender, "recipient": recipient, "type":
        message_type,
        "content": content, "timestamp": datetime.now()
    }
    self.message_history.append(message)
    self.active_conversations[f"{sender}_{recipient}"].
    append(message["id"])
```

Consider a research scenario where multiple agents collaborate to analyze market trends. A researcher agent might send a message like: "I've identified three key trends in renewable energy adoption rates. Analysis agent, please evaluate the statistical significance of the growth patterns I've found. Writer agent, prepare to synthesize this with the competitive landscape analysis when it's complete." This message contains explicit task delegation, clear deliverable expectations, and coordination information that helps recipients understand their role in the broader workflow.

The message structure must accommodate different types of agent communication: direct task delegation where one agent assigns specific work to another, information sharing where agents broadcast discoveries that might be relevant to multiple colleagues, status updates that help maintain overall project awareness, and coordination requests where agents negotiate resource usage or timing constraints.

Dynamic Task Delegation Strategies

Task delegation in multi-agent systems requires sophisticated understanding of agent capabilities, current workloads, and task requirements. Rather than following predetermined assignment rules, effective systems implement dynamic delegation that considers real-time factors like agent availability, expertise relevance, and overall system load distribution.

Production-ready task coordination systems implement sophisticated assignment logic:

```python
def assign_task(self, task_id: str, agent_name: str) -> bool:
    if not self._check_dependencies(task):
        return False  # Wait for prerequisites

    task["assigned_agent"] = agent_name
    task["status"] = "in_progress"
    self.agent_workload[agent_name].append(task_id)
    return True
```

Smart delegation systems maintain models of each agent's capabilities, specializations, and performance history. When a complex task emerges, the delegation system analyzes task requirements, identifies agents with relevant capabilities, assesses current workloads and availability, and makes assignment decisions that optimize both individual task success and overall system performance.

The delegation process must also handle task decomposition—breaking complex assignments into smaller, manageable units that can be distributed across multiple agents or executed in parallel. This decomposition requires understanding task dependencies, identifying parallelization opportunities, and ensuring that distributed subtasks can be effectively coordinated and integrated.

Coordination Without Central Control

While centralized coordination might seem simpler to implement, distributed coordination systems often prove more robust and scalable in practice. Peer-to-peer coordination protocols enable agents to self-organize around tasks, negotiate resource allocation, and adapt to changing conditions without relying on central orchestration that could become a bottleneck or single point of failure.

Distributed coordination requires agents to maintain awareness of system-wide goals while making local decisions about their activities. This balance between local autonomy and global coherence is achieved through shared goal representation, regular status broadcasting, and negotiation protocols that resolve conflicts and resource contentions as they arise.

Effective peer-to-peer coordination also implements emergent leadership patterns where different agents take coordinating roles for different types of tasks based on their expertise and current context. A research agent might coordinate information gathering phases, while an analysis agent leads data processing activities, and a quality assurance agent oversees final validation processes.

Specialized Roles: Research, Analysis, Synthesis, and Verification

The power of multi-agent collaboration becomes apparent when agents specialize in particular types of cognitive work, developing deep expertise in specific domains while contributing to broader collaborative goals. This specialization enables more sophisticated analysis than generalist approaches while providing natural quality assurance through diverse perspectives and expertise.

Research Agents: Information Discovery and Gathering

Research agents specialize in finding, evaluating, and organizing information from diverse sources. These agents develop sophisticated search strategies, maintain awareness of high-quality information sources, and excel at navigating complex information landscapes to find relevant materials. Research agents understand how to formulate effective search queries, evaluate source credibility, and organize findings in ways that support subsequent analysis and synthesis.

Specialized research agents often develop domain expertise that enables them to identify subtle patterns, recognize authoritative sources, and understand the significance of particular findings within broader knowledge contexts. A research agent focused on financial markets might develop deep understanding of economic indicators, regulatory patterns, and market dynamics that enables more effective information discovery than generalist approaches.

Research agents also serve as the primary interface between the multi-agent system and external information sources, managing API credentials, respecting rate limits, and implementing search strategies that maximize information quality while minimizing costs and resource consumption.

Analysis Agents: Pattern Recognition and Insight Generation

Analysis agents specialize in processing gathered information to identify patterns, generate insights, and develop understanding that goes beyond simple information aggregation. These agents excel at statistical analysis, trend identification, comparative evaluation, and logical reasoning that transforms raw information into actionable intelligence.

Effective analysis agents implement multiple analytical frameworks and can select appropriate methodologies based on data characteristics and investigation goals. They understand how to handle uncertainty, assess confidence levels, and communicate analytical limitations alongside insights. Analysis agents also recognize when their individual analytical capabilities are insufficient and can recommend collaboration with other specialists or additional data gathering.

The specialization of analysis agents enables development of sophisticated analytical capabilities that would be difficult to maintain in generalist systems. An analysis agent might develop expertise in particular statistical methods, industry-specific analytical frameworks, or specialized reasoning approaches that enhance the overall analytical capability of the multi-agent system.

Synthesis Agents: Integration and Communication

Synthesis agents specialize in combining insights from multiple sources and perspectives into coherent, comprehensive outputs that address original user queries or investigative goals. These agents excel at identifying connections between disparate findings, resolving apparent contradictions, and organizing complex information into clear, actionable presentations.

Effective synthesis requires understanding audience needs, communication preferences, and the broader context in which information will be used. Synthesis agents develop expertise in different communication formats, from executive summaries that highlight key decisions points to detailed technical reports that provide comprehensive analysis support.

Synthesis agents also serve as the primary interface between the multi-agent system and end users, translating complex collaborative work products into forms that meet user needs and expectations. This interface role requires sophisticated understanding of user contexts, communication preferences, and the appropriate level of detail for different types of decisions.

Verification Agents: Quality Assurance and Validation

Verification agents specialize in quality assurance, fact-checking, and validation processes that ensure collaborative outputs meet high standards for accuracy, completeness, and reliability. These agents implement systematic review processes, cross-reference claims against multiple sources, and identify potential errors or gaps in reasoning.

Quality assurance in multi-agent systems goes beyond simple fact-checking to include assessment of analytical methodology, evaluation of source reliability, verification of logical consistency, and assessment of conclusion appropriateness given available evidence. Verification agents understand common types of errors and biases that can emerge in collaborative work and implement systematic processes to identify and correct these issues.

Verification agents also maintain quality standards and best practices that guide the overall multi-agent system's work. They track common error patterns, identify opportunities for process improvement, and ensure that collaborative workflows maintain high standards for accuracy and reliability over time.

Conflict Resolution and Load Balancing

Multi-agent collaboration inevitably generates conflicts and resource contentions that must be resolved to maintain system effectiveness. Rather than viewing conflicts as problems to be avoided, sophisticated systems treat conflicts as opportunities for deeper analysis, diverse perspective integration, and improved decision-making quality.

Conflict Identification and Classification

Effective conflict resolution begins with systematic identification and classification of different types of conflicts that emerge in multi-agent collaboration. Information conflicts arise when different agents discover contradictory facts or data from different sources. Analytical conflicts emerge when agents apply different methodologies or frameworks and reach different conclusions. Resource conflicts occur when multiple agents need access to the same tools, APIs, or computational resources simultaneously.

Each type of conflict requires different resolution strategies and different types of evidence for resolution. Information conflicts might be resolved through source evaluation, additional data gathering, or temporal analysis to understand why sources disagree. Analytical conflicts might require methodology comparison, independent validation, or expert consultation to determine appropriate approaches.

The conflict identification system must operate continuously during collaborative processes, monitoring for potential disagreements, resource contentions, and incompatible recommendations. Early conflict identification enables proactive resolution before conflicts compromise overall system effectiveness.

Evidence-Based Conflict Resolution

When conflicts emerge, sophisticated systems implement evidence-based resolution processes that treat conflicts as opportunities for deeper investigation rather than problems to be quickly resolved. Evidence-based resolution involves gathering additional information, applying multiple analytical approaches, and systematically evaluating different perspectives to reach well-supported conclusions.

Modern conflict resolution systems implement structured approaches to handling agent disagreements:

```
def resolve_conflict(self, agent1_position: str, agent2_
position: str, context: str):
    resolution_prompt = f"""Analyze conflicting positions and
    provide:
    1. Strengths/weaknesses of each position
```

```
2. Areas of agreement and disagreement
3. Evidence-based resolution with confidence level"""

resolution = self.llm.invoke(resolution_prompt)
self.resolution_history.append(resolution)
return resolution
```

The conflict resolution process might involve bringing in additional specialized agents, consulting external experts, or implementing controlled experiments to test different hypotheses. For complex conflicts, the system generates multiple alternative conclusions with explicit confidence assessments rather than forcing premature resolution.

Evidence-based resolution also maintains detailed records of conflict resolution processes, enabling the system to learn from experience and improve resolution strategies over time. Patterns in conflict types and resolution approaches become part of the system's procedural knowledge, enhancing future collaboration effectiveness.

Dynamic Load-Balancing Strategies

Load balancing in multi-agent systems goes beyond simple task distribution to include sophisticated resource management that considers agent capabilities, current workloads, task priorities, and system-wide optimization goals. Dynamic load balancing systems monitor agent performance, resource utilization, and task progress to make real-time decisions about task assignment and resource allocation.

Effective load balancing considers both hard constraints—like API rate limits and memory requirements—and soft preferences like agent expertise and collaboration efficiency. The system might deliberately create some resource contention if it leads to better outcomes through collaboration, or it might distribute tasks to avoid bottlenecks even if it means using less specialized agents for particular sub-tasks.

Load balancing also includes provisions for dynamic task reassignment when agents become overloaded, encounter technical problems, or when priorities change during collaborative processes. The system maintains sufficient flexibility to adapt resource allocation while preserving collaborative coherence and task quality.

Shared Workspace Management

Effective multi-agent collaboration requires sophisticated shared workspace management that enables agents to coordinate their activities, share information, and maintain collective awareness of project progress and discoveries. The shared workspace serves as both a repository for collaborative work products and a coordination mechanism that enables distributed agents to work effectively together.

Collaborative Information Architecture

The shared workspace must accommodate diverse types of information generated during collaborative processes: raw research findings, intermediate analytical results, shared hypotheses and theories, collaborative documents and reports, task assignments and progress tracking, and communication histories and decision records. Each type of information has different access patterns, update requirements, and usage characteristics that the workspace architecture must support.

Effective information architecture enables both structured access—where agents can efficiently find specific types of information—and serendipitous discovery—where agents can encounter relevant information they weren't explicitly seeking. The workspace implements tagging, categorization, and search capabilities that help agents navigate complex information landscapes while maintaining awareness of relevant developments in other parts of the collaborative process.

The architecture also supports different levels of information maturity, from preliminary hypotheses and rough notes to validated findings and final recommendations. Agents understand how to assess information reliability and appropriateness for different types of uses, and the workspace provides metadata and context that supports these assessments.

Version Control and Change Management

Collaborative work products evolve through multiple revision cycles as different agents contribute insights, corrections, and enhancements. The shared workspace implements sophisticated version control that tracks document evolution, maintains histories of changes and contributions, and enables rollback when modifications prove problematic.

Version control in multi-agent systems goes beyond simple document management to include understanding of logical dependencies between different work products. When foundational research findings change, the system can identify downstream analyses and conclusions that might be affected and trigger appropriate review processes.

Change management also includes notification systems that keep relevant agents informed about updates to shared resources, collaborative discussion mechanisms for coordinating changes, and approval workflows for sensitive or critical modifications that require consensus or oversight.

Access Control and Information Security

Multi-agent collaboration often involves sensitive information that requires appropriate access controls and security measures. The shared workspace implements role-based access controls that limit access to sensitive information based on agent roles and task requirements, while maintaining sufficient transparency to enable effective collaboration.

Security considerations include protecting external API credentials, ensuring that sensitive research findings are appropriately protected, implementing audit trails that track access to sensitive information, and providing mechanisms for secure communication between agents when necessary.

The access control system balances security requirements with collaboration needs, enabling agents to share information effectively while protecting sensitive materials and maintaining appropriate confidentiality boundaries.

Persistent Knowledge Integration

The shared workspace serves not just as a temporary coordination mechanism but as a repository for persistent knowledge that can inform future collaborative efforts. The workspace captures successful collaboration patterns, effective research strategies, validated analytical methodologies, and insights about agent specialization and collaboration effectiveness.

This persistent knowledge enables the multi-agent system to improve over time, learning from successful collaborations and avoiding patterns that proved problematic. The workspace becomes a form of organizational memory that preserves valuable insights and approaches beyond individual collaborative episodes.

Knowledge integration also includes capabilities for cross-project learning, where insights and approaches from one collaborative effort can inform subsequent projects with similar characteristics or requirements.

The sophistication of multi-agent collaboration represents a significant evolution in autonomous information systems, enabling capabilities that approach and sometimes exceed human team performance in complex analytical tasks. However, this sophistication also requires careful attention to coordination mechanisms, quality assurance processes, and system management approaches that ensure collaborative benefits while avoiding the complexity overhead that can make systems difficult to understand, debug, and maintain.

The implementation frameworks we'll explore next provide concrete approaches for building these sophisticated collaborative capabilities, demonstrating how theoretical concepts translate into practical, deployable systems that can handle real-world complexity while maintaining reliability and effectiveness.

Multi-agent collaboration patterns and coordination mechanisms are extensively demonstrated in Notebook: Multi-Agent Collaboration with CrewAI, which provides hands-on implementation of specialized agent roles, conflict resolution, and shared workspace management.

Implementation Frameworks

The theoretical foundations of agentic RAG systems must ultimately translate into practical, deployable implementations that can handle real-world complexity while maintaining reliability and performance. This section explores the leading frameworks that enable developers to build sophisticated agentic systems without reimplementing core coordination, communication, and execution patterns from scratch.

Modern implementation frameworks have evolved to address the fundamental challenges of agentic system development: orchestrating complex multistep reasoning, managing tool integration and external API interactions, coordinating multi-agent collaboration, and providing production-ready monitoring and error handling. The frameworks we examine—LangChain Agents and CrewAI—represent complementary approaches to these challenges, each optimized for different use cases and deployment scenarios.

LangChain Agents: Executors, Tools, and Custom Implementations

LangChain Agents provide a comprehensive foundation for building individual autonomous agents with sophisticated reasoning capabilities, extensive tool integration, and flexible customization options. The framework emphasizes composability, enabling developers to combine pre-built components with custom implementations to create agents tailored to specific requirements.

Agent Executors and Core Architecture

The LangChain Agent architecture centers around the concept of an AgentExecutor—a runtime environment that orchestrates the interaction between language models, tools, and memory systems. The executor implements the core reasoning loop that enables agents to plan actions, execute tools, and integrate results into ongoing reasoning processes.

```
# Core agent creation pattern
agent = create_react_agent(llm=llm, tools=tools,
prompt=react_prompt)
executor = AgentExecutor(
    agent=agent, tools=tools, memory=memory,
    max_iterations=10, handle_parsing_errors=True
)
```

The executor handles the complexities of parsing agent outputs, routing tool calls, managing execution timeouts, and providing fallback strategies when parsing fails. This abstraction enables developers to focus on agent logic and tool design rather than low-level execution management.

Agent execution follows the pattern: *parse agent output* ➤ *validate tool selection* ➤ *execute tool call* ➤ *integrate result* ➤ *continue reasoning.* The executor maintains conversation state, tracks intermediate steps, and provides comprehensive logging for debugging and monitoring.

Tool Integration Patterns

LangChain's tool system provides standardized interfaces for integrating external capabilities into agent workflows. Tools encapsulate everything from simple calculations to complex API interactions, presenting a uniform interface that agents can discover and use dynamically.

```
@tool
def enhanced_search_tool(query: str) -> str:
    """Search for current information with error handling"""
    try:
        result = external_search_api(query)
        return f"Search results: {result}"
    except Exception as e:
        return f"Search failed: {str(e)}"
```

Tool design follows key principles: clear, descriptive names that agents can understand; comprehensive docstrings that explain tool capabilities and usage; robust error handling that provides meaningful feedback; and consistent return formats that agents can reliably process.

The framework supports tool composition, where complex capabilities are built by combining simpler tools, and dynamic tool selection, where agents choose appropriate tools based on current context and requirements. Tool orchestration enables agents to use multiple tools in sequence or in parallel to accomplish complex tasks.

Memory Integration and Context Management

LangChain provides multiple memory implementations that enable agents to maintain context across interactions and learn from experience. Memory systems range from simple conversation buffers to sophisticated summarization approaches that compress historical context while preserving essential information.

The memory integration pattern: *capture interaction* ➤ *update memory state* ➤ *retrieve relevant context* ➤ *inject into prompt*. Different memory types serve different purposes:

- **ConversationBufferMemory**: Maintains complete conversation history up to token limits

- **ConversationSummaryMemory**: Compresses older interactions while preserving recent details

- **ConversationTokenBufferMemory**: Manages memory based on token consumption rather than message count

- **VectorStoreRetrieverMemory**: Uses semantic search to retrieve relevant historical context

Memory systems can be combined and customized to create sophisticated context management that balances information retention with computational efficiency.

Custom Agent Implementation Patterns

Beyond pre-built agent types, LangChain enables custom agent implementations that can incorporate domain-specific reasoning patterns, specialized tool orchestration strategies, and unique memory management approaches.

Custom agent development follows the pattern: *define reasoning logic* ➤ *implement prompt templates* ➤ *specify tool selection strategy* ➤ *handle output parsing* ➤ *integrate memory management.* The framework provides base classes and interfaces that simplify custom development while maintaining compatibility with existing tools and infrastructure.

```
class CustomAgentImplementation:
    def plan(self, input: str, intermediate_steps: List) ->
    AgentAction:
        # Custom planning logic based on input and
        previous steps
        return AgentAction(tool="selected_tool", tool_
        input="processed_input")

    def format_prompt(self, input: str, tools: List, memory:
    str) -> str:
        # Custom prompt formatting for specific reasoning
        patterns
        return f"Custom prompt with {input}, {tools}, {memory}"
```

Custom implementations can incorporate specialized reasoning approaches like chain-of-thought for complex analytical tasks, tree-of-thoughts for exploring multiple solution paths, or domain-specific heuristics that improve performance for particular problem types.

Production Deployment Considerations

LangChain agents require careful configuration for production deployment, including timeout management, error handling strategies, cost monitoring, and performance optimization. Production deployments typically implement monitoring callbacks, resource limiting, and graceful degradation strategies.

The production deployment pattern: *configure resource limits* ➤ *implement monitoring* ➤ *add error recovery* ➤ *optimize performance* ➤ *establish maintenance procedures.* Key considerations include API rate limiting, memory usage optimization, execution timeout handling, and comprehensive logging for debugging and performance analysis.

CrewAI Multi-Agent Systems: Crew Composition and Task Coordination

CrewAI specializes in multi-agent collaboration, providing frameworks for creating teams of specialized agents that work together on complex projects. The platform emphasizes role-based specialization, coordinated task execution, and emergent collaboration patterns that arise from agent interactions.

Crew Architecture and Agent Roles

CrewAI organizes agents into crews—cohesive teams where each agent has specialized roles, capabilities, and responsibilities. The crew architecture enables sophisticated task decomposition where complex projects are broken into subtasks that match agent specializations.

```
# Specialized agent creation with distinct roles
researcher = Agent(
    role="Senior Research Analyst",
    goal="Conduct comprehensive research and gather relevant
    information",
    backstory="Expert researcher with 10+ years experience...",
    tools=[research_tool, web_search_tool]
)
```

```
analyst = Agent(
    role="Strategic Analyst",
    goal="Analyze data and provide strategic insights",
    backstory="Strategic analyst with deep expertise...",
    tools=[analysis_tool, data_tool]
)

crew = Crew(agents=[researcher, analyst], tasks=tasks,
process=Process.sequential)
```

Agent specialization goes beyond simple tool assignment to include distinct reasoning patterns, communication styles, and quality standards appropriate for each role. Researchers focus on information discovery and validation, analysts emphasize pattern recognition and insight generation, and writers concentrate on clear communication and synthesis.

Task Coordination and Workflow Management

CrewAI implements sophisticated task coordination that manages dependencies, enables parallel execution where appropriate, and ensures that agent outputs integrate effectively into cohesive project outcomes.

Task coordination follows the pattern: *decompose project* ➤ *assign tasks* ➤ *manage dependencies* ➤ *coordinate execution* ➤ *integrate results.* The framework handles task scheduling, resource allocation, and workflow optimization automatically while providing visibility into execution progress.

```
research_task = Task(
    description="Conduct comprehensive research on market trends",
    agent=researcher,
    expected_output="Detailed research report with key findings"
)
```

```
analysis_task = Task(
    description="Analyze research findings and provide
    strategic insights",
    agent=analyst,
    expected_output="Strategic analysis with recommendations"
)
```

The task system supports complex dependency relationships, conditional execution based on intermediate results, and dynamic task generation where agents can create additional subtasks as projects evolve.

Process Types and Execution Strategies

CrewAI supports multiple execution processes that optimize coordination for different types of collaborative work:

- **Sequential Process**: Tasks execute in defined order with full context passing between agents

- **Hierarchical Process**: Manager agent coordinates worker agents with dynamic task assignment

- **Parallel Process**: Independent tasks execute simultaneously with synchronized integration

Process selection depends on project characteristics: sequential for building analytical narratives, hierarchical for complex projects requiring adaptive coordination, and parallel for independent research streams that benefit from simultaneous execution.

Inter-Agent Communication and Knowledge Sharing

CrewAI implements sophisticated communication mechanisms that enable agents to share discoveries, coordinate activities, and build upon each other's work. Communication goes beyond simple task handoffs to include collaborative discussions, knowledge sharing, and peer review processes.

The communication pattern: *identify collaboration need* ➤ *select appropriate agents* ➤ *facilitate discussion* ➤ *integrate insights* ➤ *update shared knowledge*. Agents can request assistance from colleagues, share interesting discoveries, and engage in structured peer review to improve output quality.

Quality Assurance and Collaborative Review

Multi-agent systems naturally enable sophisticated quality assurance through peer review, diverse perspectives, and collaborative validation. CrewAI leverages these benefits through built-in review processes and quality checkpoints.

Quality assurance follows the pattern: *complete initial work* ➤ *identify reviewers* ➤ *conduct peer review* ➤ *integrate feedback* ➤ *validate final output*. The framework can automatically assign review tasks, facilitate collaborative improvement processes, and ensure that final outputs meet established quality standards.

Advanced Patterns: Self-improvement and Goal-Oriented Behavior

Beyond basic agent capabilities, advanced patterns enable systems that learn from experience, pursue long-term objectives, and demonstrate increasingly sophisticated autonomous behavior. These patterns represent the current frontier of agentic system development.

Self-improvement and Learning from Experience

Self-improving agents analyze their own performance, identify areas for enhancement, and adapt their strategies based on experience. This capability enables systems that become more effective over time without requiring explicit retraining.

```
def execute_with_learning(self, query: str, feedback:
Optional[str] = None):
    result = self.base_agent.invoke({"input": query})

    # Analyze performance and identify improvements
    analysis = self._analyze_performance(query, result,
    feedback)
    improvements = analysis.get("improvements", [])

    # Apply high-priority improvements immediately
    for improvement in improvements:
        if improvement["priority"] == "high":
            self._apply_improvement(improvement)

    return result
```

Self-improvement patterns include: *execute task* ➤ *analyze performance* ➤ *identify improvement opportunities* ➤ *apply optimizations* ➤ *validate improvements.* The system maintains performance history, tracks successful strategies, and learns from both successes and failures.

Learning mechanisms include strategy optimization based on success rates, tool selection refinement based on effectiveness patterns, prompt engineering improvements based on output quality, and resource allocation optimization based on efficiency metrics.

Goal-Oriented Long-Term Behavior

Goal-oriented agents maintain persistent objectives that span multiple interactions and extended time periods. These agents can break long-term goals into manageable subtasks, maintain progress toward objectives across sessions, and adapt strategies based on changing circumstances.

```python
def set_goal(self, description: str, priority: int, deadline:
datetime) -> str:
    goal = {
        "id": generate_goal_id(),
        "description": description,
        "priority": priority,
        "deadline": deadline,
        "strategy": self._generate_strategy(description),
        "progress": 0.0
    }
    self.active_goals.append(goal)
    return goal["id"]
```

Goal-oriented behavior follows the pattern: *define objectives* ➤ *develop strategies* ➤ *execute systematically* ➤ *monitor progress* ➤ *adapt approaches*. The system maintains goal hierarchies, tracks progress metrics, and can pursue multiple objectives simultaneously while managing resource allocation and priority balancing.

Adaptive Strategy Development

Advanced agents develop and refine strategies based on experience, environmental feedback, and changing requirements. Strategy adaptation enables systems to handle novel situations and improve performance in dynamic environments.

Strategy development includes: *analyze current approaches* ➤ *identify optimization opportunities* ➤ *test alternative strategies* ➤ *measure effectiveness* ➤ *adopt improvements*. The system maintains strategy libraries, conducts controlled experiments, and learns from both successful and unsuccessful approaches.

Meta-Cognition and Performance Monitoring

Sophisticated agents implement meta-cognitive capabilities that enable them to reason about their own reasoning processes, monitor their performance, and identify opportunities for improvement. Meta-cognition represents a crucial step toward truly autonomous systems.

Meta-cognitive patterns include: *monitor reasoning processes* ➤ *identify cognitive biases* ➤ *assess confidence levels* ➤ *validate reasoning chains* ➤ *improve decision-making.* The system maintains awareness of its own capabilities and limitations, can request assistance when needed, and continuously works to improve its cognitive processes.

Emergent Collaborative Behaviors

When multiple advanced agents work together, emergent behaviors often arise that exceed the capabilities of individual agents. These emergent patterns include collaborative problem-solving strategies, distributed cognition approaches, and collective intelligence phenomena.

Emergent collaboration follows the pattern: *individual agent development* ➤ *multi-agent interaction* ➤ *pattern recognition* ➤ *collaboration optimization* ➤ *emergent capability development.* The system learns effective collaboration patterns, develops specialized roles dynamically, and can adapt team composition based on project requirements.

Integration with External Learning Systems

Advanced agentic systems can integrate with external learning systems, knowledge bases, and human expertise to continuously expand their capabilities. This integration enables systems that benefit from both autonomous learning and external knowledge sources.

Integration patterns include: *identify knowledge gaps* ➤ *access external resources* ➤ *integrate new knowledge* ➤ *validate learning* ➤ *apply insights.* The system can automatically seek out relevant learning opportunities, integrate external expertise, and adapt its knowledge base based on new information.

The implementation frameworks explored in this section provide the foundation for building sophisticated agentic RAG systems that can handle real-world complexity while maintaining reliability and performance. However, successful deployment requires careful attention to evaluation, monitoring, and production considerations that ensure these powerful capabilities translate into practical value.

Implementation patterns for LangChain Agents are demonstrated in Notebook 13.1: Agent Fundamentals and Tool Integration, while CrewAI multi-agent systems are extensively covered in Notebook: Multi-Agent Collaboration with CrewAI. Advanced patterns including self-improvement and goal-oriented behavior are implemented in Notebook 13.4: Production Deployment and Advanced Patterns.

Evaluation and Production Deployment

The transition from prototype agentic systems to production-ready deployments requires comprehensive evaluation frameworks, robust monitoring systems, and careful attention to scalability, safety, and operational requirements. Unlike traditional software systems with predictable behavior patterns, agentic systems present unique evaluation challenges due to their autonomous decision-making, dynamic tool usage, and emergent behaviors that arise from complex interactions.

This section explores the methodologies, metrics, and infrastructure patterns necessary to evaluate agentic RAG systems effectively and deploy them reliably in production environments. The approaches we examine address both technical performance metrics and business value indicators, ensuring that sophisticated autonomous capabilities translate into measurable improvements in real-world applications.

Performance Metrics: Task Completion, Planning Efficiency, and User Satisfaction

Evaluating agentic systems requires multi-dimensional metrics that capture both technical performance and practical effectiveness. Traditional software metrics like response time and error rates remain important, but agentic systems demand additional measures that assess reasoning quality, planning effectiveness, and autonomous decision-making capabilities.

Task Completion Metrics

Task completion evaluation goes beyond simple success/failure binary measures to include assessment of completeness, accuracy, and efficiency Effective evaluation frameworks measure whether agents accomplish stated objectives, assess the quality and thoroughness of solutions provided, evaluate the appropriateness of approaches taken, and determine the efficiency of resource utilization.

```
def evaluate_task_completion(self, task_result: Dict, expected_
outcome: Dict) -> Dict:
    """Comprehensive task completion evaluation"""
    return {
        "success_rate": task_result["completed"] / task_
        result["attempted"],
        "accuracy_score": self._calculate_accuracy(task_
        result["output"], expected_outcome),
        "completeness_score": self._assess_completeness(task_
        result, expected_outcome["requirements"]),
```

```
  "efficiency_rating": self._evaluate_efficiency(task_
  result["steps"], task_result["time"])
}
```

Task completion metrics include: *success rate across different task types* ➤ *accuracy of final outputs* ➤ *completeness relative to requirements* ➤ *efficiency of execution paths.* These metrics must account for task complexity variations, different success criteria for different domains, and the trade-offs between thoroughness and efficiency that agents navigate.

Sophisticated evaluation systems maintain task taxonomies that enable performance comparison across similar problems, track performance trends over time, and identify patterns in agent strengths and weaknesses that inform system improvements.

Planning Efficiency Assessment

Planning efficiency measures how effectively agents decompose complex problems, select appropriate strategies, and execute multistep investigations. These metrics assess both the quality of planning decisions and the effectiveness of plan execution.

Planning evaluation includes: *strategy appropriateness for task characteristics* ➤ *resource allocation efficiency* ➤ *adaptation effectiveness when plans encounter obstacles* ➤ *coordination quality in multi-agent scenarios.* The evaluation system tracks planning patterns, measures deviation from optimal paths, and assesses agents' ability to recover from planning failures.

```
def assess_planning_efficiency(self, execution_trace:
List[Dict]) -> Dict:
    """Evaluate planning quality and execution efficiency"""
    optimal_steps = self._calculate_optimal_path(execution_
    trace[0]["goal"])
    actual_steps = len(execution_trace)
```

```
return {
    "planning_optimality": optimal_steps / actual_steps,
    "adaptation_count": sum(1 for step in execution_trace
    if step["type"] == "replan"),
    "resource_efficiency": self._calculate_resource_
    utilization(execution_trace)
}
```

Effective planning metrics distinguish between different types of inefficiency: suboptimal initial planning that could be improved through better algorithms, unavoidable adaptation due to changing circumstances or new information, and execution failures that indicate implementation rather than planning problems.

User Satisfaction and Experience Metrics

User satisfaction represents the ultimate measure of agentic system success, but measuring satisfaction with autonomous systems requires specialized approaches that account for user expectations, trust development, and perceived value delivery.

User experience evaluation includes: *perceived usefulness of autonomous assistance* ➤ *trust in agent decision-making* ➤ *satisfaction with communication clarity* ➤ *confidence in result accuracy*. These metrics require both quantitative measures (response times, completion rates) and qualitative assessments (user feedback, satisfaction surveys).

User satisfaction patterns often reveal important insights about system design: agents that explain their reasoning clearly tend to generate higher trust, systems that acknowledge uncertainty appropriately build more realistic user expectations, and agents that adapt to user preferences over time demonstrate higher long-term satisfaction.

Monitoring and Observability: Decision Tracking and Error Analysis

Production agentic systems require sophisticated observability infrastructure that provides insight into autonomous decision-making processes, enables rapid problem diagnosis, and supports continuous system improvement. Unlike traditional systems where behavior is predetermined, agentic systems demand monitoring approaches that can capture and analyze emergent behaviors.

Decision Tracking and Reasoning Transparency

Decision tracking systems capture the reasoning processes that lead to agent actions, enabling both debugging and performance analysis. Effective tracking maintains complete audit trails of agent reasoning, records tool selection rationale, and documents adaptation decisions when plans change.

```
class DecisionTracker:
    """Track agent decision-making for analysis and debugging"""

    def record_decision(self, agent_id: str, decision_type:
    str, context: Dict,
                    reasoning: str, outcome: Any):
        decision_record = {
            "timestamp": datetime.now(),
            "agent": agent_id,
            "type": decision_type,
            "context": context,
            "reasoning": reasoning,
            "outcome": outcome,
            "confidence": self._extract_confidence(reasoning)
        }
        self.decision_history.append(decision_record)
```

Decision tracking enables: *post-hoc analysis of agent reasoning* ➤ *identification of decision patterns and biases* ➤ *debugging of unexpected behaviors* ➤ *performance optimization based on decision effectiveness.* The tracking system maintains decision taxonomies, correlates decisions with outcomes, and identifies patterns that inform system improvements.

Transparency mechanisms help users understand agent decision-making, build appropriate trust levels, and provide feedback that improves system performance. Effective transparency balances comprehensiveness with usability, providing appropriate detail levels for different user types and contexts.

Error Analysis and Failure Pattern Recognition

Error analysis in agentic systems requires sophisticated approaches that distinguish between different failure types, understand failure propagation through multistep processes, and identify systemic issues that might not be apparent from individual error instances.

Error classification includes: *tool execution failures* ➤ *reasoning chain breakdowns* ➤ *planning inadequacies* ➤ *coordination failures in multi-agent scenarios.* Each error type requires different analysis approaches and different remediation strategies.

```
def analyze_error_patterns(self, error_history: List[Dict])
-> Dict:
    """Analyze error patterns for systematic issues"""
    pattern_analysis = {
        "error_types": self._classify_errors(error_history),
        "failure_correlations": self._find_error_
        correlations(error_history),
        "temporal_patterns": self._analyze_temporal_error_
        distribution(error_history),
```

```
    "remediation_suggestions": self._suggest_
    improvements(error_history)
}
return pattern_analysis
```

Failure pattern recognition identifies: *recurring error types that indicate systematic problems* ➤ *correlation patterns between different failure modes* ➤ *temporal patterns that suggest environmental or load-related issues* ➤ *cascade failures where initial problems propagate through system components.* This analysis informs both immediate fixes and longer-term system architecture improvements.

Performance Monitoring and Alerting

Production monitoring systems track both technical performance metrics and business-relevant indicators, providing early warning of problems and enabling proactive system management. Monitoring must account for the variable and sometimes unpredictable nature of agentic system behavior.

Performance monitoring includes: *response time distribution across different task types* ➤ *resource utilization patterns* ➤ *success rate trends* ➤ *user satisfaction indicators.* The monitoring system establishes baseline performance expectations, identifies anomalous behavior patterns, and triggers appropriate alerts when performance degrades.

Alerting strategies must balance sensitivity with specificity, avoiding alert fatigue while ensuring that significant problems receive prompt attention. Effective alerting systems understand normal performance variations, distinguish between temporary fluctuations and systematic problems, and provide actionable information for problem resolution.

Scalability: Resource Management and Cost Optimization

Agentic systems present unique scalability challenges due to their dynamic resource usage patterns, unpredictable execution paths, and complex interactions with external services. Production deployments must implement sophisticated resource management that maintains performance while controlling costs.

Dynamic Resource Allocation

Resource allocation for agentic systems must accommodate highly variable workloads, unpredictable execution patterns, and complex dependencies between system components. Unlike traditional web services with predictable traffic patterns, agentic systems may experience sudden spikes in computational requirements or API usage.

```python
class ResourceManager:
    """Manage dynamic resource allocation for agentic systems"""

    def allocate_resources(self, task_requirements: Dict,
    current_load: Dict) -> Dict:
        allocation = {
            "compute_nodes": self._calculate_compute_
            needs(task_requirements),
            "memory_allocation": self._estimate_memory_
            usage(task_requirements),
            "api_rate_limits": self._distribute_api_
            quotas(task_requirements),
            "priority_level": self._determine_priority(task_
            requirements)
        }
        return self._optimize_allocation(allocation,
        current_load)
```

Resource allocation patterns include: *predictive scaling based on task complexity* ➤ *dynamic load balancing across agent instances* ➤ *intelligent API quota management* ➤ *priority-based resource distribution.* The allocation system learns from historical patterns, anticipates resource needs, and adapts to changing usage patterns.

Effective resource management also implements graceful degradation strategies that maintain essential functionality when resources are constrained, intelligent caching that reduces redundant computations, and resource pooling that enables efficient sharing across multiple agent instances.

Cost Monitoring and Optimization

Cost management for agentic systems requires sophisticated tracking of various expense categories, including language model API calls, external service usage, computational resources, and storage requirements. Cost optimization must balance expense control with system effectiveness.

Cost optimization strategies include: *intelligent caching to reduce redundant API calls* ➤ *efficient tool selection to minimize expensive operations* ➤ *batch processing where possible* ➤ *usage-based scaling that aligns costs with value delivery.* The cost management system tracks expenses across different categories, identifies optimization opportunities, and implements automatic cost controls.

```
def optimize_costs(self, usage_patterns: Dict, cost_targets:
Dict) -> Dict:
    """Implement cost optimization strategies"""
    optimizations = {
        "api_call_reduction": self._identify_cacheable_
        operations(usage_patterns),
        "resource_rightsizing": self._optimize_instance_
        allocation(usage_patterns),
```

```
    "scheduling_optimization": self._optimize_task_
    scheduling(usage_patterns)
}
return self._apply_optimizations(optimizations, cost_
targets)
```

Cost optimization requires understanding the relationship between resource usage and value delivery, implementing cost-aware decision-making in agent planning, and maintaining transparency about cost implications of different operational choices.

Horizontal Scaling Patterns

Scaling agentic systems horizontally presents unique challenges due to state management requirements, coordination needs between agent instances, and the difficulty of partitioning autonomous reasoning processes across multiple nodes.

Horizontal scaling strategies include: *stateless agent design that enables easy replication* ➤ *intelligent work distribution across agent instances* ➤ *shared state management for coordination* ➤ *load balancing that considers task affinity*. The scaling system must maintain coordination capabilities while distributing workload effectively.

Scaling patterns must also account for external service dependencies, implement appropriate circuit breakers and fallback strategies, and maintain performance consistency across different load levels. Effective horizontal scaling preserves the collaborative capabilities that make multi-agent systems powerful while enabling the system to handle increased workloads.

Safety: Action Boundaries, Human Oversight, and Audit Trails

Production agentic systems require comprehensive safety mechanisms that prevent harmful actions, maintain appropriate human oversight, and provide complete audit trails for accountability and compliance. Safety considerations become particularly important as systems gain more autonomy and interact with critical business processes.

Action Boundaries and Constraint Systems

Safety boundaries define what actions agentic systems can take autonomously and what operations require human approval or oversight. These boundaries must be carefully designed to maintain system usefulness while preventing potentially harmful autonomous actions.

```
(Geoffrey Hinton)-[:AUTHORED {year: 2006}]->(Paper: "A Fast
Learning Algorithm for Deep Belief Nets")
(Paper: "A Fast Learning Algorithm for Deep Belief Nets")-
[:INTRODUCES]->(Concept: "Deep Belief Networks")
(Geoffrey Hinton)-[:AFFILIATED_WITH {since: 1987}]->(University
of Toronto)
```

Safety boundary implementation includes: *resource usage limits to prevent system abuse* ➤ *action type restrictions for sensitive operations* ➤ *approval workflows for high-risk activities* ➤ *audit requirements for compliance.* The boundary system must balance safety with operational efficiency, providing clear guidelines for autonomous operation while maintaining necessary controls.

Constraint systems also implement escalation procedures when agents encounter boundary limits, provide clear communication about safety restrictions, and maintain flexibility to adapt boundaries based on operational experience and changing requirements.

Human Oversight and Intervention Mechanisms

Human oversight systems provide mechanisms for monitoring autonomous operations, intervening when necessary, and maintaining appropriate human control over critical decisions. Effective oversight balances automation benefits with human judgment and accountability.

Oversight mechanisms include: *real-time monitoring dashboards* ➤ *alert systems for unusual behaviors* ➤ *manual intervention capabilities* ➤ *approval workflows for sensitive operations*. The oversight system must provide sufficient visibility into autonomous operations while avoiding information overload that reduces oversight effectiveness.

Human-in-the-loop patterns enable seamless collaboration between autonomous agents and human operators, with clear handoff procedures, appropriate escalation triggers, and effective communication interfaces that support human decision-making when intervention is required.

Comprehensive Audit Trails and Compliance

Audit trail systems maintain complete records of autonomous system operations, enabling accountability, compliance verification, and post-incident analysis. Audit trails must capture both technical operations and business-relevant decisions that autonomous systems make.

```python
def create_audit_entry(self, action: Dict, decision_context:
Dict, outcome: Dict):
    """Create comprehensive audit trail entry"""
    audit_entry = {
        "timestamp": datetime.now(),
        "action_type": action["type"],
        "decision_rationale": decision_context["reasoning"],
        "outcome": outcome,
        "compliance_flags": self._check_compliance_
        requirements(action),
```

```
    "human_oversight": decision_context.get("human_
    approval", False)
}
self.audit_log.append(audit_entry)
```

Audit trail implementation includes: *complete decision tracking* ➤ *outcome documentation* ➤ *compliance flag management* ➤ *retention policy enforcement.* The audit system must provide searchable records, support compliance reporting, and enable forensic analysis when problems occur.

Comprehensive audit trails also support continuous improvement by providing data for performance analysis, identifying patterns in system behavior, and enabling evidence-based optimization of safety boundaries and oversight procedures.

The evaluation and production deployment considerations explored in this section represent critical capabilities that distinguish robust, production-ready agentic systems from research prototypes. These capabilities ensure that sophisticated autonomous behaviors translate into reliable, safe, and valuable real-world applications that organizations can deploy with confidence.

Production deployment patterns, monitoring systems, and safety mechanisms are comprehensively demonstrated in Notebook: Production Deployment and Advanced Patterns, which provides practical implementations of evaluation frameworks, monitoring infrastructure, and safety controls.

Summary and Next Steps

Agentic RAG represents a fundamental shift from passive information retrieval to autonomous reasoning systems that can plan, adapt, and collaborate. This evolution transforms static knowledge bases into dynamic reasoning partners capable of complex investigation and analysis.

Key Takeaways

Key Implementation Principles

Start Simple: Begin with single-agent systems before advancing to multi-agent collaboration. Master basic planning and tool integration first.

Design for Observability: Implement monitoring and decision tracking from day one. You cannot optimize what you cannot measure.

Balance Autonomy with Control: Provide sufficient freedom for effectiveness while maintaining safety boundaries and human oversight for critical decisions.

Leverage Collaboration: Use specialized agent roles and peer review for built-in quality assurance.

Next Ses

The Path Forward

The frameworks and patterns explored in this chapter—from ReAct agents to CrewAI collaboration—provide the foundation for building production-ready autonomous information systems. Success requires combining solid engineering practices with thoughtful attention to safety, monitoring, and human cooperation.

Agentic RAG is not about replacing human intelligence but about creating collaborative partnerships in autonomous agents handle routine reasoning while humans focus on strategic decisions and complex judgment calls.

The future of information work lies in this collaboration, transforming how we discover, analyze, and act upon knowledge in an increasingly complex world.

PART IV

Production and Evaluation

## CHAPTER 14

RAG Evaluation: Measuring Quality and Performance

Building a RAG system is only half the battle. The other half, arguably the more critical half, is understanding whether your system works as intended. Throughout this book, we've explored how to construct sophisticated RAG architectures, from basic retrieval-augmented pipelines to advanced agentic systems. But without proper evaluation, you're essentially flying blind, unable to distinguish between a system that occasionally gets lucky and one that consistently delivers reliable, accurate results.

Consider this scenario: Your RAG-powered customer support system appears to be working well based on casual testing. Users seem satisfied, and the responses look reasonable. But beneath the surface, the system might be hallucinating facts 15% of the time, missing critical context in 30% of queries, or providing irrelevant answers that happen to sound authoritative. Without systematic evaluation, these problems remain invisible until they cause real damage—lost customers, misinformed decisions, or worse.

This is the central challenge of RAG evaluation: **traditional software testing approaches don't apply to systems that generate natural language responses based on retrieved context**. You can't simply write unit tests that check if a function returns the expected output. Instead, you need frameworks assessing the nuanced interplay between retrieval quality, contextual relevance, factual accuracy, and response coherence.

Why RAG Evaluation Is Uniquely Complex

RAG systems present evaluation challenges that don't exist in traditional software applications or even in stand-alone language models. These challenges stem from the multistage nature of RAG architectures:

The Retrieval Challenge: Your system must first identify and retrieve the most relevant information from potentially millions of documents. Traditional information retrieval metrics like precision and recall provide some insight, but they don't capture whether the retrieved context enables good generation. A document might be technically relevant to a query but fail to contain the specific information needed to answer it accurately.

The Generation Challenge: Even with perfect retrieval, the language model must synthesize retrieved information into coherent, accurate, and helpful responses. The model might ignore crucial retrieved context, hallucinate information not present in the documents, or fail to integrate multiple sources effectively. Traditional text generation metrics like BLEU or ROUGE capture surface-level similarity but miss deeper issues of factual accuracy and contextual faithfulness.

The End-to-End Challenge: The most insidious problems in RAG systems emerge from the interaction between retrieval and generation components. Poor retrieval can lead to hallucinations, while overly conservative retrieval might result in responses that are accurate but incomplete. The system's overall performance depends on how well these components work together, not just how well they perform individually.

The Ground Truth Problem: Unlike traditional machine learning tasks, where you have clear input–output pairs, RAG evaluation often lacks definitive "correct" answers. Multiple valid responses might exist for the same query, and the quality of a response depends on subjective factors like helpfulness, clarity, and completeness.

The Evolution of RAG Evaluation

Early RAG systems were evaluated using ad hoc methods—manual inspection of sample outputs, basic keyword matching, or repurposed metrics from other NLP tasks. These approaches quickly proved inadequate as RAG systems became more sophisticated and were deployed in production environments.

The breakthrough came with the recognition that large language models themselves could serve as sophisticated evaluators. This "LLM-as-a-judge" paradigm leverages the reasoning capabilities of advanced models to assess aspects of RAG performance that traditional metrics couldn't capture. Instead of counting word overlaps or measuring vector similarities, LLM evaluators can assess whether a response is factually grounded in the provided context, fully addresses the user's question, and whether the retrieved context is relevant to the query.

This evolution has culminated in frameworks like RAGAS (Retrieval-Augmented Generation Assessment), which provide standardized, automated approaches to evaluating RAG systems across multiple dimensions of quality. These frameworks don't replace human judgment but provide scalable, consistent evaluation that can guide system development and monitor production performance.

Modern Evaluation Approaches: A Multifaceted Strategy

Today's best practices for RAG evaluation recognize that no single metric or method provides a complete picture of system performance. Instead, effective evaluation strategies combine multiple complementary approaches:

Automated Metrics: Frameworks like RAGAS assess key quality dimensions—faithfulness, relevance, precision, and recall. These metrics enable rapid iteration during development and continuous monitoring in production.

Traditional IR Metrics: Classic information retrieval metrics remain valuable for assessing the retrieval component in isolation. Understanding whether your system retrieves relevant documents is foundational to diagnosing performance issues.

Human Evaluation: Despite advances in automated evaluation, human judgment remains the gold standard for assessing subjective qualities like helpfulness, clarity, and user satisfaction. Well-designed human evaluation studies provide crucial validation of automated metrics.

Production Monitoring: Real-world RAG systems require continuous evaluation as data distributions shift, user needs evolve, and system components degrade. Automated monitoring pipelines can detect performance drift and alert teams to emerging issues.

What This Chapter Will Teach You

This chapter will equip you with a comprehensive toolkit for evaluating RAG systems effectively. You'll learn to

1. **Implement RAGAS evaluation**: Master the leading framework for automated RAG assessment, understanding when and how to apply its core metrics—faithfulness, answer relevancy, context precision, and context recall.

2. **Design comprehensive evaluation strategies**: Combine automated metrics, traditional IR approaches, and human evaluation to create robust assessment frameworks tailored to your specific use case.

3. **Build production evaluation pipelines**: Create automated systems that continuously monitor RAG performance, detect drift, and alert teams to quality issues before they impact users.

4. **Avoid common evaluation pitfalls**: Learn from the mistakes that trip up even experienced practitioners, from dataset bias to metric gaming to evaluation–production mismatches.

By the end of this chapter, you'll have the knowledge and tools to ensure your RAG systems don't just work—they work reliably, accurately, and in ways you can measure and improve over time. The era of "hope and pray" RAG deployment is over. It's time to build trustworthy systems because you know exactly how well they perform.

The RAGAS Framework

RAGAS Core Philosophy

RAGAS (Retrieval-Augmented Generation Assessment) represents a paradigm shift in how we evaluate RAG systems. Born from the recognition that traditional NLP metrics fail to capture the nuanced requirements of retrieval-augmented generation, RAGAS introduces a fundamentally different approach: using large language models as sophisticated judges of RAG system performance.

The LLM-As-a-Judge Revolution

The core insight behind RAGAS is deceptively simple yet profound: **if large language models are sophisticated enough to power RAG systems, they're sophisticated enough to evaluate them**. This "LLM-as-a-judge" paradigm leverages the reasoning capabilities of advanced models like GPT-4 or Claude to assess aspects of RAG performance that traditional metrics simply cannot capture.

Consider the challenge of evaluating whether a RAG system's response is "faithful" to its retrieved context. Traditional approaches might use keyword overlap or semantic similarity scores, but these miss crucial nuances. A response might use words completely different from the source document while remaining perfectly faithful to its meaning. Conversely, a response might copy exact phrases from the source while completely misrepresenting their context or significance.

An LLM evaluator, however, can read both the source context and the generated response, understand their semantic relationship, and make nuanced judgments about faithfulness that mirror human reasoning. It can detect subtle hallucinations, identify when responses go beyond the context's support, and recognize when accurate information is presented in misleading ways.

Why Traditional Metrics Fall Short for RAG

To understand RAGAS's revolutionary approach, we need to examine why traditional evaluation metrics prove inadequate for RAG systems:

Surface-Level Similarity Isn't Enough: Metrics like BLEU and ROUGE measure n-gram overlap between generated and reference texts. While useful for tasks like machine translation where there's a clear "correct" output, they fail for RAG systems where multiple valid responses exist and where semantic accuracy matters more than lexical similarity.

For example, consider these two responses to the question "What causes diabetes?":

Response A: "Diabetes is caused by insufficient insulin production or insulin resistance." *Response B*: "The pancreas fails to produce adequate insulin, or cells become resistant to insulin's effects, leading to diabetes."

Traditional metrics would score these very differently despite both being equally accurate. RAGAS evaluators can recognize their semantic equivalence.

Context Awareness is Missing: Traditional metrics evaluate generated text in isolation, ignoring the retrieved context that should ground RAG responses. They can't assess whether a response appropriately uses its source material or whether it introduces information not present in the retrieval results.

Retrieval Quality Gets Lost: Most traditional approaches focus solely on the final generated output, providing no insight into whether the retrieval component performed well. A RAG system might generate a perfect response based on irrelevant retrieved documents—a concerning failure mode that traditional metrics would miss entirely.

The Hallucination Blind Spot: Perhaps most critically, traditional metrics cannot reliably detect hallucinations—instances where the model generates factual-sounding but incorrect information. A response might score highly on ROUGE while containing dangerous misinformation.

RAGAS's Philosophical Foundations

RAGAS is built on several key philosophical principles that differentiate it from traditional evaluation approaches:

Holistic Assessment: Rather than evaluating generation in isolation, RAGAS assesses the entire RAG pipeline. It considers how well the system retrieves relevant information, how effectively it uses that information in generation, and how well the final output serves the user's needs.

Multidimensional Quality: RAGAS recognizes that RAG quality isn't a single number but a multifaceted concept. A response might be highly faithful to its sources but irrelevant to the user's question, or perfectly relevant but based on poorly retrieved context. RAGAS provides separate metrics for different quality dimensions.

Semantic Understanding Over Lexical Matching: By leveraging LLM reasoning capabilities, RAGAS evaluates meaning rather than surface forms. This enables more nuanced and accurate quality assessment that aligns better with human judgment.

Automated Yet Sophisticated: While RAGAS provides automated evaluation, it maintains the sophistication typically associated with human assessment. This combination enables both rapid iteration during development and scalable monitoring in production.

Reference-Free Evaluation: Unlike traditional metrics that require "gold standard" reference answers, RAGAS can evaluate RAG systems using only the query, retrieved context, and generated response. This is crucial because creating comprehensive reference datasets for RAG applications is often impractical.

The Four Pillars of RAGAS Evaluation

RAGAS structures its evaluation around four core dimensions that together provide a comprehensive view of RAG system performance:

Faithfulness: Does the generated response accurately reflect the information in the retrieved context? This metric detects hallucinations and ensures that responses are grounded in the provided source material.

Answer Relevancy: How well does the generated response address the user's specific question? This assesses whether the system understood the query and provided appropriate information.

Context Precision: Among the retrieved documents, how many of the top-ranked results are actually relevant to the query? This evaluates the quality of the retrieval ranking.

Context Recall: Of all the relevant information available in the knowledge base, how much was successfully retrieved? This measures the completeness of the retrieval process.

Together, these four metrics provide visibility into every stage of the RAG pipeline, from initial retrieval through final generation.

Practical Implications of the RAGAS Approach

The LLM-as-a-judge paradigm introduced by RAGAS has profound practical implications for RAG system development:

Faster Iteration Cycles: Developers can rapidly test changes to their RAG systems without waiting for human evaluation studies. RAGAS provides immediate, sophisticated feedback that guides optimization efforts.

Consistent Evaluation Standards: Unlike human evaluation, which can vary between annotators and over time, RAGAS provides consistent assessment criteria. This enables reliable comparison between different system configurations and tracking of performance over time.

Scalable Quality Assurance: RAGAS can evaluate thousands of query–response pairs automatically, enabling comprehensive testing that would be prohibitively expensive with human evaluators.

Production Monitoring: The automated nature of RAGAS makes it suitable for continuous monitoring of production RAG systems, detecting performance degradation or drift in real time.

However, the LLM-as-a-judge approach also introduces new considerations. The quality of RAGAS evaluation depends on the sophistication of the judge model, and different LLMs might provide varying assessments. Additionally, LLM evaluators can have their own biases and limitations, making it important to validate RAGAS scores against human judgment in critical applications.

Looking Ahead: RAGAS in Practice

Understanding RAGAS's philosophical foundations prepares us for the practical implementation details we'll explore in the following sections. The framework's emphasis on semantic understanding, multidimensional assessment, and automated sophistication makes it an invaluable tool for RAG system development and deployment.

In the next section, we'll dive deep into each of RAGAS's four core metrics, exploring how they work, when to use them, and how to interpret their results. We'll see how these abstract concepts translate into concrete tools for building better RAG systems.

Essential RAGAS Metrics

The power of RAGAS lies in its four core metrics, each designed to evaluate a specific dimension of RAG system performance. These metrics work together to provide comprehensive coverage of the entire RAG pipeline, from retrieval quality through generation accuracy.

Basic RAGAS Setup

RAGAS evaluation follows a straightforward pattern that enables rapid assessment of all four metrics:

```
from ragas import evaluate
from ragas.metrics import faithfulness, answer_relevancy,
context_precision, context_recall

result = evaluate(dataset, metrics=[faithfulness,
answer_relevancy, context_precision, context_recall])
```

This single function call provides comprehensive RAG evaluation across all dimensions. The real value lies in understanding what each metric reveals about your system's performance.

Faithfulness: The Foundation of Trust

Faithfulness is arguably the most critical RAGAS metric because it directly addresses hallucination—one of the most serious problems in RAG systems. This metric evaluates whether the generated response accurately reflects information in the retrieved context, without adding, distorting, or misrepresenting facts.

How Faithfulness Works: The evaluation process involves three sophisticated LLM-powered steps. First, the evaluator analyzes the generated response and identifies all factual claims or assertions it contains, going beyond simple sentence parsing to understand semantic claims that might span multiple sentences. Second, each extracted claim is systematically checked against the provided context documents to determine if it can be directly supported. Finally, the faithfulness score is calculated as the ratio of verified claims to total claims.

```
faithfulness_score = result['faithfulness']
# Score interpretation
if faithfulness_score >= 0.9:
    status = "Excellent: Highly faithful to context"
elif faithfulness_score >= 0.7:
    status = "Good: Mostly grounded responses"
else:
    status = "Concern: Potential hallucination issues"
```

Score Interpretation: Faithfulness scores range from 0.0 to 1.0. Scores above 0.9 indicate excellent faithfulness with nearly every claim supported by context. Scores between 0.7 and 0.8 represent good faithfulness with minimal unsupported assertions. Scores in the 0.5–0.6 range suggest moderate faithfulness with significant hallucination risks, while scores below 0.5 indicate serious reliability problems requiring immediate attention.

Common Faithfulness Issues: The most frequent problems include inference overreach where the model makes logical leaps beyond what the context supports, precision errors where responses are more specific than the context warrants, and causal attribution where the model claims causation when the context only establishes correlation.

Real-World Example: Consider a question about diabetes treatment. A response stating "Insulin therapy helps manage blood sugar levels in diabetic patients" would score high on faithfulness if the context contains information about insulin's role. However, adding "Insulin therapy cures diabetes completely" would lower the faithfulness score significantly, even if everything else is accurate.

Answer Relevancy: Measuring Response Appropriateness

Answer relevancy evaluates how well the generated response addresses the specific question asked. A response can be factually accurate and well-grounded in context but still receive a low relevancy score if it doesn't actually answer the user's question.

The Reverse-Engineering Approach: RAGAS uses an innovative method for measuring relevancy. Instead of directly judging whether an answer addresses a question (which can be subjective), it reverse-engineers the process. Given the generated answer, the evaluator LLM generates multiple questions that this answer would appropriately address. The original question is then compared with these generated questions using semantic similarity measures. Higher similarity indicates better relevancy.

```
relevancy_score = result['answer_relevancy']
# Diagnostic pattern for common issues
if faithfulness_score > 0.8 and relevancy_score < 0.6:
    issue = "Accurate but off-topic responses"
    solution = "Improve query understanding"
```

This approach is particularly clever because it provides more objective and consistent evaluation. Rather than asking "Does this answer address the question?" it asks "What questions would this answer appropriately address?" and measures the overlap.

Score Ranges and Interpretation: Scores above 0.9 indicate excellent relevancy where the answer directly and completely addresses the question. Scores between 0.7 and 0.8 represent good relevancy with minor tangents or incomplete coverage. Moderate relevancy (0.5–0.6) suggests the answer is partially relevant but misses key aspects, while scores below 0.5 indicate the response addresses a different question or topic entirely.

Common Relevancy Issues: Frequent problems include scope mismatch (answering a narrow question with broad information or vice versa), focus drift (starting relevant but moving to tangential topics), partial coverage (addressing only part of a multipart question), and wrong granularity (providing detail levels that don't match what the question seeks).

Context Precision: Evaluating Retrieval Quality

Context precision evaluates the quality of the retrieval component by measuring how much of the retrieved context is actually relevant to answering the question. High precision means the retrieval system efficiently finds useful information without cluttering the context with irrelevant documents.

Beyond Simple Relevance Counting: Context precision goes beyond just counting relevant documents because the order of retrieved documents significantly impacts RAG performance. Most generation models give more attention to earlier context, so having relevant information ranked higher is crucial for good generation quality. The evaluation process analyzes both the relevance of individual documents and whether relevant documents are properly ranked above irrelevant ones.

```
precision_score = result['context_precision']
recall_score = result['context_recall']
# Trade-off analysis
if precision_score < 0.7 and recall_score > 0.8:
    strategy = "Implement relevance filtering"
elif precision_score > 0.8 and recall_score < 0.7:
    strategy = "Expand retrieval parameters"
```

Balancing Signal and Noise: The metric addresses a fundamental challenge in RAG systems: the tension between comprehensive retrieval and focused generation. Too much irrelevant context can confuse the generation model and lead to poor responses, while too little context might miss crucial information. Context precision helps identify when retrieval is introducing too much noise into the generation process.

Optimization Strategies: Improving context precision typically involves better embedding models (especially domain-specific ones), enhanced query processing and expansion techniques, sophisticated re-ranking algorithms based on multiple signals, and relevance filtering before presenting context to the generation model.

Context Recall: Measuring Retrieval Completeness

Context recall evaluates whether the retrieval system successfully found all the information needed to answer the question completely. While precision focuses on avoiding irrelevant information, recall focuses on not missing relevant information.

The Ground Truth Requirement: Unlike other RAGAS metrics, context recall requires reference answers to determine what information should have been retrieved. Using the ground truth answer, the evaluator LLM identifies key information necessary to answer the question properly, then checks whether this necessary information is present in the retrieved context documents. The score represents the proportion of required information that was successfully retrieved.

```
recall_score = result['context_recall']
# Calculate completeness
completeness = f"Retrieved {recall_score:.1%} of required
information"
```

Application-Specific Trade-offs: The balance between precision and recall depends heavily on your specific application. High-stakes applications (medical, legal) might prioritize recall to avoid missing critical information, accepting some noise in the process. User-facing applications might prioritize precision to provide cleaner, more focused responses. Exploratory applications might accept lower precision for comprehensive coverage.

The Four-Metric Diagnostic Framework

The true power of RAGAS emerges when all four metrics are considered together, as different combinations reveal specific system issues:

```
def diagnose_system(faithfulness, relevancy, precision,
recall):
    if faithfulness < 0.7:
        return "Priority: Address hallucination"
    elif relevancy < 0.7:
        return "Priority: Improve query understanding"
    elif precision < 0.7:
        return "Priority: Enhance retrieval ranking"
    elif recall < 0.7:
        return "Priority: Expand knowledge coverage"
    return "System performing well"
```

High Faithfulness + Low Relevancy indicates the system is accurate but doesn't understand what users are asking, suggesting improvements needed in query understanding and response filtering. **High Relevancy + Low Faithfulness** means the system understands questions but hallucinates answers, requiring focus on generation prompting and fact-checking. **High Precision + Low Recall** suggests retrieval is accurate but incomplete, while **High Recall + Low Precision** indicates retrieval is comprehensive but noisy.

Optimization Priority Framework: Address issues in order of impact: Faithfulness (trust) comes first because unreliable systems are dangerous regardless of other qualities. Relevancy (usefulness) comes second because irrelevant responses frustrate users. Precision (efficiency) and Recall (completeness) balance based on application requirements.

Practical Implementation Considerations

Computational Efficiency: RAGAS metrics require LLM calls for evaluation, introducing both latency and cost considerations. For large-scale evaluation, use efficient models like GPT-3.5 rather than GPT-4 for routine monitoring, implement sampling strategies for very large datasets, and consider caching evaluation results for repeated tests.

```
# Cost-efficient evaluation strategy
if len(dataset) > 1000:
    sample_size = min(200, len(dataset) // 5)
    eval_dataset = dataset.shuffle().select(range(sample_size))
else:
    eval_dataset = dataset
```

Consistency and Validation: LLM-based evaluation can have some variability, so ensure reliable results by using temperature=0 for deterministic evaluation, running multiple evaluation passes for critical assessments, and validating automated scores against human judgment periodically.

Domain Adaptation: While RAGAS provides excellent general-purpose evaluation, consider domain-specific adaptations by adjusting evaluation prompts for technical domains, adding domain-specific criteria to the evaluation framework, and validating that automated scores align with domain expert judgment.

The four RAGAS metrics together provide unprecedented insight into RAG system performance, enabling data-driven optimization and reliable quality assurance. Each metric serves as a lens for understanding different aspects of system behavior, and their combination creates a comprehensive framework for building trustworthy, effective RAG applications.

Implementation and Custom Metrics

While RAGAS provides excellent general-purpose evaluation metrics, real-world RAG applications often require specialized assessment criteria tailored to specific domains, use cases, or organizational requirements. This section explores how to implement RAGAS evaluation pipelines effectively and create custom metrics that address unique evaluation needs beyond the standard four metrics.

Setting Up RAGAS Evaluation Pipelines

Creating a robust RAGAS evaluation pipeline involves more than simply calling the evaluate function. Production-ready pipelines need to handle data preparation, error management, result interpretation, and integration with existing development workflows.

Data Preparation and Validation: The foundation of reliable evaluation is properly formatted data. RAGAS expects specific data structures, and ensuring your evaluation dataset meets these requirements and prevents runtime errors and invalid results.

```
dataset = Dataset.from_dict({
    'question': questions,
    'contexts': contexts,  # List of lists
```

```
'answer': answers,
'ground_truth': ground_truths or [""] * len(questions)
})
```

The contexts field requires special attention—it should be a list of lists, where each inner list contains the retrieved documents for that specific question. Proper validation ensures data consistency and provides default values where needed.

Efficient Evaluation Execution: For production use, evaluation pipelines need to balance thoroughness with computational efficiency. Large datasets can result in significant API costs and processing time, making sampling strategies essential.

```
# Sample for large datasets to control costs
eval_dataset = dataset.shuffle().select(range(min(sample_size,
len(dataset))))
result = evaluate(eval_dataset, metrics=[faithfulness, answer_
relevancy, context_precision, context_recall])
```

Error Handling and Robustness: Real-world evaluation pipelines must gracefully handle various failure modes, from API rate limits to malformed data. The most common issues include context lists that are too long (exceeding LLM token limits), malformed questions or answers, and API timeouts during evaluation. Implementing retry logic, data validation, and graceful degradation helps maintain pipeline reliability.

Result Aggregation and Reporting: Raw RAGAS scores need to be transformed into actionable insights. This involves calculating aggregate statistics, identifying patterns across different question types, and generating reports that guide optimization efforts.

```
scores = {metric: np.mean(result[metric]) for metric in
result.keys()}
# Pseudocode: if score < threshold → add to issues list
# Pseudocode: generate_optimization_recommendations(scores)
```

Creating Domain-Specific Custom Metrics

While RAGAS's four core metrics provide comprehensive general evaluation, many applications benefit from additional metrics that capture domain-specific quality dimensions. Custom metrics can address specialized requirements like medical accuracy, legal compliance, financial regulation adherence, or technical precision.

The Custom Metric Framework: RAGAS custom metrics follow a consistent pattern—they take the same inputs as standard metrics (question, context, answer, and optionally ground truth) and return numerical scores. The key is designing evaluation prompts that capture your specific quality criteria.

Medical Domain Example: Healthcare applications require evaluation beyond general accuracy—they need to assess medical correctness, safety implications, appropriate disclaimers, and scope limitations.

```
medical_prompt = """
Evaluate medical response for accuracy and safety:
Question: {question}, Context: {context}, Answer: {answer}
Rate 0.0-1.0 on: Medical accuracy, Safety, Disclaimers
Format: Score: X.X
"""

def medical_accuracy(row):
    response = llm.predict(medical_prompt.format(**row))
    return float(response.split('Score:')[1].strip())
```

This example demonstrates the basic pattern for custom metrics: define evaluation criteria, create a structured prompt, and implement scoring logic. The prompt explicitly guides the LLM evaluator to consider domain-specific factors that general metrics might miss.

Legal Domain Considerations: Legal applications require metrics that assess legal accuracy, jurisdiction awareness, appropriate disclaimers about legal advice, and clarity in explaining complex legal concepts. A legal precision metric might evaluate whether responses correctly explain legal principles while appropriately advising readers to consult qualified attorneys.

Financial Compliance Metrics: Financial applications need metrics that evaluate regulatory compliance, risk disclosure adequacy, accuracy of financial calculations, and appropriate investment disclaimers. These metrics help ensure responses meet fiduciary standards and regulatory requirements.

Technical Accuracy Metrics: Engineering and technical applications might require metrics that assess code correctness, algorithm accuracy, best practice adherence, and security considerations.

Advanced Custom Metric Patterns

Multidimensional Scoring: Some applications benefit from metrics that evaluate multiple dimensions simultaneously rather than single scores. Multidimensional metrics can provide granular feedback on different quality aspects.

```
def multi_dimensional_evaluator(row):
    scores = {}
    # Pseudocode: for each criterion → evaluate_criterion(row,
    criterion)
    # Pseudocode: calculate weighted_overall_score(scores,
    weights)
    return {'overall': overall_score, 'breakdown': scores}
```

Comparative Metrics: Some evaluation needs require comparing responses against multiple standards or benchmarks. Comparative metrics can assess how well responses perform relative to expert answers, competing systems, or historical performance.

Temporal Consistency Metrics: Applications that handle time-sensitive information might need metrics that evaluate temporal accuracy, currency of information, and appropriate handling of time-dependent queries.

Implementing Custom Metric Evaluation

Integration with RAGAS Framework: Custom metrics can be integrated into existing RAGAS evaluation pipelines, allowing you to combine standard and domain-specific evaluation in a single workflow.

```
# Standard RAGAS evaluation
ragas_result = evaluate(dataset, metrics=[faithfulness, answer_
relevancy, context_precision, context_recall])

# Custom metric evaluation
custom_results = {name: np.mean([func(row) for row in dataset])
                  for name, func in custom_metrics.items()}
```

Validation Against Human Judgment: Custom metrics should be validated against human expert judgment to ensure they accurately capture the intended quality dimensions. This validation process helps refine metric definitions and builds confidence in automated evaluation results.

Performance Optimization: Custom metrics that rely on LLM evaluation can be computationally expensive. Optimization strategies include using faster models for initial screening, implementing caching for repeated evaluations, and developing heuristic pre-filters that identify cases requiring detailed evaluation.

Metric Composition and Weighting

Weighted Scoring Systems: Different applications may prioritize different quality dimensions. A weighted scoring system allows you to combine multiple metrics while emphasizing the most critical aspects for your specific use case.

```
# Example weighting for high-stakes application
weights = {'faithfulness': 0.4, 'answer_relevancy': 0.2,
           'context_precision': 0.2, 'domain_accuracy': 0.2}
weighted_score = sum(scores[m] * weights[m] for m in scores) /
sum(weights.values())
```

Threshold-Based Quality Gates: Production systems often need binary quality decisions rather than continuous scores. Threshold-based systems can automatically flag responses that fall below acceptable quality levels across any evaluation dimension.

Adaptive Weighting: Advanced implementations might use adaptive weighting that adjusts metric importance based on query type, user context, or system confidence. This allows for more nuanced quality assessment that reflects the varying requirements of different interactions.

Practical Implementation Strategies

Development Workflow Integration: Effective custom metrics integrate seamlessly into existing development workflows. This includes integration with continuous integration pipelines, automatic evaluation of model updates, and integration with A/B testing frameworks for system improvements.

Monitoring and Alerting: Production custom metrics should include monitoring capabilities that track metric performance over time and alert teams to quality degradation.

```
def check_quality_thresholds(current_scores, thresholds):
    alerts = [f"{metric} below threshold" for metric, score in
    current_scores.items()
            if metric in thresholds and score <
            thresholds[metric]]
    return alerts
```

Documentation and Sharing: Custom metrics should be well-documented with clear descriptions of what they measure, how they're calculated, and what scores mean in practical terms. This documentation enables team members to effectively use and interpret custom evaluation results.

Best Practices for Custom Metric Development

Start Simple, Iterate: Begin with straightforward custom metrics that address your most pressing evaluation needs. Complex multidimensional metrics can be developed iteratively as you gain experience with simpler approaches and better understand your specific requirements.

Validate Early and Often: Regular validation against human judgment ensures custom metrics remain aligned with actual quality requirements. This validation should include edge cases and challenging examples that test metric robustness.

Consider Computational Costs: Balance metric sophistication with computational efficiency. Overly complex custom metrics can make evaluation prohibitively expensive, limiting their practical utility in development and production monitoring.

Maintain Metric Quality: Custom metrics require ongoing maintenance to ensure they remain effective as your system and requirements evolve. Regular review and refinement help maintain metric relevance and accuracy.

The combination of RAGAS's sophisticated general-purpose metrics with thoughtfully designed custom metrics creates a comprehensive evaluation framework that can assess RAG systems across all dimensions of quality relevant to your specific application. This comprehensive approach enables both rapid development iteration and confident production deployment, ensuring your RAG systems meet both general quality standards and domain-specific requirements.

Complementary Evaluation Methods

While RAGAS provides sophisticated evaluation tailored specifically for RAG systems, it's valuable to understand and selectively apply traditional evaluation approaches that can complement RAGAS metrics. These traditional methods offer different perspectives on system performance and can provide useful benchmarks, especially when comparing against established baselines or conducting academic research.

Traditional IR Metrics: When to Use Precision@K, Recall@K, NDCG

Traditional Information Retrieval (IR) metrics focus on evaluating the retrieval component of RAG systems in isolation. These metrics have decades of research behind them and provide well-understood benchmarks for retrieval quality. While they don't capture the nuanced requirements of RAG generation, they remain valuable for diagnosing retrieval performance and comparing against established baselines.

Precision@K: Measuring Retrieval Accuracy

What Precision@K Measures: Precision@K evaluates what fraction of the top K retrieved documents are actually relevant to the query. This metric focuses on the accuracy of high-ranked results, which is crucial since most RAG systems only use the top few retrieved documents for generation.

When Precision@K is Useful: This metric excels in scenarios where retrieval accuracy is more important than completeness. Applications with limited context windows, real-time response requirements, or user-facing systems that need clean, focused information often prioritize precision over recall.

```
def precision_at_k(retrieved_docs, relevant_docs, k):
    top_k = retrieved_docs[:k]
    relevant_in_top_k = sum(1 for doc in top_k if doc in
    relevant_docs)
    return relevant_in_top_k / k
```

Interpretation and Thresholds: Precision@K scores range from 0.0 to 1.0. High-performing retrieval systems typically achieve Precision@5 scores above 0.8, while scores below 0.6 suggest significant retrieval quality issues. However, these thresholds vary significantly by domain—technical documentation might achieve higher precision than general web search.

Limitations for RAG: Precision@K measures document-level relevance but can't assess whether retrieved documents contain the specific information needed to answer the query. A document might be topically relevant but lack the crucial details required for accurate generation.

Recall@K: Measuring Retrieval Completeness

What Recall@K Measures: Recall@K evaluates what fraction of all relevant documents in the collection appear among the top K retrieved results. This metric focuses on completeness—ensuring that important information isn't missed during retrieval.

When Recall@K is Critical: High-stakes applications like medical research, legal discovery, or comprehensive analysis scenarios often prioritize recall. Missing critical information can have serious consequences, making completeness more important than efficiency.

```
def recall_at_k(retrieved_docs, relevant_docs, k):
    top_k = set(retrieved_docs[:k])
    relevant_found = sum(1 for doc in relevant_docs if doc
    in top_k)
    return relevant_found / len(relevant_docs)
```

The Precision–Recall Trade-off: Increasing K typically improves recall but may hurt precision. The optimal K depends on your application's tolerance for noise vs. sensitivity to missing information. Production systems often need to balance this trade-off based on user experience requirements and computational constraints.

Challenges in RAG Context: Recall@K requires knowing all relevant documents in the collection, which is often impractical for large knowledge bases. Additionally, recall doesn't indicate whether the retrieved information is sufficient to answer the specific question posed.

NDCG: Normalized Discounted Cumulative Gain

What NDCG Measures: NDCG evaluates both relevance and ranking quality by considering the position of relevant documents in the retrieval results. Documents ranked higher contribute more to the score, reflecting the reality that users and RAG systems pay more attention to top-ranked results.

The Discounting Principle: NDCG applies logarithmic discounting to lower-ranked positions, recognizing that the value of a relevant document decreases as its rank increases. This makes NDCG particularly suitable for RAG evaluation since generation models typically give more weight to earlier context.

```
import math
def ndcg_at_k(retrieved_docs, relevance_scores, k):
    dcg = sum(rel / math.log2(i + 2) for i, rel in
    enumerate(relevance_scores[:k]))
    # Pseudocode: ideal_dcg = calculate_ideal_dcg(sorted_
    relevance_scores, k)
    return dcg / ideal_dcg if ideal_dcg > 0 else 0
```

Graded Relevance: Unlike binary precision and recall, NDCG can handle graded relevance scores (e.g., 0–3 scale) that reflect varying degrees of document usefulness. This nuanced approach better captures the reality that some documents are more helpful than others for answering specific queries.

NDCG Interpretation: NDCG scores range from 0.0 to 1.0, with 1.0 representing perfect ranking of all relevant documents. Good retrieval systems typically achieve NDCG@10 scores above 0.7, while scores below 0.5 indicate significant ranking problems.

When Traditional IR Metrics Add Value to RAG Evaluation

Baseline Establishment: Traditional metrics provide well-understood baselines for comparing retrieval performance across different systems, algorithms, or time periods. This is particularly valuable when evaluating retrieval improvements or comparing against published benchmarks.

Component-Level Diagnosis: IR metrics can isolate retrieval performance from generation quality, helping identify whether poor RAG performance stems from inadequate retrieval or generation issues. If traditional metrics are high but RAGAS scores are low, the problem likely lies in generation rather than retrieval.

Academic and Research Context: Traditional metrics remain standard in academic literature, making them essential for research comparisons and publications. They provide common ground for discussing retrieval performance across different research groups and methodologies.

Rapid Iteration: Traditional IR metrics can be computed quickly without LLM calls, enabling rapid evaluation during retrieval system development. This speed advantage makes them valuable for initial system tuning and A/B testing of retrieval parameters.

Limitations of Traditional IR Metrics for RAG

Context Insensitivity: Traditional metrics evaluate document relevance without considering how well the information enables question answering. A document might be topically relevant but lack the specific details needed for accurate generation.

Binary Thinking: Most traditional metrics treat relevance as binary (relevant/irrelevant) when RAG systems benefit from understanding degrees of usefulness, information completeness, and contextual appropriateness.

Generation Blindness: Traditional metrics can't assess whether retrieved information actually improves generation quality. Documents that score well on traditional metrics might still lead to poor RAG responses due to conflicting information, unclear explanations, or inappropriate detail levels.

Static Evaluation: Traditional metrics assume fixed relevance judgments, but RAG effectiveness often depends on dynamic factors like query complexity, user context, and the specific information needs that emerge during generation.

Best Practices for Using Traditional IR Metrics

Complement, Don't Replace: Use traditional IR metrics alongside RAGAS rather than as substitutes. Traditional metrics provide useful baseline information, while RAGAS captures RAG-specific quality dimensions.

Focus on Ranking Quality: When using traditional metrics for RAG, emphasize ranking-aware measures like NDCG rather than simple precision/recall. Ranking quality is crucial for RAG systems since generation models are sensitive to context order.

```
# Combined evaluation approach
ir_scores = {
    'precision_5': precision_at_k(retrieved, relevant, 5),
    'recall_10': recall_at_k(retrieved, relevant, 10),
    'ndcg_10': ndcg_at_k(retrieved, relevance_scores, 10)
}
# Pseudocode: ragas_scores = evaluate_with_ragas(dataset)
# Pseudocode: combined_analysis = analyze_ir_vs_ragas(ir_
scores, ragas_scores)
```

Domain-Specific Tuning: Adapt traditional metric thresholds to your specific domain and use case. Medical applications might require higher recall thresholds, while user-facing applications might prioritize precision.

Temporal Analysis: Track traditional metrics over time to identify retrieval drift, index quality degradation, or query distribution changes that might not be immediately apparent in end-to-end RAG evaluation.

Integration with Modern RAG Evaluation

Diagnostic Workflows: Traditional IR metrics work best as diagnostic tools within broader evaluation frameworks. Use them to investigate specific performance issues identified through RAGAS evaluation or user feedback.

Ablation Studies: Traditional metrics excel in ablation studies where you need to isolate the impact of specific retrieval improvements. They provide consistent baselines for measuring incremental changes to embedding models, retrieval algorithms, or indexing strategies.

Multilevel Analysis: Combine traditional metrics with RAGAS for comprehensive system understanding:

- Traditional metrics reveal retrieval component performance

- RAGAS metrics assess end-to-end RAG quality

- The combination identifies optimization priorities and validates improvements

Quality Assurance: Traditional metrics can serve as fast quality checks during development, with RAGAS providing deeper assessment for release candidates and production monitoring.

Practical Implementation Considerations

Ground Truth Requirements: Traditional IR metrics require relevance judgments that can be expensive to create and maintain. Consider the cost–benefit trade-off of creating comprehensive relevance datasets vs. investing in more sophisticated end-to-end evaluation.

Metric Selection: Choose traditional metrics based on your specific needs:

- **Precision@K** for applications prioritizing accuracy over completeness

- **Recall@K** for comprehensive coverage requirements

- **NDCG** for ranking-sensitive applications with graded relevance

Evaluation Efficiency: Traditional metrics compute quickly, making them suitable for continuous integration pipelines and rapid experimentation. Use them for initial filtering before applying more expensive RAGAS evaluation to promising candidates.

Traditional IR metrics remain valuable tools in the RAG evaluation toolkit, providing well-understood baselines and enabling component-level performance analysis. When used appropriately alongside modern RAG-specific metrics, they contribute to comprehensive system understanding and effective optimization strategies. The key is recognizing their strengths and limitations, using them to complement rather than replace sophisticated RAG evaluation approaches.

Generation Quality Metrics: BLEU, ROUGE, Semantic Similarity

Traditional generation quality metrics provide complementary perspectives on RAG system outputs by focusing on surface-level and semantic similarities between generated and reference texts. While these metrics have limitations for RAG evaluation, they offer valuable benchmarking capabilities and can quickly identify certain types of generation issues.

BLEU: Measuring N-gram Precision

What BLEU Measures: BLEU (Bilingual Evaluation Understudy) evaluates the precision of n-gram overlaps between generated text and reference answers. Originally designed for machine translation, BLEU focuses on lexical similarity and fluency indicators.

BLEU's Strengths in RAG Context: BLEU excels at detecting when RAG systems produce completely irrelevant responses or when generation quality degrades significantly. It's particularly useful for tracking consistency in factual responses where specific terminology and phrasing matter.

```
from sacrebleu import BLEU
bleu = BLEU()
score = bleu.sentence_score(prediction, [reference]).
score / 100.0
# Pseudocode: batch_scores = [calculate_bleu(pred, ref) for
pred, ref in pairs]
```

Interpretation Guidelines: BLEU scores range from 0.0 to 1.0. Scores above 0.4 indicate strong lexical similarity, while scores between 0.2 and 0.4 suggest moderate overlap. Scores below 0.1 typically indicate poor generation quality or completely different approaches to answering the question.

BLEU Limitations for RAG: BLEU cannot assess semantic equivalence, factual accuracy, or contextual appropriateness. Two semantically identical answers using different vocabulary will receive low BLEU scores despite being equally correct.

ROUGE: Evaluating Recall-Oriented Overlap

What ROUGE Measures: ROUGE (Recall-Oriented Understudy for Gisting Evaluation) focuses on recall of n-grams, capturing how much of the reference content appears in the generated response. ROUGE variants (ROUGE-1, ROUGE-2, ROUGE-L) measure different granularities of overlap.

ROUGE's Value for RAG: ROUGE effectively identifies when RAG responses miss key information present in reference answers. It's particularly valuable for evaluating completeness of factual responses and ensuring important details aren't omitted.

```
from rouge_score import rouge_scorer
scorer = rouge_scorer.RougeScorer(['rouge1', 'rouge2', 'rougeL'])
scores = scorer.score(reference, prediction)
rouge_l_score = scores['rougeL'].fmeasure
```

Understanding ROUGE Variants:

- **ROUGE-1**: Unigram recall, measuring coverage of individual words

- **ROUGE-2**: Bigram recall, capturing phrase-level overlap

- **ROUGE-L**: Longest common subsequence, emphasizing sequential similarity

ROUGE Interpretation: ROUGE scores range from 0.0 to 1.0. Good RAG systems typically achieve ROUGE-L scores above 0.3, while scores below 0.15 suggest significant content gaps or topic drift.

ROUGE Limitations: Like BLEU, ROUGE focuses on surface forms rather than meaning. It cannot detect paraphrasing, semantic equivalence, or assess whether the generated content is actually helpful for answering the question.

Semantic Similarity: Beyond Surface Forms

What Semantic Similarity Measures: Semantic similarity metrics use neural embeddings to assess meaning-level similarity between generated and reference texts. These metrics can capture paraphrasing, conceptual overlap, and semantic equivalence that traditional metrics miss.

Implementation with Sentence Transformers: Modern semantic similarity evaluation typically uses pre-trained sentence embedding models that map text to high-dimensional vector spaces where cosine similarity reflects semantic relatedness.

```
from sentence_transformers import SentenceTransformer
model = SentenceTransformer('all-MiniLM-L6-v2')
embeddings = model.encode([prediction, reference])
similarity = cosine_similarity([embeddings[0]],
[embeddings[1]])[0][0]
```

Advantages for RAG Evaluation: Semantic similarity can identify when RAG responses convey the same information as references despite using different words. This makes it more suitable for evaluating paraphrased answers and responses that restructure reference information.

Score Interpretation: Semantic similarity scores range from -1.0 to 1.0, though practical scores typically fall between 0.0 and 1.0. Scores above 0.8 indicate strong semantic alignment, while scores between 0.6 and 0.8 suggest moderate similarity. Scores below 0.4 typically indicate semantic divergence.

Limitations: Semantic similarity metrics can miss factual errors, hallucinations, and nuanced quality differences. High semantic similarity doesn't guarantee factual accuracy or appropriateness.

When Traditional Generation Metrics Provide Value

Rapid Quality Assessment: Traditional metrics compute quickly without requiring LLM calls, making them suitable for rapid iteration during development and continuous integration pipelines.

Benchmark Comparison: These metrics provide standardized baselines for comparing RAG systems against published results and established benchmarks in academic literature.

Regression Detection: Traditional metrics excel at detecting significant quality regressions during system updates. Sudden drops in BLEU or ROUGE scores can quickly identify when changes negatively impact generation quality.

```
def quick_generation_assessment(predictions, references):
    bleu_avg = np.mean([calculate_bleu(p, r) for p, r in
    zip(predictions, references)])
    rouge_avg = np.mean([calculate_rouge_l(p, r) for p, r in
    zip(predictions, references)])
```

```
# Pseudocode: semantic_avg = calculate_semantic_similarity_
batch(predictions, references)
return {'bleu': bleu_avg, 'rouge_l': rouge_avg, 'semantic':
semantic_avg}
```

Component Isolation: Traditional metrics can isolate generation quality from retrieval performance, helping diagnose whether poor RAG performance stems from generation issues rather than retrieval problems.

Comparative Analysis: Traditional vs. RAGAS Metrics

Complementary Strengths: Traditional metrics and RAGAS serve different purposes and should be used together rather than as alternatives. Traditional metrics provide quick quality checks and benchmark comparisons, while RAGAS offers deep RAG-specific assessment.

Speed vs. Depth Trade-off: Traditional metrics compute in milliseconds, while RAGAS metrics require LLM calls that can take seconds. Use traditional metrics for rapid feedback and RAGAS for thorough evaluation.

Coverage Differences: Traditional metrics focus on text similarity, while RAGAS evaluates faithfulness, relevance, and context usage. This difference makes them complementary rather than competitive.

Practical Implementation Strategies

Multimetric Dashboard: Combine traditional and RAGAS metrics in evaluation dashboards that provide both immediate feedback and deep analysis capabilities.

```
def comprehensive_evaluation_suite(dataset):
    # Fast traditional metrics for immediate feedback
    traditional_scores = quick_generation_
    assessment(predictions, references)
```

729

```
# Detailed RAGAS analysis for deep insights
# Pseudocode: ragas_scores = evaluate_with_ragas(dataset)
# Pseudocode: combined_report = generate_analysis_
report(traditional_scores, ragas_scores)

return traditional_scores, ragas_scores
```

Threshold-Based Filtering: Use traditional metrics as first-pass filters to identify potentially problematic responses before applying more expensive RAGAS evaluation.

Development Workflow Integration: Integrate traditional metrics into continuous integration pipelines for immediate feedback, with periodic RAGAS evaluation for comprehensive assessment.

Limitations and Cautions

Surface-Level Focus: Traditional generation metrics emphasize lexical and shallow semantic similarity rather than the deep quality dimensions most important for RAG applications.

Reference Dependency: All traditional generation metrics require reference answers, which can be expensive to create and may not capture the full range of valid responses to open-ended questions.

Context Blindness: Traditional metrics cannot assess whether generated text appropriately uses retrieved context or whether it introduces information not present in the source documents.

Misleading Scores: High traditional metric scores don't guarantee good RAG performance, and low scores don't necessarily indicate poor quality if the response uses different but equally valid phrasing.

Best Practices for Traditional Generation Metrics

Use as Supplements: Apply traditional metrics alongside RAGAS rather than as primary evaluation tools. They provide useful supplementary information but shouldn't drive main optimization decisions.

Focus on Trends: Traditional metrics are most valuable for tracking trends over time rather than absolute score interpretation. Look for sudden changes that might indicate system regressions.

Domain Calibration: Calibrate traditional metric thresholds for your specific domain and use case. Technical domains might have different score distributions than general conversation.

Multireference Evaluation: When possible, use multiple reference answers to capture the range of valid responses and reduce bias toward specific phrasings.

Traditional generation quality metrics remain useful tools for rapid assessment and benchmark comparison in RAG evaluation. When combined thoughtfully with RAGAS metrics, they contribute to comprehensive system understanding while providing the speed advantages needed for development workflows. The key is understanding their limitations and using them appropriately within a broader evaluation strategy that prioritizes RAG-specific quality dimensions.

Human Evaluation: Expert Assessment and User Satisfaction Studies

Despite advances in automated evaluation metrics, human judgment remains the gold standard for assessing RAG system quality. Human evaluation captures subjective quality dimensions that automated metrics miss, validates the real-world effectiveness of RAG systems, and provides crucial insights for system improvement. This section explores how to design and conduct effective human evaluation studies for RAG systems.

731

The Irreplaceable Value of Human Judgment

Nuanced Quality Assessment: Humans excel at evaluating complex quality dimensions that resist automated measurement—helpfulness, clarity, appropriateness, trustworthiness, and user satisfaction. These subjective qualities often determine real-world system success more than technical metrics.

Context-Sensitive Evaluation: Human evaluators can assess whether responses are appropriate for specific contexts, user types, and situations. They understand implicit requirements, cultural nuances, and domain-specific expectations that automated metrics cannot capture.

Error Detection Beyond Metrics: Humans identify subtle issues like misleading implications, inappropriate tone, confusing explanations, and missing context that automated systems might miss. They can detect when technically accurate responses fail to address user needs effectively.

Validation of Automated Metrics: Human evaluation serves as the ultimate validation for automated assessment approaches. Correlation between human judgment and automated metrics indicates whether technical evaluation aligns with real-world quality requirements.

Types of Human Evaluation for RAG Systems

Expert Assessment Studies: Domain experts evaluate RAG responses for accuracy, completeness, and appropriateness within their field of expertise. Medical professionals might assess healthcare RAG systems, while legal experts evaluate legal document systems.

Expert assessment provides authoritative quality judgment but can be expensive and time-consuming. Experts offer deep domain knowledge but may have different perspectives than typical users. These studies work best for high-stakes applications where accuracy is paramount.

User Satisfaction Studies: End users evaluate RAG systems based on their actual needs and preferences. These studies capture real-world usability, helpfulness, and satisfaction metrics that directly relate to system adoption and success.

User satisfaction studies provide authentic feedback but may miss technical quality issues that users cannot detect. They excel at identifying user experience problems and feature gaps but may not catch subtle accuracy issues.

Comparative Evaluation: Human evaluators compare multiple RAG systems or different versions of the same system. Comparative studies help rank system quality and identify relative strengths and weaknesses.

Comparative evaluation reduces absolute judgment variability by focusing on relative quality differences. However, it requires careful experimental design to avoid bias and ensure fair comparison conditions.

Task-Based Evaluation: Evaluators complete realistic tasks using RAG systems and assess how well the systems support task completion. This approach captures the end-to-end user experience and system effectiveness in real scenarios.

Task-based evaluation provides practical insights but requires careful task design and can be resource-intensive. It excels at identifying workflow issues and usability problems that other approaches might miss.

Designing Effective Human Evaluation Studies

Clear Evaluation Criteria: Successful human evaluation requires well-defined, specific criteria that evaluators can apply consistently. Vague instructions like "rate overall quality" lead to inconsistent results and poor inter-rater reliability.

Effective criteria might include: "Rate factual accuracy on a 1–5 scale where 5 means all facts are correct and verifiable, and 1 means multiple factual errors are present." Specific anchors and examples help evaluators understand expectations and apply consistent standards.

Representative Sampling: Evaluation datasets should represent the full range of queries, difficulty levels, and edge cases that production systems encounter. Biased sampling toward easy questions or specific topics provides misleading quality estimates.

Stratified sampling across query types, difficulty levels, user demographics, and domain areas ensures comprehensive evaluation coverage. Include both typical use cases and challenging edge cases that stress system capabilities.

Inter-Rater Reliability: Multiple evaluators should assess the same examples to measure agreement and identify inconsistent criteria application. High inter-rater reliability indicates that evaluation criteria are clear and consistently applicable.

Cohen's kappa or intraclass correlation coefficients can quantify agreement levels. Low reliability suggests that criteria need refinement or that evaluators need additional training.

Bias Mitigation: Human evaluation is susceptible to various biases, including order effects, halo effects, and evaluator preferences. Careful experimental design helps minimize these biases and improves result validity.

Randomize evaluation order, blind evaluators to system identities when possible, and include control examples with known quality levels. Consider using multiple evaluation rounds to capture consistency and reduce random variation.

Evaluation Dimensions for RAG Systems

Factual Accuracy: Evaluators assess whether generated responses contain correct, verifiable information. This dimension requires domain expertise and access to authoritative sources for fact-checking.

Completeness: Evaluators determine whether responses adequately address all aspects of the user's question. Incomplete responses that miss important information reduce user satisfaction and system effectiveness.

Relevance and Focus: Evaluators assess whether responses stay on-topic and address the specific question asked. Responses that drift to tangential topics or provide excessive irrelevant detail receive lower relevance scores.

Clarity and Comprehensibility: Evaluators judge whether responses are clearly written, well-organized, and appropriate for the intended audience. Technical accuracy means little if users cannot understand the information.

Helpfulness and Usefulness: Evaluators assess whether responses actually help users accomplish their goals. This pragmatic dimension captures the ultimate value of RAG system outputs.

Appropriateness and Safety: Evaluators check whether responses are appropriate for the context and avoid harmful, biased, or inappropriate content. This dimension is crucial for user-facing applications.

Scaling Human Evaluation

Crowdsourcing Approaches: Platforms like Amazon Mechanical Turk or specialized evaluation services can provide scalable human evaluation for large datasets. Crowdsourcing offers cost advantages but requires careful quality control.

Design clear instructions, provide training examples, and implement quality checks to ensure reliable crowdsourced evaluation. Use qualification tests to identify capable evaluators and monitor response quality continuously.

Expert Panel Methods: Smaller groups of domain experts can provide high-quality evaluation for critical applications. Expert panels offer authoritative judgment but are more expensive and less scalable than crowdsourcing.

Recruit experts with relevant credentials and experience, provide clear evaluation protocols, and facilitate discussion to resolve disagreements and refine criteria.

Hybrid Approaches: Combine automated pre-filtering with human evaluation to balance cost and quality. Use automated metrics to identify potentially problematic responses for human review, focusing human attention where it's most valuable.

Longitudinal Studies: Track human evaluation results over time to identify trends, validate system improvements, and detect quality drift. Regular evaluation provides ongoing quality assurance and guides system evolution.

Integrating Human Evaluation with Automated Metrics

Correlation Analysis: Compare human judgment with automated metric scores to validate technical evaluation approaches. Strong correlation suggests that automated metrics capture human-relevant quality dimensions.

Weak correlation indicates that automated metrics miss important quality aspects or that human evaluation criteria need refinement. Use correlation analysis to improve both automated and human evaluation approaches.

Validation Studies: Use human evaluation to validate new automated metrics or evaluate the effectiveness of existing metrics for specific domains or applications.

Calibration Studies: Use human evaluation to set appropriate thresholds for automated metrics. Determine what automated metric scores correspond to acceptable human-judged quality levels.

Feedback Loop Creation: Use human evaluation insights to improve automated evaluation approaches, system design, and evaluation criteria. Human feedback provides valuable guidance for technical development.

Practical Considerations and Challenges

Cost and Resource Management: Human evaluation is significantly more expensive than automated assessment. Budget for appropriate evaluator compensation, platform costs, and analysis time when planning evaluation studies.

Evaluator Training and Management: Provide clear instructions, training materials, and ongoing support to ensure consistent, high-quality evaluation. Monitor evaluator performance and provide feedback to maintain standards.

Result Analysis and Interpretation: Human evaluation generates rich but complex data that requires careful analysis. Use appropriate statistical methods to identify significant differences and meaningful patterns.

Ethical Considerations: Ensure that human evaluation studies treat participants fairly, protect privacy, and follow appropriate ethical guidelines. Consider the burden on evaluators and design studies that respect their time and expertise.

Best Practices for RAG Human Evaluation

Start with Pilot Studies: Conduct small-scale pilot studies to refine evaluation criteria, identify potential issues, and estimate resource requirements before launching full-scale evaluation.

Document Everything: Maintain detailed documentation of evaluation procedures, criteria, and results to enable replication and future comparison studies.

Plan for Iteration: Design evaluation studies that can be repeated over time to track system improvement and identify emerging issues.

Balance Scope and Depth: Choose evaluation scope and depth based on available resources and decision requirements. Comprehensive studies provide rich insights but require significant investment.

Validate Findings: Cross-validate human evaluation results using multiple approaches, evaluator groups, or evaluation rounds to ensure robust and reliable conclusions.

Human evaluation remains essential for comprehensive RAG system assessment despite the sophistication of automated metrics. When designed and conducted thoughtfully, human evaluation studies provide irreplaceable insights into system quality, user satisfaction, and real-world effectiveness. The key is balancing the depth and authority of human judgment with the practical constraints of cost, time, and scalability.

Production Evaluation and Monitoring

Moving from development to production requires robust, automated evaluation systems that can continuously assess RAG performance without manual intervention. Production evaluation faces unique challenges: scale, reliability, cost management, and the need for real-time insights. This section explores how to build evaluation pipelines that operate effectively in production environments.

Automated Evaluation Pipelines

Production RAG systems require evaluation pipelines that run automatically, handle failures gracefully, and provide actionable insights at scale. Unlike development evaluation, production pipelines must balance thoroughness with computational efficiency, cost constraints, and operational reliability.

Core Pipeline Architecture

Pipeline Components: Effective automated evaluation pipelines consist of several interconnected components that work together to provide comprehensive system assessment. The data collection component samples queries and responses from production traffic, the evaluation engine applies RAGAS and custom metrics, the analysis component interprets results and generates insights, and the alerting system notifies teams of quality issues.

```python
class ProductionEvaluationPipeline:
    def __init__(self, config):
        self.llm = ChatOpenAI(model="gpt-3.5-turbo",
        temperature=0)
        self.alert_thresholds = config.get('alert_
        thresholds', {
            'faithfulness': 0.7, 'answer_relevancy': 0.7
        })
        self.results_history = []
```

Data Source Management: Production pipelines must handle multiple data sources with different characteristics and availability patterns. Production logs provide real-world query distributions but may lack ground truth answers. Curated test cases offer consistent evaluation baselines but may not reflect current user needs. Synthetic datasets enable controlled testing but might miss edge cases encountered in production.

Error Handling and Recovery: Production evaluation pipelines must handle various failure modes gracefully. API rate limits, network timeouts, malformed data, and LLM service disruptions are common issues that can disrupt evaluation. Robust pipelines implement retry logic, graceful degradation, and comprehensive error logging.

```python
def run_evaluation(self, dataset):
    try:
        result = evaluate(dataset, metrics=[faithfulness,
        answer_relevancy, context_precision, context_recall])
        scores = {metric: float(np.mean(result[metric])) for
        metric in result.keys()}
        # Pseudocode: store_results(scores), check_alerts(scores)
        return {'scores': scores, 'timestamp': datetime.now().
        isoformat()}
```

```
except Exception as e:
    # Pseudocode: log_error(e), return_default_result()
    return {'error': str(e), 'scores': None}
```

Continuous Evaluation Frameworks

Scheduling and Orchestration: Continuous evaluation requires sophisticated scheduling that balances evaluation frequency with computational costs. High-frequency monitoring (every few minutes) provides rapid feedback but consumes significant resources. Deep evaluation (daily or weekly) offers comprehensive assessment but may miss rapidly developing issues.

Background Processing: Production evaluation pipelines typically run as background services that don't interfere with user-facing operations. This requires careful resource management, queue processing, and coordination with other system components.

```
class ContinuousEvaluationFramework:
    def start_continuous_evaluation(self, interval_hours=6):
        schedule.every(interval_hours).hours.do(self._run_
        scheduled_evaluation)
        # Pseudocode: start_background_thread(), run_initial_
        evaluation()

    def _run_scheduled_evaluation(self):
        # Pseudocode: sample_production_data(), run_
        evaluation(),
        # generate_report(), send_alerts_if_needed()
```

Resource Management: Continuous evaluation must manage computational resources carefully to avoid impacting production systems. This includes rate limiting API calls, managing memory usage for large datasets, and implementing circuit breakers to prevent cascade failures.

Result Storage and Retrieval: Production evaluation generates substantial amounts of data that must be stored efficiently and retrieved quickly for analysis. Time-series databases work well for metric trends, while document stores handle detailed evaluation results and metadata.

Quality Gates and Thresholds

Threshold-Setting Strategy: Effective quality gates require carefully calibrated thresholds that balance sensitivity with specificity. Thresholds set too low generate excessive false alarms, while thresholds set too high miss genuine quality issues. The optimal approach involves analyzing historical performance data to establish baseline ranges and setting thresholds based on business impact rather than arbitrary numbers.

Multilevel Alerting: Production systems benefit from graduated alerting that matches alert severity to response urgency. Critical alerts (faithfulness below 0.6) might trigger immediate escalation, while warning alerts (slight performance degradation) generate notifications for investigation during business hours.

```
def check_quality_alerts(self, scores):
    alerts = []
    for metric, score in scores.items():
        threshold = self.alert_thresholds.get(metric, 0.5)
        if score < threshold:
            severity = "CRITICAL" if score < threshold * 0.8
            else "WARNING"
            alerts.append(f"{severity}: {metric} below
            threshold")
    return alerts
```

Contextual Thresholds: Advanced pipelines use contextual thresholds that adjust based on query type, user segment, or system load. Medical queries might have higher faithfulness requirements than general conversation, while complex technical queries might accept lower relevancy scores.

Data Sampling Strategies

Cost-Effective Sampling: Production evaluation must balance comprehensive coverage with computational costs. Random sampling provides unbiased coverage but may miss important edge cases. Stratified sampling ensures representation across query types and user segments. Intelligent sampling focuses evaluation resources on queries most likely to reveal quality issues.

Temporal Considerations: Production systems experience varying query patterns throughout the day, week, and year. Evaluation sampling must account for these temporal patterns to ensure representative coverage. Peak usage periods might require higher sampling rates, while off-peak times allow for more comprehensive evaluation.

Quality-Driven Sampling: Advanced pipelines implement quality-driven sampling that increases evaluation frequency for queries exhibiting potential quality issues. Queries with high response times, user dissatisfaction signals, or unusual patterns receive priority for detailed evaluation.

Integration with Development Workflows

CI/CD Pipeline Integration: Automated evaluation pipelines integrate with continuous integration and deployment workflows to provide quality gates for system updates. Pre-deployment evaluation ensures that new versions maintain quality standards, while post-deployment monitoring detects any quality regressions.

Feature Flag Coordination: Modern deployment strategies use feature flags to gradually roll out changes. Evaluation pipelines can coordinate with feature flag systems to assess the impact of specific features on system quality, enabling rapid rollback if quality degrades.

Version Comparison: Production pipelines enable systematic comparison between different system versions, configurations, or models. This capability supports data-driven decisions about system improvements and helps identify the root causes of quality changes.

Performance Optimization

Evaluation Efficiency: Production evaluation must optimize for speed and cost without sacrificing accuracy. This includes using smaller, faster models for initial screening, implementing result caching for repeated evaluations, and optimizing batch processing for large datasets.

Parallel Processing: Large-scale evaluation benefits from parallel processing that distributes evaluation tasks across multiple workers. This requires careful coordination to ensure result consistency and proper error handling across distributed components.

Resource Scaling: Production evaluation pipelines must scale resources dynamically based on evaluation load. Auto-scaling mechanisms ensure adequate capacity during peak evaluation periods while minimizing costs during quiet periods.

Monitoring and Observability

Pipeline Health Monitoring: Evaluation of pipelines themselves requires monitoring to ensure they operate reliably. Key metrics include evaluation completion rates, processing latency, error rates, and resource utilization. Pipeline health dashboards provide operators with visibility into evaluation system performance.

Evaluation Quality Metrics: Production pipelines track meta-metrics that assess the quality of the evaluation process itself. These include inter-evaluator consistency, evaluation coverage across different query types, and correlation between automated metrics and user satisfaction.

Alerting on Pipeline Issues: Pipeline monitoring includes alerting on evaluation system failures, performance degradation, or anomalous patterns. Pipeline alerts help operations teams maintain evaluation system reliability and ensure continuous quality oversight.

Cost Management

Budget Controls: Production evaluation can consume significant computational resources through LLM API calls and infrastructure usage. Effective pipelines implement budget controls that prevent runaway costs while maintaining adequate evaluation coverage.

Cost–Quality Trade-offs: Different evaluation approaches offer varying cost–quality trade-offs. Lightweight heuristics provide fast, cheap quality signals but miss nuanced issues. Comprehensive LLM-based evaluation offers sophisticated assessment but consumes substantial resources. Optimal pipelines combine multiple approaches based on query importance and available budget.

Usage Optimization: Production pipelines optimize resource usage through intelligent scheduling, result caching, and progressive evaluation strategies. High-confidence evaluations might skip detailed analysis, while uncertain cases receive comprehensive assessment.

Best Practices for Production Deployment

Start Simple, Scale Gradually: Begin with basic evaluation pipelines that cover essential quality metrics and gradually add sophistication as experience and confidence grow. This approach minimizes deployment risk while building operational expertise.

Comprehensive Testing: Evaluation pipelines require thorough testing in staging environments that mirror production conditions. Testing should cover normal operations, failure scenarios, and edge cases to ensure reliable production performance.

Documentation and Runbooks: Production evaluation systems require comprehensive documentation and operational runbooks that enable teams to operate, troubleshoot, and improve the systems effectively. This includes architecture documentation, configuration guides, and incident response procedures.

Team Training: Successful production evaluation requires teams trained in interpreting results, responding to alerts, and improving system performance based on evaluation insights. Investment in team capability building ensures effective utilization of evaluation capabilities.

Automated evaluation pipelines form the foundation of production RAG quality assurance, enabling continuous monitoring and improvement of system performance. When designed and implemented thoughtfully, these pipelines provide the insights and confidence needed to operate RAG systems successfully at scale.

Real-Time Quality Monitoring

While automated evaluation pipelines provide periodic comprehensive assessment, real-time monitoring enables immediate detection of quality issues as they occur. Production RAG systems face dynamic challenges— traffic spikes, infrastructure issues, model degradation, and data quality problems—that require continuous vigilance and rapid response.

Real-Time Monitoring Architecture

Stream Processing for Continuous Assessment: Real-time monitoring processes each user interaction as it occurs, extracting quality signals and maintaining running statistics. Unlike batch evaluation, stream processing provides immediate feedback but must operate within strict latency constraints to avoid impacting user experience.

```
class RealTimeQualityMonitor:
    def __init__(self, alert_config):
        self.metrics_buffer = []
        self.alert_thresholds = alert_config['alert_
        thresholds']
        self.baseline_metrics = None

    def record_interaction(self, interaction_data):
        interaction_data['timestamp'] = datetime.now()
        self.metrics_buffer.append(interaction_data)
        # Pseudocode: check_immediate_alerts(), update_running_
        stats(),
        # trigger_drift_detection_if_needed()
```

Lightweight Quality Signals: Real-time monitoring relies on quality signals that can be computed quickly without expensive LLM calls. Response time, error rates, user feedback signals, and basic text metrics provide immediate indicators of system health. These lightweight signals complement deeper periodic evaluation rather than replacing it.

Circular Buffer Management: Real-time systems maintain sliding windows of recent interactions to compute trends and detect changes. Circular buffers provide memory-efficient storage for recent data while automatically discarding older observations to maintain constant memory usage.

Immediate Alert Detection

Response Time Monitoring: Performance degradation often manifests first as increased response times. Real-time monitoring tracks response latency at multiple percentiles (median, 95th, 99th) to identify both systematic slowdowns and outlier performance issues.

Error Rate Tracking: System errors provide immediate quality signals that don't require sophisticated evaluation. Tracking error rates, error types, and error patterns enables rapid detection of infrastructure issues, model failures, or data quality problems.

```python
def _check_immediate_alerts(self, interaction):
    alerts = []
    if interaction['response_time'] > self.alert_
    thresholds['response_time']:
        alerts.append(f"SLOW RESPONSE: {interaction['response_
        time']:.2f}s")

    if 'quality_scores' in interaction:
        for metric, score in interaction['quality_scores'].
        items():
            if score < self.alert_thresholds.get(metric, 0.5):
                alerts.append(f"LOW QUALITY: {metric} =
                {score:.3f}")
    return alerts
```

User Feedback Integration: Real-time systems can incorporate immediate user feedback—thumbs up/down ratings, follow-up questions, session abandonment—as quality signals. While not comprehensive, user feedback provides valuable real-world quality indicators that complement automated metrics.

Quality Score Thresholds: When lightweight quality assessment is available (simplified faithfulness checks, basic relevance scoring), real-time systems can apply threshold-based alerting. These simplified metrics sacrifice some accuracy for speed but provide valuable early warning signals.

Performance Drift Detection

Statistical Change Detection: Real-time drift detection compares current performance against established baselines using statistical methods. Moving averages, cumulative sum (CUSUM) algorithms, and exponential smoothing help identify gradual performance changes that might not trigger immediate threshold alerts.

Baseline Establishment and Updates: Effective drift detection requires representative baselines that reflect normal system performance. Initial baselines establish normal operating ranges, while adaptive baselines update gradually to account for legitimate system evolution while detecting abnormal changes.

```
def _check_performance_drift(self):
    if not self.baseline_metrics:
        self._set_baseline_metrics()
        return []

    current_metrics = self._calculate_aggregate_metrics(self.
    metrics_buffer[-50:])
    drift_alerts = []

    for metric, current_value in current_metrics.items():
        baseline_value = self.baseline_metrics.get(metric)
        if baseline_value and abs(current_value - baseline_
        value) / baseline_value > 0.1:
            # Pseudocode: calculate_drift_significance(),
            format_drift_alert()
            drift_alerts.append(f"DRIFT DETECTED: {metric}")
    return drift_alerts
```

Multidimensional Drift Analysis: Production systems experience drift across multiple dimensions simultaneously—query distribution changes, user behavior evolution, seasonal patterns, and gradual model degradation. Sophisticated drift detection analyzes multiple metrics jointly to distinguish between normal variation and concerning degradation.

Temporal Pattern Recognition: Real-time monitoring must distinguish between temporary fluctuations and persistent changes. Traffic spikes, daily usage patterns, and seasonal variations create temporary metric changes that shouldn't trigger drift alerts. Pattern recognition helps separate noise from signal in performance monitoring.

User Feedback Integration

Implicit Feedback Signals: User behavior provides valuable quality signals without explicit feedback collection. Session duration, follow-up query patterns, result clicks, and task completion rates indicate user satisfaction levels and system effectiveness.

Explicit Feedback Processing: Direct user ratings, reported issues, and satisfaction surveys provide authoritative quality signals but require careful processing to account for selection bias and response patterns. Users more likely to provide feedback may not represent the overall user population.

Feedback Aggregation and Weighting: Real-time systems must aggregate diverse feedback signals into actionable quality metrics. Different feedback types carry different weights—explicit negative feedback might trigger immediate investigation, while implicit signals contribute to longer-term trend analysis.

Rapid Response to Critical Feedback: Certain user feedback demands immediate response—reports of dangerous medical misinformation, legal inaccuracies, or harmful content. Real-time systems implement priority queues and escalation procedures for critical feedback that requires urgent attention.

Alert Systems for Quality Degradation

Graduated Alert Severity: Real-time alerting systems implement multiple severity levels that trigger appropriate responses. Critical alerts require immediate attention and may trigger automatic failover, while warning alerts notify teams for investigation during business hours.

Alert Fatigue Prevention: Poorly configured alerting generates excessive false positives that lead to alert fatigue and reduced responsiveness. Effective alerting systems use hysteresis (different thresholds for triggering and clearing alerts), alert aggregation, and intelligent filtering to maintain signal quality.

```
def _trigger_alerts(self, alerts):
    print(f"\n QUALITY ALERTS ({datetime.now()}):")
    for alert in alerts:
        print(f"  {alert}")
    # Pseudocode: send_to_slack(), create_incident_ticket(),
    # notify_on_call_engineer(), update_dashboard()
```

Alert Routing and Escalation: Different alert types require different response teams and procedures. Response time alerts might go to infrastructure teams, while content quality alerts route to product teams. Escalation procedures ensure that unresolved alerts receive appropriate attention and resources.

Alert Context and Enrichment: Effective alerts provide sufficient context for rapid triage and response. Alert messages include affected metrics, severity levels, potential impact, and suggested investigation steps. Enriched alerts reduce time-to-resolution and improve response effectiveness.

Monitoring Dashboard Design

Real-Time Visualization: Monitoring dashboards provide operators with immediate visibility into system health through real-time charts, metrics displays, and status indicators. Effective dashboards balance information density with clarity, highlighting the most critical information while maintaining comprehensive coverage.

Multilevel Detail: Dashboard design follows a drill-down pattern—high-level system health at the top level, component-specific metrics at intermediate levels, and detailed diagnostic information at the bottom level. This hierarchy enables rapid triage and focused investigation.

Historical Context: Real-time displays benefit from historical context that shows current performance relative to past behavior. Time-series charts, trend indicators, and comparison views help operators distinguish between normal variation and unusual patterns.

Collaborative Features: Modern monitoring dashboards support collaborative investigation through shared views, annotation capabilities, and communication integration. Teams can share dashboard views, add contextual notes, and coordinate response efforts directly within the monitoring interface.

Integration with Monitoring Infrastructure

Metrics Export and Aggregation: Real-time RAG monitoring integrates with enterprise monitoring platforms (Prometheus, Grafana, DataDog) through standardized metrics export. This integration enables unified operational dashboards and leverages existing alerting infrastructure.

Log Aggregation and Analysis: RAG quality monitoring generates substantial log data that requires aggregation and analysis. Integration with log management platforms (ELK stack, Splunk) enables sophisticated query analysis and correlation with other system metrics.

```
def get_monitoring_dashboard_data(self):
    recent_metrics = self._calculate_aggregate_metrics(self.
    metrics_buffer[-50:])
    return {
        'current_metrics': recent_metrics,
        'baseline_metrics': self.baseline_metrics,
        'time_series': {
            'timestamps': [i['timestamp'].isoformat() for i in
            self.metrics_buffer],
            'response_times': [i.get('response_time', 0) for i
            in self.metrics_buffer]
        },
        'last_updated': datetime.now().isoformat()
    }
```

Incident Management Integration: Quality alerts integrate with incident management systems (PagerDuty, Opsgenie) to ensure appropriate response and tracking. Integration enables automatic incident creation, escalation management, and resolution tracking.

Communication Platform Integration: Real-time alerts integrate with team communication platforms (Slack, Microsoft Teams) to provide immediate notification and enable collaborative response. Bot integration allows teams to query system status and acknowledge alerts directly from chat platforms.

Performance Considerations

Low-Latency Processing: Real-time monitoring must process interactions with minimal latency to avoid impacting user experience. This requires efficient algorithms, optimized data structures, and careful resource management to maintain sub-millisecond processing times.

Memory Management: Continuous operation requires careful memory management to prevent memory leaks and resource exhaustion. Circular buffers, periodic cleanup, and memory pooling help maintain stable memory usage over extended operation periods.

Scalability Planning: Real-time monitoring systems must scale with production traffic growth. Horizontal scaling, load balancing, and distributed processing enable monitoring systems to handle increasing interaction volumes without degrading performance.

Quality Assurance for Monitoring Systems

Monitoring the Monitors: Monitoring systems themselves require oversight to ensure reliability. Meta-monitoring tracks monitoring system health, processing latency, alert delivery success, and data quality to ensure the monitoring infrastructure operates correctly.

Testing and Validation: Real-time monitoring systems require comprehensive testing under various conditions—normal operation, traffic spikes, partial failures, and edge cases. Synthetic traffic generation and chaos engineering help validate monitoring system robustness.

Operational Runbooks: Monitoring systems require detailed operational procedures that guide teams through common scenarios— alert investigation, system maintenance, configuration changes, and incident response. Well-documented procedures ensure consistent and effective monitoring operations.

Cost and Resource Optimization

Sampling and Filtering: Real-time monitoring can process massive interaction volumes, requiring intelligent sampling and filtering to manage computational costs. Statistical sampling maintains representative coverage while reducing processing overhead.

Tiered Processing: Monitoring systems benefit from tiered processing that applies increasingly sophisticated analysis based on initial quality signals. Lightweight screening identifies potential issues, while detailed analysis focuses on concerning interactions.

Resource Budgeting: Real-time monitoring systems implement resource budgets that prevent runaway costs during traffic spikes or system anomalies. Circuit breakers, rate limiting, and priority queues help maintain system stability within resource constraints.

Real-time quality monitoring provides the immediate feedback and rapid response capabilities essential for production RAG systems. When implemented effectively, these systems enable teams to maintain high quality standards, respond quickly to emerging issues, and build user confidence through reliable system performance. The key is balancing comprehensive coverage with operational efficiency, ensuring that monitoring systems enhance rather than hinder production operations.

Evaluation Best Practices

Effective RAG evaluation depends not only on sophisticated metrics and monitoring systems but also on the quality and representativeness of evaluation datasets. Poor dataset design can lead to misleading evaluation results, hidden biases, and false confidence in system performance. This section explores how to design robust evaluation datasets and mitigate various forms of bias that can compromise evaluation validity.

Dataset Design and Bias Mitigation

The foundation of reliable RAG evaluation lies in carefully designed datasets that accurately represent real-world usage patterns while avoiding systematic biases that could skew results. Dataset design involves balancing multiple competing objectives: comprehensiveness vs. manageability, realism vs. control, and diversity vs. focus.

Comprehensive Dataset Coverage Analysis

Multidimensional Coverage Assessment: Effective evaluation datasets must cover the full spectrum of system usage across multiple dimensions simultaneously. Query complexity ranges from simple factual questions to complex analytical requests. Domain coverage spans the breadth of knowledge areas the system handles. User types include novices seeking basic information and experts requiring detailed analysis. Temporal patterns capture both current events and timeless knowledge.

```
class EvaluationDatasetDesigner:
    def analyze_dataset_coverage(self, dataset):
        analysis = {
            'size': len(dataset),
            'question_analysis': self._analyze_question_
            types(dataset['question']),
            'domain_distribution': self._analyze_
            domains(dataset['question']),
            'bias_indicators': self._detect_biases(dataset)
        }
        return analysis

    def _detect_biases(self, dataset):
        biases = []
        # Pseudocode: check_length_variance(), check_
        repetition_patterns(),
        # analyze_domain_balance(), assess_question_type_
        distribution()
        return biases
```

Usage Pattern Representation: Production evaluation datasets should reflect actual system usage patterns rather than idealized test scenarios. Query frequency distributions, seasonal variations, user behavior patterns, and failure modes from production systems provide essential input for dataset design. This realism ensures that evaluation results predict actual system performance.

Edge Case and Stress Testing: Comprehensive datasets include challenging edge cases that stress system capabilities—ambiguous queries, conflicting information requests, queries outside the system's knowledge domain, and adversarial inputs. While edge cases represent small fractions of normal usage, they often reveal the most significant system weaknesses.

Question Type and Complexity Stratification

Systematic Question Classification: Effective dataset design requires systematic classification of questions across multiple dimensions. Factual questions (who, what, when, where) test basic information retrieval. Analytical questions (how, why) require synthesis and reasoning. Comparative questions assess the system's ability to contrast different concepts or approaches. Open-ended questions evaluate creativity and comprehensive coverage.

Complexity Gradient Management: Questions span complexity gradients that should be represented proportionally in evaluation datasets. Simple questions with clear answers test basic system functionality. Moderate complexity questions requiring multistep reasoning evaluate integration capabilities. Complex questions involving nuanced judgment test the system's sophisticated reasoning abilities.

Domain-Specific Question Patterns: Different domains exhibit characteristic question patterns that evaluation datasets must capture. Medical queries often seek diagnostic information and treatment options. Legal questions require precise interpretation and jurisdictional awareness. Technical queries demand accuracy and current best practices. Financial questions need risk awareness and regulatory compliance.

Bias Detection and Analysis

Statistical Bias Identification: Systematic analysis reveals various forms of dataset bias that can compromise evaluation validity. Length bias occurs when questions or answers cluster around specific lengths, potentially favoring systems optimized for particular response sizes. Repetition bias emerges when similar questions appear multiple times, artificially inflating performance metrics. Domain bias manifests when certain knowledge areas dominate the dataset.

```
def _analyze_question_types(self, questions):
    question_types = {'what': 0, 'how': 0, 'why': 0,
    'other': 0}
    for question in questions:
        q_lower = question.lower().strip()
        # Pseudocode: classify_question_type(), update_
        counters()
    return question_types

def _detect_biases(self, dataset):
    biases = []
    if self._check_length_variance(dataset) > 0.8:
        biases.append("High length variance detected")
    if self._check_domain_balance(dataset) < 0.3:
        biases.append("Domain imbalance detected")
    return biases
```

Demographic and Cultural Bias: RAG systems often exhibit biases related to demographics, culture, and language patterns present in training data and evaluation datasets. Evaluation datasets should include diverse perspectives, cultural contexts, and linguistic patterns to ensure fair assessment across different user populations.

Temporal Bias: Many datasets suffer from temporal bias, overrepresenting certain time periods or failing to account for evolving knowledge. Current events, changing regulations, scientific discoveries, and technological developments create temporal dependencies that evaluation datasets must address.

Selection Bias: The process of choosing evaluation examples can introduce selection bias that favors certain types of questions or domains. Random sampling helps reduce selection bias, but purely random approaches may miss important edge cases or underrepresent critical use cases.

Balanced Dataset Creation Strategies

Stratified Sampling Approaches: Balanced dataset creation employs stratified sampling that ensures appropriate representation across all relevant dimensions. Equal stratification assigns equal weight to each category, while proportional stratification reflects real-world usage patterns. Minimum representation ensures that even rare but important categories receive adequate coverage.

```
def create_balanced_dataset(self, raw_data, balance_criteria):
    groups = self._group_data_by_criteria(raw_data, balance_
    criteria)
    balanced_data = self._sample_balanced_groups(groups,
    balance_criteria)

    dataset = Dataset.from_dict({
        'question': [item['question'] for item in
        balanced_data],
        'contexts': [item['contexts'] for item in
        balanced_data],
        'answer': [item['answer'] for item in balanced_data]
    })
```

```
# Pseudocode: verify_balance(), generate_coverage_report()
return dataset
```

Multi-Criteria Balancing: Sophisticated dataset design balances multiple criteria simultaneously rather than optimizing single dimensions. This requires careful consideration of trade-offs and priorities—domain balance vs. complexity distribution, current events vs. timeless knowledge, common queries vs. edge cases.

Iterative Refinement: Dataset design benefits from iterative refinement based on evaluation results and production feedback. Initial datasets reveal gaps and biases that subsequent iterations address. Continuous improvement ensures that evaluation datasets evolve with system capabilities and user needs.

Representative Sampling Techniques

Production-Driven Sampling: The most representative evaluation datasets derive from actual production usage patterns. Production log analysis reveals real query distributions, user behavior patterns, and failure modes that artificial datasets miss. However, production sampling requires careful privacy protection and may lack ground truth answers for evaluation.

Synthetic Data Generation: Controlled synthetic data generation enables systematic coverage of specific scenarios while maintaining dataset balance. Template-based generation creates variations on core patterns, while AI-assisted generation produces more natural variations. Synthetic approaches ensure comprehensive coverage but may miss real-world nuances.

Hybrid Sampling Strategies: Effective dataset design combines multiple sampling approaches to balance representativeness with control. Core evaluation sets use production sampling for realism, supplemented by synthetic data for systematic coverage and curated examples for critical edge cases.

Ground Truth and Reference Answer Quality

Multiple Reference Standards: High-quality evaluation requires authoritative reference answers that reflect the range of acceptable responses. Single reference answers may be too restrictive, while multiple references capture valid variation in response styles and content. Expert-generated references provide authority, while crowd-sourced references offer diversity.

Reference Answer Validation: Reference answers themselves require validation to ensure accuracy and appropriateness. Expert review, fact-checking against authoritative sources, and consistency verification help maintain reference quality. Outdated or incorrect references can severely compromise evaluation validity.

Acceptable Response Ranges: Many questions have multiple valid answers that evaluation datasets should acknowledge. Factual questions may have single correct answers, while analytical questions permit various valid interpretations. Evaluation frameworks must account for this response diversity rather than penalizing legitimate variation.

Temporal Considerations and Dataset Freshness

Knowledge Currency: RAG systems must handle both timeless knowledge and rapidly evolving information. Evaluation datasets require regular updates to include current events, policy changes, scientific discoveries, and technological developments. Stale datasets may overestimate system performance on outdated information while missing contemporary challenges.

Seasonal and Cyclical Patterns: User queries exhibit seasonal patterns—tax questions peak during tax season, educational queries align with academic calendars, and product queries surge during shopping seasons. Evaluation datasets should capture these temporal patterns to ensure year-round performance assessment.

Historical Perspective: While currency is important, evaluation datasets also need historical depth to assess system performance on established knowledge. Balancing current and historical content ensures comprehensive evaluation across the knowledge spectrum.

Domain-Specific Considerations

Regulatory and Compliance Requirements: Certain domains impose specific requirements on evaluation datasets. Medical evaluation must avoid patient privacy violations while ensuring clinical relevance. Financial evaluation requires regulatory compliance and risk awareness. Legal evaluation needs jurisdictional accuracy and current case law.

Expert Knowledge Requirements: Domain-specific evaluation often requires subject matter expertise for both question formulation and answer validation. Medical questions need clinical accuracy, legal questions require jurisprudential precision, and technical questions demand current best practices.

Domain-Specific Bias Patterns: Different domains exhibit characteristic bias patterns that evaluation datasets must address. Technical domains may overrepresent certain technologies or approaches. Medical domains might reflect demographic biases in clinical research. Academic domains could emphasize certain theoretical perspectives.

Evaluation Dataset Maintenance

Continuous Quality Monitoring: Evaluation datasets require ongoing maintenance to ensure continued relevance and accuracy. Automated checks can identify inconsistencies, outdated information, and emerging bias patterns. Regular human review validates automated assessments and identifies subtle quality issues.

Version Control and Change Tracking: Dataset evolution requires careful version control that tracks changes, rationale, and impact on evaluation results. Change logs enable teams to understand evaluation trends and validate system improvements against consistent baselines.

```python
def generate_dataset_report(self, dataset):
    analysis = self.analyze_dataset_coverage(dataset)

    report = ["EVALUATION DATASET ANALYSIS REPORT",
              f"Dataset Size: {analysis['size']} examples",
              f"Question Types: {analysis['question_analysis']
              ['question_types']}",
              f"Domain Distribution: {analysis['domain_
              distribution']}"]

    if analysis['bias_indicators']:
        report.append("POTENTIAL BIASES DETECTED:")
        # Pseudocode: format_bias_warnings(), add_mitigation_
        suggestions()

    return "\n".join(report)
```

Community and Collaborative Maintenance: Large-scale evaluation datasets benefit from community contributions and collaborative maintenance. Open-source dataset initiatives leverage distributed expertise while shared maintenance reduces individual organization burden. However, collaborative approaches require governance frameworks and quality standards.

Practical Implementation Guidelines

Start with Production Data: Begin dataset design with analysis of production usage patterns to ensure baseline representativeness. Production data provides authentic query distributions and user behavior patterns that artificial datasets struggle to replicate.

Implement Systematic Coverage: Use systematic approaches to ensure comprehensive coverage across all relevant dimensions. Checklists, coverage matrices, and automated analysis help identify gaps and biases that manual approaches might miss.

Plan for Evolution: Design datasets with evolution in mind, including version control, change tracking, and update procedures. Regular refresh cycles ensure continued relevance while preserving historical comparisons.

Validate Through Multiple Lenses: Assess dataset quality through multiple perspectives—statistical analysis, domain expert review, and correlation with production performance. Multifaceted validation provides confidence in dataset quality and evaluation validity.

Well-designed evaluation datasets form the foundation of reliable RAG assessment, enabling accurate performance measurement and meaningful system improvement. The investment in thoughtful dataset design pays dividends through more reliable evaluation results, better system optimization decisions, and increased confidence in production performance. The key is balancing comprehensiveness with practicality while systematically addressing the various forms of bias that can compromise evaluation validity.

Common Pitfalls and How to Avoid Them

Even experienced practitioners fall into evaluation traps that can lead to overconfident assessments, missed quality issues, and poor system optimization decisions. Understanding these common pitfalls and implementing strategies to avoid them is crucial for building reliable evaluation practices.

Metric Gaming and Optimization Myopia

The Goodhart's Law Problem: When evaluation metrics become targets for optimization, they often cease to be good measures of quality. Teams may inadvertently optimize systems to perform well on specific metrics while degrading overall user experience. For example, optimizing solely for RAGAS faithfulness scores might lead to overly conservative responses that technically avoid hallucination but fail to provide helpful information.

Single-Metric Tunnel Vision: Focusing on individual metrics in isolation can create blind spots that miss critical quality dimensions. A system might achieve excellent faithfulness scores while producing irrelevant responses, or maintain high relevancy while hallucinating facts. Comprehensive evaluation requires balanced attention across multiple quality dimensions.

Evaluation–Production Mismatch: Systems optimized for evaluation datasets may not perform well in production environments. Evaluation datasets, no matter how carefully designed, cannot capture the full complexity of real-world usage. This mismatch becomes particularly problematic when evaluation datasets are small, outdated, or biased toward specific use cases.

Short-Term Optimization Bias: Pressure to show rapid improvement can lead to optimizations that improve immediate metrics while creating long-term problems. Quick fixes that boost evaluation scores may introduce subtle biases, reduce system robustness, or create maintenance burdens that compound over time.

Evaluation Dataset Limitations

Sample Size Inadequacy: Small evaluation datasets often provide misleading confidence in system performance. Statistical significance requires adequate sample sizes, and many evaluation efforts underestimate the number of examples needed for reliable assessment. A system might appear to perform well on 50 examples but fail when exposed to broader usage patterns.

Distribution Shift Blindness: Evaluation datasets that don't reflect current production distributions can provide false confidence. User behavior evolves, query patterns change, and new use cases emerge, making static evaluation datasets progressively less representative. This drift can be particularly rapid in dynamic domains like current events or emerging technologies.

Difficulty Calibration Errors: Evaluation datasets that skew too easy or too difficult provide distorted performance pictures. Easy datasets inflate confidence and miss system limitations, while overly difficult datasets discourage improvement efforts and mask genuine progress. Balanced difficulty distributions better reflect real-world performance ranges.

Context Length Mismatches: Many evaluation datasets use simplified contexts that don't reflect production complexity. Real RAG systems often work with longer, more complex, and noisier retrieved contexts. Evaluation on clean, short contexts may overestimate system robustness and miss important failure modes.

Automated Metric Overreliance

Human Judgment Substitution: Automated metrics provide valuable scalable assessment but cannot fully replace human judgment. Over-relying on automated evaluation can miss nuanced quality issues that humans readily detect—inappropriate tone, cultural insensitivity, misleading implications, or response unhelpfulness despite technical accuracy.

LLM Evaluator Limitations: Even sophisticated LLM-based evaluation frameworks like RAGAS have limitations and biases. LLM evaluators may exhibit consistency issues, cultural biases, or systematic blind spots. Treating LLM evaluation as ground truth without validation against human judgment can perpetuate and amplify these limitations.

Correlation Assumptions: Assuming that automated metrics perfectly correlate with user satisfaction can lead to misguided optimization efforts. While metrics like faithfulness and relevancy generally align with user preferences, the correlation is imperfect and context-dependent. Regular validation against user feedback ensures metric relevance.

Evaluation Speed Prioritization: The convenience and speed of automated evaluation can lead to insufficient investment in human evaluation studies. While automated metrics enable rapid iteration, periodic human evaluation provides essential validation and insight that automated approaches cannot match.

Statistical and Analytical Errors

Significance Testing Neglect: Many evaluation efforts report metric improvements without proper statistical significance testing. Small apparent improvements may represent random variation rather than genuine system enhancement. Proper statistical analysis includes confidence intervals, significance tests, and effect size calculations.

Multiple Comparison Problems: When evaluating many metrics simultaneously, the probability of finding spurious "significant" results increases. Multiple comparison corrections (Bonferroni, Benjamini-Hochberg) help maintain appropriate statistical rigor when conducting comprehensive evaluations.

Baseline Comparison Failures: Meaningful evaluation requires appropriate baselines for comparison. Comparing against weak baselines inflates apparent performance, while missing important baseline comparisons can obscure relative system positioning. Strong baselines include previous system versions, competitive systems, and theoretical upper bounds.

Cherry-Picking and Selection Bias: Selectively reporting favorable results while ignoring unfavorable ones creates misleading performance pictures. This bias can occur unconsciously through focusing on successful use cases while downplaying failures, or systematically through deliberate result selection.

Production-Evaluation Disconnects

Evaluation Environment Idealization: Evaluation environments often idealize conditions that don't reflect production realities. Clean data, optimal infrastructure, and controlled conditions may not represent the noisy, resource-constrained, and dynamic production environment where systems actually operate.

User Behavior Modeling Gaps: Evaluation scenarios may not accurately model real user behavior patterns. Users ask follow-up questions, provide partial information, make typos, and interact with systems in ways that static evaluation datasets cannot capture. These behavioral nuances significantly impact system performance.

Scale and Performance Interactions: System performance often degrades under production scale and load conditions that evaluation environments don't replicate. Latency constraints, resource limitations, and concurrent usage patterns can reveal quality issues that don't appear in controlled evaluation settings.

Temporal Dynamics Ignorance: Production systems operate in dynamic environments where query patterns, user expectations, and information currency constantly evolve. Static evaluation approaches may miss performance degradation that occurs over time or fail to assess system adaptability to changing conditions.

Choosing the Right Evaluation Strategy

Effective RAG evaluation requires matching evaluation approaches to specific goals, constraints, and contexts. No single evaluation strategy suits all situations, and optimal approaches depend on system maturity, available resources, user requirements, and organizational priorities.

Context-Driven Strategy Selection

Development Stage Considerations: Early-stage systems benefit from rapid, broad evaluation that identifies major issues and guides development priorities. Mature systems require more sophisticated evaluation that detects subtle quality differences and validates incremental improvements. The evaluation sophistication should match system development maturity.

Application Domain Requirements: Different domains demand different evaluation emphases. High-stakes applications (medical, legal, financial) prioritize accuracy and safety, requiring rigorous evaluation with expert validation. Consumer applications may emphasize user experience and engagement, calling for different evaluation approaches and metrics.

Resource and Constraint Analysis: Evaluation strategy must align with available resources—budget, time, expertise, and infrastructure. Resource-constrained environments may rely more heavily on automated evaluation, while resource-rich settings can invest in comprehensive human evaluation studies. Understanding constraints enables realistic strategy selection.

Risk Tolerance and Consequences: Systems with severe failure consequences require more comprehensive evaluation than those with limited impact. The potential costs of quality failures should guide evaluation investment levels and sophistication. High-risk applications justify expensive evaluation approaches that would be inappropriate for low-risk scenarios.

Multimodal Evaluation Frameworks

Layered Evaluation Approaches: Effective evaluation employs multiple layers that provide different perspectives on system quality. Automated metrics enable rapid assessment and continuous monitoring. Human evaluation provides authoritative quality judgment and user perspective validation. A/B testing reveals real-world performance differences under actual usage conditions.

Temporal Evaluation Strategies: Comprehensive evaluation spans multiple time horizons with different focuses. Real-time monitoring detects immediate issues and enables rapid response. Periodic deep evaluation provides comprehensive assessment and long-term trend analysis. Longitudinal studies track system evolution and user satisfaction over extended periods.

Stakeholder-Specific Evaluation: Different stakeholders require different evaluation perspectives and metrics. Technical teams need detailed performance metrics and diagnostic information. Product teams focus on user experience and business impact metrics. Executives require high-level quality summaries and trend analysis. Effective evaluation serves multiple stakeholder needs simultaneously.

Complementary Method Integration: The most robust evaluation strategies combine complementary methods that address different quality dimensions. Automated metrics provide scalable assessment, human evaluation offers nuanced judgment, user feedback captures real-world satisfaction, and production monitoring ensures ongoing quality assurance.

Cost-Benefit Optimization

Evaluation ROI Analysis: Evaluation efforts should demonstrate clear return on investment through improved system quality, reduced failure costs, or enhanced user satisfaction. Expensive evaluation approaches must provide proportional benefits, while cost-effective methods should be maximized before investing in premium approaches.

Progressive Evaluation Sophistication: Start with basic evaluation approaches and progressively add sophistication as systems mature and resources permit. Initial evaluation might rely primarily on automated metrics, with human evaluation and advanced analytics added as systems prove their value and justify additional investment.

Sampling and Coverage Strategies: Optimize evaluation coverage through intelligent sampling that maximizes quality insight per evaluation dollar. Focus comprehensive evaluation on high-impact use cases while using lighter evaluation for routine scenarios. Risk-based sampling concentrates evaluation resources where quality issues have the greatest potential impact.

Automation Investment Priorities: Invest in evaluation automation that provides the highest value for ongoing operations. Automated evaluation pipelines, monitoring dashboards, and alert systems reduce long-term evaluation costs while maintaining quality oversight. However, automation investment should complement rather than replace human evaluation entirely.

Evaluation Strategy Evolution

Adaptive Strategy Development: Evaluation strategies should evolve with system maturity, user needs, and organizational learning. Initial strategies focus on basic functionality and major issues, while mature strategies address subtle quality differences and optimization opportunities. Regular strategy review ensures continued relevance and effectiveness.

Learning-Driven Improvement: Use evaluation results to improve evaluation approaches themselves. Metrics that don't correlate with user satisfaction should be de-emphasized, while evaluation gaps revealed through production issues should be addressed. Evaluation strategies should be self-improving systems that become more effective over time.

Community and Industry Integration: Leverage industry best practices, academic research, and community knowledge to improve evaluation approaches. Participate in evaluation benchmarks, share insights with the community, and adopt proven practices from other organizations. External perspectives often reveal evaluation blind spots and improvement opportunities.

Future-Proofing Considerations: Design evaluation strategies that can adapt to evolving RAG capabilities, changing user expectations, and emerging quality requirements. Flexible evaluation frameworks accommodate new metrics, evaluation methods, and quality dimensions without requiring complete strategy overhauls.

Implementation Guidelines

Start Simple, Scale Systematically: Begin with straightforward evaluation approaches that provide immediate value and build confidence. Basic RAGAS evaluation, simple monitoring, and periodic human assessment provide foundational quality insight. Add sophistication systematically as experience and needs grow.

Balance Automation and Human Insight: Combine automated evaluation for scale and consistency with human evaluation for nuance and validation. Automated metrics enable continuous assessment and rapid iteration, while human evaluation provides authoritative quality judgment and user perspective. The optimal balance depends on resources, requirements, and risk tolerance.

Integrate with Development Workflows: Embed evaluation into development and deployment processes rather than treating it as a separate activity. Quality gates in CI/CD pipelines, automated evaluation on code changes, and production monitoring integration ensure that evaluation insights influence system development effectively.

Plan for Continuous Improvement: Design evaluation systems that improve over time through learning, feedback, and systematic enhancement. Regular evaluation strategy reviews, metric validation studies, and process improvement initiatives ensure that evaluation capabilities evolve with system sophistication and organizational maturity.

Avoiding common pitfalls and choosing appropriate evaluation strategies are crucial for building reliable RAG assessment capabilities. Success comes from understanding the limitations and trade-offs inherent in different evaluation approaches, matching strategies to specific contexts and requirements, and continuously improving evaluation practices based on experience and results. The goal is not perfect evaluation but rather evaluation that provides sufficient insight to build, deploy, and maintain high-quality RAG systems with confidence.

Summary and Next Steps

RAG evaluation has evolved from an afterthought to a critical discipline that determines the success of production systems. This chapter has equipped you with the tools and knowledge needed to evaluate RAG systems effectively, from development through production deployment.

Key Takeaways

- RAGAS provides sophisticated automated evaluation that captures the unique challenges of RAG systems—faithfulness, relevancy, precision, and recall.

- Traditional metrics remain valuable for benchmarking and rapid assessment, while human evaluation offers the gold standard for nuanced quality judgment.

- Combining these approaches creates comprehensive evaluation frameworks to ensure system reliability.

Production Readiness

Successful RAG deployment requires

- Automated evaluation pipelines
- Real-time monitoring
- Robust alerting systems

These capabilities enable continuous quality assurance, rapid issue detection, and confident system operation at scale.

Avoiding Pitfalls

Common evaluation mistakes—metric gaming, dataset bias, overreliance on automation—can be mitigated by

- Balancing multiple evaluation methods
- Designing unbiased datasets
- Validating results against human judgment and user feedback

Next Steps

1. **Implement RAGAS**: Start with core metrics (faithfulness, answer_relevancy) and expand to custom metrics for domain-specific needs.

2. **Build monitoring pipelines**: Integrate evaluation into CI/CD workflows and production systems.

3. **Iterate**: Use evaluation insights to refine retrieval, generation, and user interaction components.

The future of RAG systems depends on rigorous evaluation practices. With the frameworks in this chapter, you're equipped to build systems you can trust—because you can measure and improve their performance systematically.

CHAPTER 15

Production Deployment and Scaling Strategies

Building a RAG system that works in development is fundamentally different from deploying one that serves thousands of users reliably. Production flips every assumption: thousands of concurrent users, messy real-world data, sub-second response requirements, and zero tolerance for downtime.

Scale Changes Everything: Your development system handling ten users becomes a different beast serving thousands. Vector searches that previously took 200ms with small indexes now take over 2 seconds with production data.

Users Are Unpredictable: Real users make typos, ask ambiguous questions, submit edge cases you never considered, and expect instant responses regardless of complexity.

Production Requirements: Sub-second responses under heavy load, 24/7 uptime with automatic failure recovery, enterprise-grade security, comprehensive monitoring, and predictable costs.

© Ranajoy Bose 2025
R. Bose, *Mastering Retrieval-Augmented Generation*,
https://doi.org/10.1007/979-8-8688-1808-0_15

Scalable Architecture Patterns

Microservices Decomposition

RAG systems naturally decompose into services with different scaling characteristics:

```
services = {
    'query-service': 'Handle requests, orchestrate pipeline',
    'retrieval-service': 'Vector search and document retrieval',
    'generation-service': 'LLM inference and response generation',
    'embedding-service': 'Text-to-vector conversion'
}
```

Each service scales independently run multiple generation instances during peak hours while keeping retrieval services constant.

Load Balancing and Caching

Multilayer caching reduces expensive operations:

- **Query-level cache**: Store complete responses for identical questions

- **Retrieval cache**: Cache vector search results for similar queries

- **Embedding cache**: Reuse embeddings for repeated text chunks

```
@cache_with_ttl(hours=24)
def get_embeddings(text):
    return embedding_model.encode(text)
@cache_with_similarity_threshold(0.95)
def retrieve_documents(query_embedding):
    return vector_db.similarity_search(query_embedding)
```

Intelligent Routing: Route embedding requests to GPU-optimized instances, direct simple queries to lightweight inference servers.

Infrastructure and DevOps

Containerization Essentials

```
FROM python:3.11-slim
COPY requirements.txt .
RUN pip install --no-cache-dir -r requirements.txt
HEALTHCHECK --interval=30s --timeout=10s \
  CMD curl -f http://localhost:8000/health
CMD ["uvicorn", "app:app", "--host", "0.0.0.0"]
```

Configure resource limits appropriately embedding services need GPU access, vector databases need substantial memory.

CI/CD for RAG Systems

RAG deployments involve models, embeddings, and data indices versioned together:

```
stages:
  - test: Run unit tests, integration tests
  - build: Build containers, validate models
  - evaluate: Run RAGAS evaluation suite
  - deploy: Rolling deployment to production
```

Model Versioning: Track model versions alongside code. Maintain deployment manifests specifying compatible component versions.

Performance Optimization

Vector Database Tuning

Choose the right index type—HNSW for fast approximate search, IVF for memory efficiency:

```
index_configs = {
    'production': {
        'type': 'hnsw',
        'M': 16,   # Connections per node
        'ef_search': 100   # Query-time search width
    }
}
```

LLM Inference Optimization

Dynamic Batching balances latency with throughput wait briefly to accumulate requests, then processes together:

```
class BatchProcessor:
    def __init__(self, max_batch_size=8, max_wait_ms=50):
        self.batch_queue = []
        # Pseudocode: implement batching logic
```

End-to-End Latency Reduction

Make independent operations concurrent:

```
async def process_query(query):
    embedding_task = asyncio.create_task(get_embedding(query))
    embedding = await embedding_task
```

```
docs_task = asyncio.create_task(retrieve_docs(embedding))
docs = await docs_task
return await generate_response(query, docs)
```

Stream responses to users as they're generated rather than waiting for completion.

Security and Monitoring

Essential Security

- **Data Protection**: Encrypt all data at rest and in transit.

- **Input Validation**: Prevent prompt injection attacks.

- **Access Control**: Role-based access restricting users to appropriate documents.

- **Audit Logging**: Log all interactions for compliance.

```
def validate_query(query):
    if len(query) > MAX_QUERY_LENGTH:
        raise ValidationError("Query too long")
    if detect_injection_patterns(query):
        raise SecurityError("Suspicious query detected")
    return sanitize_input(query)
```

Production Monitoring

Track metrics at every level:

```
monitoring_layers = {
    'infrastructure': ['CPU', 'memory', 'GPU utilization'],
    'application': ['response_time', 'error_rate'],
```

```
'rag_specific': ['retrieval_accuracy', 'generation_
quality'],
'business': ['user_satisfaction', 'task_completion_rate']
}
```

Configure alerts for critical issues:

```
alerts:
  high_error_rate:
    condition: error_rate > 5%
    severity: critical
  slow_responses:
    condition: p95_latency > 2s
    severity: warning
```

Cost Management

Track costs at the query level:

```
def track_query_cost(query_id):
    costs = {
        'embedding_api': calculate_embedding_cost(tokens),
        'vector_search': calculate_search_cost(operations),
        'llm_inference': calculate_generation_cost(tokens)
    }
    log_cost_metrics(query_id, sum(costs.values()))
```

Deployment Best Practices
Safe Deployment Patterns

Canary Releases: Gradually roll out changes, start with 5% traffic, monitor metrics, then increase to 25%, 50%, and 100%.

```
def route_request(user_id):
    if user_id % 100 < canary_percentage:
        return route_to_canary_version()
    return route_to_stable_version()
```

Health Checks: Verify all components are functional:

```
@app.get("/health")
async def health_check():
    checks = {
        'vector_db': await check_vector_db_connection(),
        'llm_service': await check_llm_availability(),
        'embedding_model': await check_embedding_service()
    }
    if all(checks.values()):
        return {"status": "healthy"}
    raise HTTPException(503, {"status": "unhealthy"})
```

Operational Excellence

Runbooks: Document common scenarios and solutions. When vector search is slow, when embeddings fail, when costs spike, have clear procedures ready.

Incident Response: Establish clear escalation paths and communication channels. Learn from every incident without blame.

Summary and Next Steps

Production RAG deployment transforms experimental systems into reliable, scalable services. Success requires systematic architecture design, performance optimization, robust infrastructure, and comprehensive monitoring.

781

Key Takeaways

Architecture First: Decompose RAG into scalable microservices—query, retrieval, generation, and embedding services that scale independently based on demand.

Performance Matters: Users expect sub-second responses. Optimize through multilayer caching, dynamic batching, and streaming responses. Monitor actual bottlenecks, not assumptions.

Deploy Safely: Use canary releases, health checks, and rollback procedures. Plan for failure with comprehensive monitoring and incident response.

Production Readiness Checklist

- Load testing under realistic conditions
- Security review and monitoring are configured
- Cost controls and budget alerts implemented
- Rollback procedures tested
- Team training on operations completed

Next Steps

Week 1–2: Basic infrastructure and monitoring; **Week 3–4**: Performance optimization and gradual rollout **Month 2+**: Scale based on real usage data and continuous improvement

Start simple, scale systematically, optimize based on data. Production RAG is a journey of continuous improvement, not a destination.

## CHAPTER 16

Security, Privacy, and Ethical Considerations in Enterprise RAG

Enterprise RAG systems present unique security challenges that traditional frameworks weren't designed to handle. Unlike conventional software, RAG systems ingest vast unstructured data, perform complex AI operations, and can expose sensitive information in unexpected ways.

Expanded Attack Surface: RAG systems blur input/output boundaries—user queries trigger complex retrieval across multiple data sources, and responses might combine information from different security contexts.

AI-Specific Vulnerabilities: Prompt injection, model poisoning, and adversarial inputs represent new threat categories requiring specialized security approaches.

Security-First Design: Embed security into every architectural decision—least privilege access, defense in depth, zero trust verification, and comprehensive audit trails.

© Ranajoy Bose 2025
R. Bose, *Mastering Retrieval-Augmented Generation*,
https://doi.org/10.1007/979-8-8688-1808-0_16

```python
class SecureRAGPipeline:
    def process_query(self, user, query):
        # Authenticate, validate input, filter by permissions
        sanitized_query = self.validate_and_sanitize(query)
        accessible_docs = self.filter_by_permissions(user.
        permissions)
        results = self.retrieve_with_audit(sanitized_query,
        accessible_docs)
        return self.generate_and_filter(sanitized_query,
        results)
```

Data Security and Privacy

Encryption Strategies

Data at Rest: Encrypt vector databases, document stores, and model caches. Use key management systems with role-based access to encryption keys.

Data in Transit: TLS 1.3 for all API communications, encrypted connections to vector databases, and secure model serving endpoints.

```python
encryption_config = {
    'vector_db': {'encryption': 'AES-256', 'key_rotation':
    'monthly'},
    'document_store': {'encryption': 'AES-256', 'field_
    level': True},
    'api_transport': {'tls_version': '1.3', 'cert_
    validation': True}
}
```

PII Detection and Anonymization

Implement automated PII detection before document ingestion and response generation:

```
def sanitize_document(content):
    # Detect and mask PII patterns
    patterns = {
        'ssn': r'\d{3}-\d{2}-\d{4}',
        'email': r'\S+@\S+\.\S+',
        'phone': r'\d{3}-\d{3}-\d{4}'
    }
    for pii_type, pattern in patterns.items():
        content = re.sub(pattern, f'[{pii_type.upper()}_
        REDACTED]', content)
    return content
```

Access Controls and Data Lineage

Document-Level Security: Tag documents with access levels, departments, or classification levels. Filter retrieval results based on user permissions.

Data Lineage Tracking: Maintain complete audit trails showing which documents contributed to each response for compliance and forensic analysis.

Authentication and Access Control

Role-Based Access Control (RBAC)

Implement granular permissions that control document access, feature usage, and system operations:

```python
class RAGPermissions:
    def __init__(self, user_role):
        self.permissions = {
            'analyst': ['read_public', 'read_internal'],
            'manager': ['read_public', 'read_internal', 'read_
            confidential'],
            'admin': ['read_all', 'system_config', 'audit_access']
        }[user_role]
    def can_access_document(self, doc_classification):
        return f'read_{doc_classification}' in self.permissions
```

API Security

Rate Limiting: Prevent abuse and ensure fair resource allocation across users and applications.

Authentication Tokens: Use short-lived tokens with refresh mechanisms. Implement token scoping for different access levels.

```python
# Rate limiting configuration
rate_limits = {
    'free_tier': {'requests_per_minute': 10, 'queries_per_
    hour': 100},
    'enterprise': {'requests_per_minute': 100, 'queries_per_
    hour': 10000},
    'admin': {'requests_per_minute': 'unlimited'}
}
```

Prompt Security and Input Validation

Prompt Injection Defense

Prompt injection attacks attempt to manipulate system behavior through crafted inputs. Implement multiple defense layers:

```
def detect_prompt_injection(query):
    suspicious_patterns = [
        r'ignore.*(previous|above|prior).*(instruction|prompt)',
        r'system.*(role|prompt|instruction)',
        r'(pretend|act).*(as|like).*(admin|root|system)'
    ]
    for pattern in suspicious_patterns:
        if re.search(pattern, query.lower()):
            return True
    return False
def validate_input(query):
    if len(query) > MAX_QUERY_LENGTH:
        raise ValidationError("Query too long")
    if detect_prompt_injection(query):
        raise SecurityError("Potential prompt injection
        detected")
    return sanitize_query(query)
```

Output Filtering

Filter generated responses to prevent information leakage:

```
def filter_output(response, user_permissions):
    # Remove PII and unauthorized content
    filtered_response = remove_pii_from_response(response)
    # Check for information disclosure
```

```
if contains_restricted_info(filtered_response, user_
permissions):
    return "I cannot provide information on that topic."
return filtered_response
```

Ethical AI and Bias Mitigation

Bias Detection and Measurement

Monitor RAG responses for bias across different user groups and
content types:

```
def assess_response_bias(query, response, user_demographics):
    bias_metrics = {
        'gender_bias': analyze_gendered_language(response),
        'cultural_bias': check_cultural_assumptions(response),
        'demographic_bias': evaluate_representation(response,
        user_demographics)
    }
    return bias_metrics
```

Fairness and Transparency

Equal Access: Ensure all authorized users receive comparable response
quality regardless of demographics or query patterns.

Explainability: Provide source attribution and confidence indicators
to help users understand response reliability.

Bias Auditing: Regularly evaluate system outputs across different user
segments and content areas to identify and address systematic biases.

Compliance and Regulatory Framework

Data Protection Laws

GDPR Compliance: Implement right to erasure, data portability, and consent management for EU users.

CCPA Requirements: Provide transparency about data collection and enable user data deletion requests.

```python
class ComplianceManager:
    def handle_deletion_request(self, user_id):
        # Remove user data from all systems
        self.delete_user_queries(user_id)
        self.remove_from_training_data(user_id)
        self.update_vector_indices()
        self.log_deletion_completion(user_id)
```

Industry-Specific Compliance

HIPAA (Healthcare): Encrypt PHI, implement access controls, maintain audit logs, and ensure business associate agreements cover AI vendors.

SOX (Financial): Maintain data integrity controls, implement change management, and ensure accurate financial reporting capabilities.

PCI-DSS (Payment): Secure payment-related information with appropriate controls and regular security assessments.

Security Monitoring and Operations

Threat Detection

Monitor for suspicious patterns and security incidents:

```python
def detect_security_anomalies(user_activity):
    alerts = []
```

```
if user_activity['queries_per_hour'] > normal_
threshold * 3:
    alerts.append('Unusual query volume detected')
if user_activity['failed_auth_attempts'] > 5:
    alerts.append('Potential brute force attack')
if suspicious_query_patterns(user_activity['recent_
queries']):
    alerts.append('Potential prompt injection attempts')
return alerts
```

Incident Response

Rapid Response: Automated detection and alerting for security incidents with clear escalation procedures.

Forensic Analysis: Maintain detailed logs enabling post-incident analysis and system improvement.

Recovery Procedures: Document and test procedures for containing breaches and restoring normal operations.

Summary and Next Steps

Enterprise RAG security requires comprehensive approaches addressing data protection, access control, input validation, bias mitigation, and regulatory compliance. Success comes from embedding security into system design rather than adding it as an afterthought.

Security Checklist

- Encryption implemented for data at rest and in transit

- PII detection and anonymization processes are active

- Role-based access controls enforcing document-level permissions

- Prompt injection detection and input validation are deployed

- Output filtering prevents information leakage

- Bias monitoring and mitigation processes established

- Compliance requirements mapped and implemented

- Security monitoring and incident response procedures tested

Building Security Culture

Team Training: Educate development and operations teams on RAG-specific security threats and mitigation strategies.

Regular Audits: Conduct periodic security assessments and penetration testing focused on AI-specific vulnerabilities.

Continuous Improvement: Learn from security incidents and emerging threats to strengthen defenses over time.

Security, privacy, and ethics aren't constraints on RAG systems—they're enablers that build user trust and enable broader enterprise adoption. The organizations that get security right from the beginning will have sustainable competitive advantages as RAG technology becomes mainstream.

Index

A

Ablation studies, 724
A/B testing
  controlled variable testing,
    405, 406
  setting up, 404
  significance testing, 406
Access controls, 248, 286, 287,
  665, 785
Accuracy, 30, 42, 45, 50, 51,
  61, 258
Accuracy metrics, 274
Adaptability, 30, 33, 43, 44
Adaptation strategies, 649
Adaptive learning systems, 29, 30
Adaptive strategy development,
  676, 770
Adaptive weighting, 716
Adversarial self-critique, 419, 420
Agent communication, 655
Agentic RAG
  architecture, 615–617
  autonomy adds value *vs.*
    complexity, 614, 615
  communication and
    coordination, 621, 622
  components, 617, 618
  concepts, 611, 612

decision tracking and reasoning
  transparency, 682, 683
design patterns, 618–621
error analysis and failure
  pattern recognition, 683
implementation, 666–678, 691
  error handling and graceful
    degradation, 623
  observability and
    debugging, 623
  security and safety
    boundaries, 623, 624
memory and context
  management, 643–653
multi-agent
  collaboration, 653–666
performance metrics
  planning efficiency, 680, 681
  task completion, 680
  user satisfaction and
    experience metrics, 681
performance monitoring and
  alerting, 684
planning and multistep
  reasoning, 624–633
safety
  action boundaries and
    constraint systems, 688

© Ranajoy Bose 2025
R. Bose, *Mastering Retrieval-Augmented Generation*,
https://doi.org/10.1007/979-8-8688-1808-0

F

H

M

Q